PA

# Heating Technology
**Principles, Equipment, and Application**

# Heating Technology
## Principles, Equipment, and Application

**S. Don Swenson**

North Seattle Community College

**Breton Publishers**

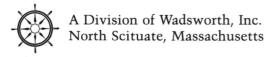

A Division of Wadsworth, Inc.
North Scituate, Massachusetts

Breton Publishers
A Division of Wadsworth, Inc.

**Library of Congress Cataloging in Publication Data**

Swenson, S. Don, 1928–
  Heating technology.

  Includes index.
  1. Heating.  I. Title.
TH7222.S94  1983      697      82–20792
ISBN 0–534–01481–X

Printed in the United States of America
  2  3  4  5  6  7  8  9–87  86  85  84

*Heating Technology: Principles, Equipment, and Application* was
prepared for publication by the following people:
  Sponsoring editor: Jay P. Bartlett
  Production editor: Linda F. Kluz
  Art coordinator: Ellie Connolly
  Copy editor: Beverly Miller
  Interior designer: Joanna Prudden Drummond
  Cover designer: Trisha Hanlon
Illustrations were done by Arvak. The book was typeset in Trump
Mediaeval by Modern Graphics, Inc.; it was printed and bound by
The Maple-Vail Book Manufacturing Group.

# Contents

**Chapter 15
Selection and Sizing
of Heating
Equipment**—186

**Chapter 16
Warm Air
Distribution
and Duct Sizing**—203

**Chapter 20
Hydronic System
Design and
Component Sizing—**

270

**Chapter 21
Troubleshooting
a Warm Air
Furnace—293**

# Preface

The production and management of heat is one of the most important functions of a total comfort system for a residential or commercial building. However, unless there is a complete failure or drastic loss of efficiency, most people take the comfortable atmosphere in which they live and work for granted. They seldom realize that the level of comfort depends on the knowledge and skills of trained heating technicians. Heating technicians must understand the basic processes of heat production, the operation of heating equipment, and the principles of system design. They must be able to calculate the heating requirements of a building, choose the correct equipment for the application, design an efficient distribution system, and troubleshoot the entire system once it is in place.

*Heating Technology: Principles, Equipment, and Application* provides a practical and technically sound textbook and reference book on heating for both practicing and future heating technicians. While many books on various aspects of heat and heating technology have been published, most focus on only one aspect of heating technology. They cover either the principles or the application of heating technology; they cover only one type of heating system; or they provide only servicing information. *Heating Technology: Principles, Equipment, and Application*, however, covers fundamental principles and skills for all common heating systems in a single volume.

The first four chapters cover the basics—physics, fuels, combustion, and heating mediums. The next six chapters introduce the important types of warm air and hydronic heating equipment. Chapters 11 and 12 describe unit and system controls. Chapters 13 and 14 provide detailed coverage of heat load calculations, and Chapter 15 covers equipment selection. The next five chapters cover warm air and hydronic system design. Chapter 21 presents troubleshooting of a warm air system, and Chapter 22 introduces the heat pump. Most of the information in this text is related to small heating units and systems; however, the same principles apply to larger units and systems.

Readers will find several features in this book that make both the first reading and later review or reference easy. New words and technical terms are defined in the margin of the page where they first appear. Each chapter concludes with a summary of important points and numerous self-study questions. Finally, reference materials that are frequently used on the job are gathered in Appendix Tables at the end of the book.

## ACKNOWLEDGMENTS

The author would like to express his thanks to Russell C. Stevens, Jr., Ferris State College, and Dean Harlow, Indiana Vocational-Technical College, for reviewing the manuscript of *Heating Technology: Principles, Equipment, and Application.*

The author would also like to thank the following manufacturers and photographers who provided photographs and illustrations for the preparation of the book:

Adams Manufacturing Company
Air Conditioning and Refrigeration Institute
Air Conditioning Contractors of America
Airtherm Manufacturing Company
Alco Controls Division of Emerson Electric
    Company
Alnor Instrument Company
American Gas Association
American Society of Heating, Refrigeration and
    Air-Conditioning Engineers
ARCO Comfort Products Company
Bacharach Instrument Company
Barry Blower, a Marley Company
R.W. Beckett Corporation
Bell and Gossett ITT
Bryant, Day & Night, Payne
Burnham Corporation, Hydronics Division
Dwyer Instruments, Inc.
Emerson Electric
Farr Filter Company
Friedrich Air Conditioning and Refrigeration
    Company, Room Air Conditioning Division
General Electric Company
Honeywell, Inc.
ITT Fluid Handling Division
Johnson Controls, Inc., Control Products Division
Kewanee Boiler Corporation
Lennox Industries, Inc.

Lima Register Company
National Environmental Systems Contractors
  Association
Pearless of America, Inc.
Perfection Products Company
Rheem Manufacturing Company
Robertshaw Controls Company
Taylor Instrument/Sybron Corporation
York Division of Borg-Warner Corporation
United Technologies, Bacharach

## 1.1 INTRODUCTION

This text is written primarily for students interested in heat as it is used to keep buildings comfortable. The physical composition of materials is important to an understanding of heat and temperature. In this chapter, physical composition of material is explained by the molecular theory. The chapter then covers heat and temperature and the relationship between them. Heat and temperature are defined, and the methods used for measuring them are covered. Heat transfer and the terms used in relation to heat transfer are also described and explained.

## 1.2 MATTER

Scientists believe that all matter is made up of extremely small particles of material called **molecules.** A molecule is described as the smallest particle that a material can be divided into and still retain the properties of the material. Molecules are made up of still smaller particles called **atoms,** and atoms are made up of smaller particles called **neutrons, protons,** and **electrons.** The atom, then, is the basic building block of material.

The neutrons and protons are the particles that make up the nucleus of the atom. Orbiting about this nucleus are other particles called electrons. Figure 1–1

# CHAPTER
# 1
# Heat

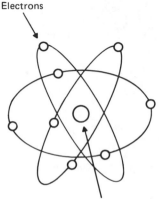

Electrons

Nucleus (neutrons and protons)

FIGURE 1-1 _____

Basic parts of an atom. Neutrons and protons are found in the nucleus, and electrons orbit the nucleus.

**Molecule***

**Atom**   The smallest particle of an element that can take part in a chemical reaction and retain its identity

**Neutron**   A basic part of the atom; particles with a neutral polarity that are part of the nucleus

**Proton**   A basic part of the atom; particles with a positive polarity that are part of the nucleus

**Electron**   A basic part of an atom; particles of energy with a negative electric charge that rotate in orbit around the nucleus

*Term defined in text.

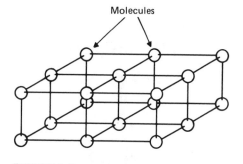

Molecules

FIGURE 1-2 _____

Molecule. Molecules combine in a lattice
arrangement to form an element or compound.

illustrates the construction of an atom. The number of
particles forming an atom and the general arrangement
of those particles determine what materials the atoms
will make as they combine to form molecules. If all of
the atoms are of the same kind, the molecules and the
material are called an **element.** If atoms of different
kinds combine, the molecules and the material are
called a **compound.** When molecules combine to form
elements or compounds, they do so in a lattice-like
arrangement. See Figure 1–2.

In a pure element, all of the atoms are alike. More
than one hundred elements are now known. Most of
these are natural elements, such as carbon, hydrogen,
and oxygen, but some—among them, nobelium, law-
rencium, and mendelevium—have been created by
physicists in their work with nuclear physics. *Created*
is used loosely here since matter cannot be created;
instead the term means that the basic parts can be
reorganized to form new elements.

In cases where atoms of two or more kinds combine
to form molecules, the resulting material is a compound. An example of a compound is table salt. Table
salt is made up of atoms of sodium that have formed
molecules and atoms of chlorine that have formed
molecules. The molecules of the two materials combine to form sodium chloride, or common table salt.
See Figure 1–3.

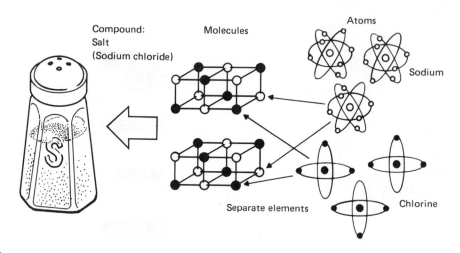

Compound:
Salt
(Sodium chloride)

Molecules

Atoms

Sodium

Separate elements

Chlorine

FIGURE 1-3 _____

Compound. Atoms of different elements combine to form compounds.

By definition, then, an element is a material in which all of the molecules are the same, and the material cannot be divided into particles smaller than an atom without the material's losing its identity as an element. A compound is a material made up of molecules of two or more elements; the material cannot be subdivided into particles smaller than molecules without losing its identity as a compound. In other words, salt can be divided into the elements of sodium and chlorine, but the result will no longer be salt.

**Element***

**Compound***

**Cohesion***

**Heat***

## 1.3  DEFINITION OF HEAT

In the composition of matter, molecules combine in lattice-like arrangements. These molecules are held together by mutual attraction, friction, and pressure, which we call **cohesion.** Normally we think of matter as being rigid in form and usually solid, which would lead us to think that the arrangement of molecules is static or unmoving. We would be wrong in our belief, however. Although the molecules are united, they move a great amount. Even iron and steel, which feel so solid, are made up of molecules that are constantly in motion. The molecules have enough cohesion to keep the materials within certain boundaries that define the shape of the material, but the molecules do move within the material.

The amount of motion of the molecules indicates the amount of heat in the material. When any material is heated, the relative motion of the molecules increases. Thus the amount of motion of the molecules indicates the amount of heat that has been added. See Figure 1–4.

As an example, when heat is added to ice, the ice changes into water. When more heat is added, the water changes to steam. Here we observe the change from solid to liquid and to gas. The change is brought about by the addition of heat, which increases the amount of motion of the molecules. See Figure 1–5 for an illustration of this law of physics.

**Heat** can be defined as the measurement of the amount of motion of the molecules in matter. This definition is not a very practical one for application, but it does help to explain what is observed.

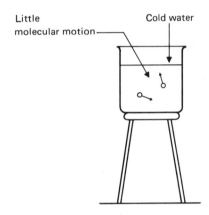

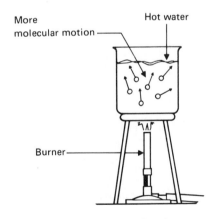

FIGURE 1-4 _____

Molecular motion. Relative motion of molecules in a material indicates the amount of heat in the material.

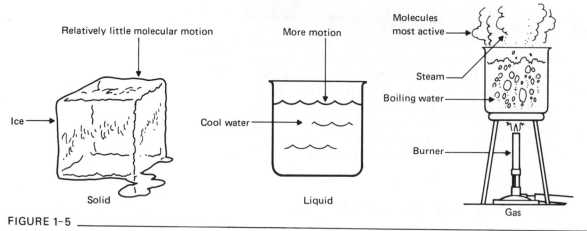

FIGURE 1-5

Temperature and state of a material. Both are related to the speed at which the molecules are moving.

## Heat and State of Matter

Matter exists in one or more of three forms or states: solid, liquid, or gas.

In the solid **state,** the motion of the molecules is somewhat limited. The molecules will stay within certain boundaries without special containment. An example of a substance in the solid state is ice, the solid form of water.

In the liquid state, the motion of the molecules has increased, and they will no longer stay within definite boundaries unless they are in a container that limits their motion. In a liquid, gravity exerts enough force downward on the molecules to prevent them from moving up, with the exception of those that escape from the surface by evaporation. A container with sides and bottom is necessary to prevent the molecules from moving down or out to the sides. Figure 1–6 illustrates molecular action in a liquid. Water in a glass or pitcher is a good example of the behavior of molecules when a material is in a liquid state.

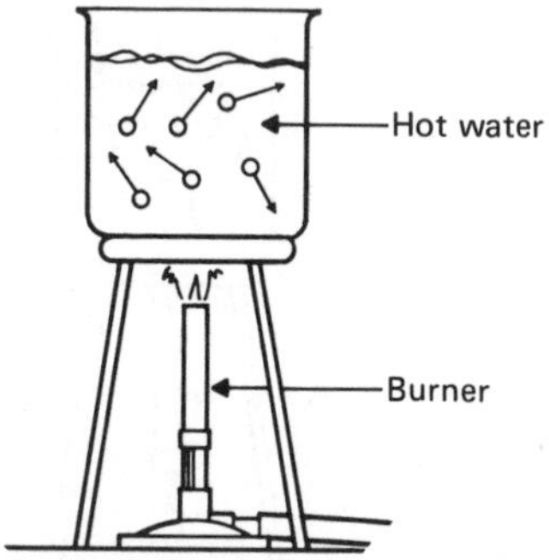

In the gaseous state, molecules are moving so rapidly that they have a tendency to dissipate in all directions. Consequently a gas will fill all the space in any container in which it is placed. Steam from boiling water is a good example of a gas. The steam will fill the space over boiling water, such as the top of a tea-

FIGURE 1-6

Molecular action in a liquid. The molecules have enough speed to allow the material to move out and down, but not enough to break surface film.

kettle, and will exert equal pressure on all surfaces of the container, including the surface of the water.

The speed at which the molecules in any material move is increased when heat is applied. Consequently, heat added to ice changes the ice to water; more heat added to the water changes the water to steam. The change has been from solid to liquid to gas. The application of heat is necessary to increase the speed of the molecular motion. Conversely, the relative motion of the molecules indicates the heat content of the material. Figure 1–4 illustrates this principle.

At normal atmospheric pressure, water can stay in the solid state only when its temperature is 32°F or lower. If heat is added to raise the temperature above 32°F, the ice melts, or changes into the liquid state. The heat involved in this change is called **latent heat.** Heat that causes a change of state without a change in the temperature that can be measured with a thermometer is latent heat. Water between 32°F and 212°F remains in the liquid state. Heat that changes the measurable temperature of a material but does not change the state is called **sensible heat.** This term is used because the temperature difference can be sensed. It can also be measured with a thermometer. If the temperature is increased above 212°F, the water turns into steam, the gaseous state. Each increase in temperature is represented by a definite amount of heat that has been added to the water. Figure 1–7 is a graph showing the heat-temperature relationship for water in the three states.

## Transfer of Heat

To use heat as a means of controlling the temperature in a space, we normally have to move the heat from the point of generation to a point of utilization. To move the heat, we take advantage of one or more of the methods of heat transfer:

1. Conduction,
2. Convection,
3. Radiation.

Heat can be, and is, moved by all three methods.

One of the first laws of **thermodynamics** states that heat always flows from a warm body to a cooler one, *never* in the opposite direction.

**State**    Form, or physical structure, of a material, such as solid, liquid, or gas

**Latent heat***

**Sensible heat***

**Thermodynamics**    The science of heat and its related phenomena

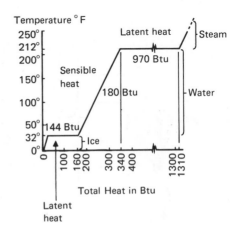

FIGURE 1-7 _____

Graph representing water in the three states and heat related to the temperature for 1 lb

**Conduction***

**Convection***

**Radiation***

**Temperature**    A term related to the intensity of heat  as measured by a thermometer

**Thermometer***

When heat is transferred by **conduction,** heat flows from one molecule to another in the material being heated. An example is a bar of iron with one end held in a flame. The end of the bar that is in the flame gets very hot; the heat travels along the bar by conduction; and gradually the other end of the bar gets hot also. Conducted heat is not transferred instantaneously because it takes time for each molecule to be heated in turn and then pass the heat along to the next molecule. Figure 1–8 illustrates another good example. Heat from a stove burner is conducted through the bottom of a pan to boil water in the pan.

When heat is transferred by **convection,** the heat is carried by the molecules in the material that is heated. An example is the heat carried by the air from a heating coil. The molecules of air carry the heat. Another example is heat carried by water that is heated and then pumped through a piping system to remote locations. In these examples, heat is being carried by the air or water. See Figure 1–9.

The third method of heat transfer, **radiation,** depends on heat waves or rays to transfer the heat. Heat waves function similarly to light waves. If two bodies of different temperatures are reasonably close to each other, heat will flow from the warmer to the cooler through any intervening space by heat waves. An example is the heat we feel from an open fire. If we stand close to a fire and feel warm on the side facing the fire, our other side will still feel cold. Radiant heat can act

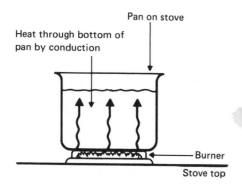

FIGURE 1-8 _____

Conduction. Heat is transferred through bottom of pan.

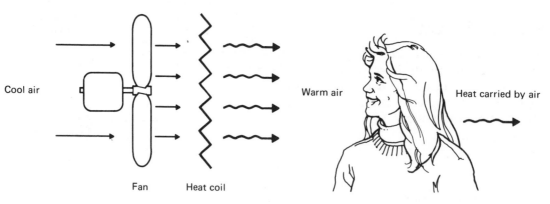

FIGURE 1-9 _____

Convection. Heat is transferred as molecules of air move.

only in direct line of sight. Any intervening body will intercept the waves and be heated by them, but the heat will not go beyond that body. Radiant heat does not warm the air directly; it warms only the objects that the waves themselves reach. Figure 1–10 shows an example of radiant heat transfer.

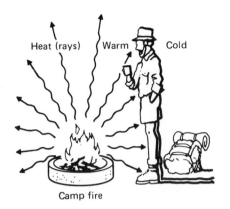

FIGURE 1-10 _____
Radiant heat transfer. Heat is carried by rays of energy.

## Measurement of Heat

It is important to keep in mind that heat is a quantity. One of the most common terms we use in relation to heat is **temperature**, which is a term related to the intensity of heat and only indirectly related to the quantity of heat. To calculate heat as a quantity, we first measure temperature.

**Temperature.**    Temperature is a measurement of the intensity of heat. The difference between temperature and heat is important and needs to be understood.

When the temperature of a room is 76°F, this reading indicates the intensity of heat in the room. If more heat is introduced into the room by the heating system, the temperature will increase, and if heat is lost from the room, the temperature will decrease. In these cases, the temperature of the air is related to the quantity of heat involved; the temperature is not a direct measurement of the quantity of heat.

**Thermometers.**    The most common device for measuring temperature is the **thermometer.** The operation of most thermometers is based upon the principle that materials expand as they are heated and contract as they cool.

The standard glass thermometer is a closed glass tube with a small-diameter bore and a glass bulb at the bottom. Figure 1–11 shows a typical glass thermometer.

The bulb and some portion of the tube contain a liquid with a low freezing point, such as alcohol or mercury. As the temperature surrounding the glass bulb increases, the alcohol or mercury expands, and the column of liquid moves up in the stem of the thermometer. If the temperature goes down, the alcohol or mercury contracts, and the column goes down because the volume of the liquid is reduced. The glass tube is

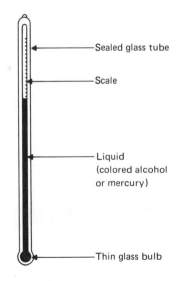

FIGURE 1-11 _____
Standard glass thermometer

**Fahrenheit***

**Celsius***

marked in equally spaced gradations over a certain length called the scale. The intensity of the heat surrounding the thermometer can be read by referring to the movement of the column on the particular scale being used.

Many types of thermometers are available, including direct-reading dial types that are actuated by bimetal springs and remote-reading dial types that are actuated by thermocouples. Figure 1–12 shows some examples.

Special thermometers are available to measure extremely high temperatures, such as those found in flames, and to measure low temperatures, such as those found in refrigeration systems.

**Temperature Scales.**    The **Fahrenheit** scale has been the standard scale for measuring temperature in the United States. This scale was originally established by using the temperature of an ice-salt mixture as 0°F, and the temperature of boiling water at sea level as 212°F. This scale places the temperature of freezing water at 32°F. Because the scale does not readily relate to any of the other systems of measurement that we use, it is rather clumsy in calculations.

The other common scale used on thermometers is called the **Celsius.** The Celsius scale is used in the metric system and is much easier to use for calculations than the Fahrenheit. The Celsius is the common scale used for temperature measurements in all countries of the world with the exception of the United States and has even been used here for scientific and engineering work. The Celsius scale is now being adopted in the United States along with the other metric values.

In the Celsius scale, the freezing point of water at

Dial-Type of Thermometer

Pocket-Type Glass Thermometer

FIGURE 1–12 _____

Two types of thermometers commonly used by heating technicians

sea-level pressure is 0°C, and the boiling-point temperature at the same pressure is 100°C. See Figure 1–13 for an illustration of the two common temperature scales.

In solving application problems, it is often necessary to convert Fahrenheit temperature readings to Celsius, and vice-versa. To do this conversion, the relationship of the Fahrenheit scale to the Celsius scale can be stated as a mathematical ratio. On the Fahrenheit scale, 180° separates the freezing point of water (32°) and the boiling point of water (212°), as found by subtracting 32° from 212°:

$$212° - 32° = 180°$$

On the Celsius scale, 100° separates the freezing point of water (0°) and the boiling point (100°), as found by subtracting 0° from 100°:

$$100° - 0° = 100°$$

The ratio of the Fahrenheit scale to the Celsius scale, then, is 180:100, or 1.8:1. This ratio gives a factor of 1.8, which is used in the conversion process. The 32° difference between the freezing point temperatures on the two scales also has to be considered.

To convert a Fahrenheit temperature reading to a Celsius, the difference between the freezing points (32°) is subtracted from the Fahrenheit reading, and the remainder is divided by 1.8 (the ratio between the two scales), as follows:

$$C° = \frac{F° - 32°}{1.8}$$

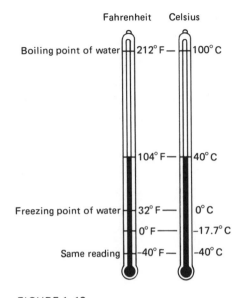

FIGURE 1-13 _____

Fahrenheit and Celsius thermometers

## EXAMPLE 1–1

If the temperature of the air leaving an air-conditioning coil is 59°F, what is the Celsius temperature?

**Solution**

$$C° = \frac{F° - 32°}{1.8}$$

so

$$C° = \frac{59° - 32°}{1.8} = 15°C$$

**Absolute zero***

**Rankine**    A temperature scale related to the Fahrenheit scale in which absolute zero (0°R) is 460° below 0°F

**Kelvin**    A temperature scale related to the Celsius scale in which absolute zero (0°K) is −273°C

To convert a Celsius temperature, multiply by 1.8 and add 32° to the product:

$$F° = (C° \times 1.8) + 32°$$

# EXAMPLE 1–2

If the temperature of the air leaving a furnace is measured at 73°C, what is the Fahrenheit temperature?

## Solution

$$F° = (C° \times 1.8) + 32°$$
$$= (73° \times 1.8) + 32° = 163.4°F$$

Technicians need to be familiar with two other temperature scales. The scales are seldom used in practical application, but are used in many calculations concerning air and temperature. Both of the scales in these systems are related to **absolute zero,** the theoretical point at which no heat exists in a substance. The names of the scales are the **Rankine** and the **Kelvin,** each named after a famous scientist. The Rankine scale is related to the Fahrenheit scale and the Kelvin to the Celsius. Figure 1–14 compares the four temperature scales.

Absolute zero can be computed. It is found to be 460° below 0°F and 273° below 0°C. Since Celsius and Fahrenheit scales have different bases, absolute zero is numerically different for each of the scales, but the numbers relate to the same actual temperature. No thermometer can read as low as absolute zero, but the temperature can be calculated.

Since there is a 460° difference between the zero points on the Fahrenheit scale and the Rankine scale, to change a Fahrenheit reading to a Rankine, we simply add 460° to the Fahrenheit reading as follows:

$$R° = F° + 460°$$

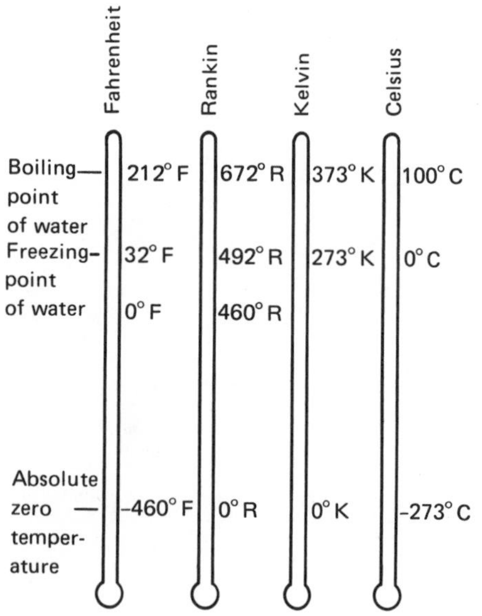

|  | Fahrenheit | Rankin | Kelvin | Celsius |
|---|---|---|---|---|
| Boiling point of water | 212°F | 672°R | 373°K | 100°C |
| Freezing point of water | 32°F | 492°R | 273°K | 0°C |
|  | 0°F | 460°R |  |  |
| Absolute zero temperature | −460°F | 0°R | 0°K | −273°C |

FIGURE 1-14 _____
Comparison of the four temperature scales

# EXAMPLE 1–3

If a Fahrenheit temperature reading is 73°F, what is the absolute reading on the Rankine scale?

**Solution**

Add 460° to 73°. The sum is the Rankine reading:

$$R° = F° + 460°$$
$$= 73° + 460° = 533°R$$

Since there is 273° difference between the zero points on the Celsius and the Kelvin scales, add 273° to a Celsius reading to change it to a Kelvin reading as follows:

$$K° = C° + 273°$$

## EXAMPLE 1–4

If a Celsius temperature reading is 22.8°C, what is the Kelvin temperature reading?

**Solution**

Add 273° to the 22.8°C. The sum of 295.8°K is the answer:

$$K° = C° + 273°$$
$$= 22.8° + 273° = 295.8°K$$

Although many types of thermometers are available, each type is calibrated in either the Fahrenheit or the Celsius scale. Each may serve one particular purpose better than another, but each does the same thing: measures the intensity of heat.

## Heat Quantity

Temperature is only one of the variables that we are concerned with in a discussion about heat. Temperature is a measure of the intensity of heat; more often, we are concerned with the quantity of heat.

To measure quantity of heat, we must consider the temperature plus the mass or volume of the material involved in the temperature change. For instance, a teakettle of boiling water at sea level pressure that holds 1 gallon (gal) has a temperature of 212°F. Because the volume of water and steam is not large, a large

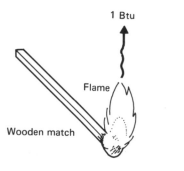

FIGURE 1-15 _____

British thermal units. When a wooden kitchen match is burned, approximately 1 Btu of heat is produced.

TABLE 1-1 _____

Specific Heat of Some Common Materials

| Material | Specific heat (Btu/lb/°F) |
|---|---|
| Aluminum | 0.22 |
| Carbon | 0.17 |
| Concrete, stone | 0.19 |
| Copper | 0.09 |
| Glass | 0.20 |
| Ice, 32°F | 0.49 |
| Silver | 0.06 |
| Water | 1.00 |
| Air, dry, 50°F | 0.24 |
| Steam, 220°F | 0.47 |
| Beef carcass | |
|   Above freezing | 0.68 |
|   Below freezing | 0.48 |

quantity of heat will not be involved. A large steam boiler that drives a turbine to generate electricity, however, may have hundreds of gallons of water. And if the temperature of the water and steam in the boiler is 212°F, there is a great amount of heat represented because of the large volume of water and steam. More heat is needed to raise the temperature of the water in the large boiler to the boiling point than to raise the temperature of the water in the small teakettle to the boiling point.

The term used to describe quantity of heat is **British thermal unit** (Btu). The definition of a Btu is the amount of heat required to raise the temperature of 1 pound (lb) water 1°F, or to be more precise, from 59°F to 60°F. A Btu is approximately the amount of heat released when a common wooden match is burned. See Figure 1–15.

**Specific heat** refers to the quantity of heat contained in a material. Specific heat is the amount of heat contained in the material as measured by temperature difference. Specific heat is always related to 1°F and is compared to water as a standard. The specific heat of water is 1.0 since 1 Btu is needed to raise the temperature of 1 lb of water 1°F. Each material has its own specific heat. The specific heat of air is 0.24 since 0.24 Btu is needed to raise the temperature of 1 lb of air 1°F. See Table 1–1 for a comparison of the specific heat of some common materials.

In the metric system, the term **calorie** is used for heat. A calorie is described as the amount of heat required to raise the temperature of 1 gram (g) of water 1°C. In this text, the Fahrenheit scale and the term Btu will be used.

Both mass and temperature are needed to measure heat. If the weight of any material is known, and that material undergoes a temperature change, the amount of heat involved in the process can be calculated mathematically.

The formula used for calculating the amount of heat in a process is:

$$Btu = ms(T_1 - T_2)$$

where:

Btu = the total heat in the process

$m$ = the weight of the material

$s$ = the specific heat of the material

$(T_1 - T_2)$ = the temperature difference of the process

To find the amount of heat involved in a process, multiply the weight of the material times the specific heat of the material times the temperature difference. The result will be the amount of heat in Btu.

**British thermal unit***

**Specific heat**    A ratio of the amount of heat required to raise the temperature of 1 lb of a substance 1°F compared to the amount required for water

**Calorie***

## EXAMPLE 1–5

If 125 lb of water is at room temperature, which is 60°F, and you want to raise the temperature to 140°F, how much heat will be required in Btu?

### Solution

Multiply 125 lb times the specific heat of water (1.0) times the temperature difference (140° − 60°):

$$\text{Btu} = ms(T_1 - T_2)$$

$$= 125 \text{ lb} \times 1.0 \times (140° - 60°) = 10,000 \text{ Btu}$$

To find the amount of heat involved if a quantity of air changes temperature, multiply the weight of the air times the specific heat of air times the temperature difference.

## EXAMPLE 1–6

If 200 lb of air is brought into a building through the fresh-air system when the temperature outside is 36°F, and it is necessary to heat the air to 75° (the inside temperature in the building), how much heat is required?

### Solution

$$\text{Btu} = ms(T_1 - T_2)$$

$$= 200 \text{ lb} \times 0.24 \text{ Btu/lb} \times (75° - 36°)$$

$$= 1872 \text{ Btu}$$

The amount of heat required to raise the temperature of the air from 36°F to 75°F is 1872 Btu.

## 1.4  SUMMARY

Heat is an important subject for heating, ventilating, and air-conditioning technicians to study because heat is the most important consideration in providing comfort in a building.

To understand how heat is produced, controlled, and applied in heating a building, it is necessary to understand how materials are formed, how heat is affected by that structure, and how heat affects the structure of materials. The molecular theory of matter makes possible an understanding of the phenomena we observe when dealing with heat. Heat, temperature, and change of state can all be explained by the molecular theory of matter and heat.

All materials are made up of molecules, small particles of matter, that are in motion. The relative speed of that motion depends upon the amount of heat in the material. The amount of heat determines the temperature of the material, and the speed of the molecules determines whether the material is in the solid, liquid, or gaseous state.

Heat is a term related to quantity, measured in British thermal units (Btu). Temperature is a term related to intensity, measured in degrees Fahrenheit or Celsius. The temperature in the theoretical condition in which no heat exists in a material is called absolute zero. Scales have been developed relative to this value for use in many of the calculations made concerning air. The Rankine scale relates the Fahrenheit scale to absolute zero, in which 0°R equals −460°F. The Kelvin scale relates the Celsius scale to absolute zero, in which 0°K equals −273°C.

## 1.5  QUESTIONS

1. All matter is made up of _____ or combinations of them.

2. When atoms combine under the right conditions, they form _____ , which in turn combine to form matter.

3. When two or more elements combine, a _____ is formed.

4. The amount of _____ in a material is an indication of the relative motion of the _____ in the material.

5. It requires the addition or removal of _____ to cause a material to change state, as indicated by relative molecular motion.

6. A material in the state in which it retains its shape without any containment is called a _____ .

7. A material that has to be contained on all sides is in the _____ state.

8. A material that has to be contained on all sides except the top is in the _____ state.

9. A change of state is always accompanied by a change of temperature. True or false?

10. Match the words in the first column with the description found in the second column.

| 1. Conduction | a. Transfers by energy waves |
| 2. Convection | b. Transfers from one molecule to another |
| 3. Radiation | c. Molecules carry the heat |

11. Radiant energy heats the air in a space directly. True or false?

12. Temperature is a measurement of _____ of heat, as read on a thermometer scale.

13. A temperature scale that reads the freezing point of water as 32°F is called a _____ scale.

14. The Celsius scale reads the boiling point of water as _____ °C and the freezing point as _____ °C.

15. Absolute zero is a theoretical temperature at which no heat will exist in a substance. Give the temperature readings that would be indicated at absolute zero on each of the following scales: (a) Fahrenheit; (b) Rankine; (c) Celsius; and (d) Kelvin.

16. If the temperature in a room as measured by a Fahrenheit thermometer was found to be 73°F, what would the Celsius temperature be in that room?

17. If an outside temperature was read on a Celsius scale to be 6°C, what would the Fahrenheit temperature be?

18. British thermal units (Btu) are used to indicate heat _____ .

19. The specific heat of materials is always given as a ratio. What is the material used as the standard for the ratio?

20. The specific heat of standard air is 0.24. That means that it requires 0.24 _BTU_ of heat to raise the temperature of one _lb_ of air 1°F.

21. In the formula, Btu = $ms(T_1 - T_2)$, define each of the terms: (a) Btu; (b) $m$; (c) $s$; and (d) $(T_1 - T_2)$.
     _mass_                              _Temp._

22. How much heat would be required to raise the temperature of 12 lb of water from 36°F to 48°F?

23. How much heat in Btu would be required to heat 561 lb of air from 72°F to 158°F?

## 2.1 INTRODUCTION

Heat in a building can be produced by two traditional methods and by several alternative methods. The first of the traditional methods is combustion of fuel under controlled conditions. The second is the use of electricity through a resistance heater. These two traditional methods are covered in this chapter.

The energy shortage and subsequent high cost of fuel have created a demand for new technology and better use of old technology in the production of heat. Some of the more common alternative methods being used in response to this demand are covered briefly in this chapter. These alternative methods include solar energy, geothermal heat, and heat pumps.

## 2.2 COMMONLY USED FUELS

The **fuels** most commonly used for the production of heat are natural gas, fuel oil, manufactured gases, and coal. Gas, oil, and coal are similar in chemical makeup, but their physical differences require that they be handled differently as fuels. Heating technicians must understand these differences to be able to understand the procurement, distribution, and use of the fuels.

### Gas Fuels

The two most commonly used gases for fuel are natural gas and manufactured or liquid petroleum gas (LPG). Some of the other gases that are used in special cases are blast furnace gas, casing head gas, oil gas, refinery gas, and sewage gas. Table 2–1 lists properties of some of the fuel gases.

Because the use of many of these gases is so limited, our discussion will deal only with natural and LPGs. An understanding of those two gases will help technicians to understand the basic use of the other gases also.

**Natural Gas.**    Although natural gas is found quite commonly, the origin of it is not actually known. Because natural gas is usually found in relation to petroleum products, scientists think that oil and natural gas

**CHAPTER**

# 2

# Sources of Heat

**Fuel**    Any material that is burned to produce heat

TABLE 2–1 _____

Properties of Some Fuel Gases

| Fuel | Source | Chemical composition | Btu value |
| --- | --- | --- | --- |
| Butane | By-product of refining process; also found at wellhead | $C_4H_{10}$ | 3200–3260 Btu per cu ft |
| Natural gas | Gas wells and found with oil | Varies; mostly $CH_4$, $C_2H_6$, and $C_3H_8$ | 950–1150 Btu per cu ft |
| Propane | By-product of refining process | $C_3H_8$ | 2500 Btu per cu ft |
| Refinery gas | By-product of refining process | Varies; mostly propane and butane | 1200–2000 Btu per cu ft |
| Sewage gas | Sewage disposal plants | Varies | 600–700 Btu per cu ft |

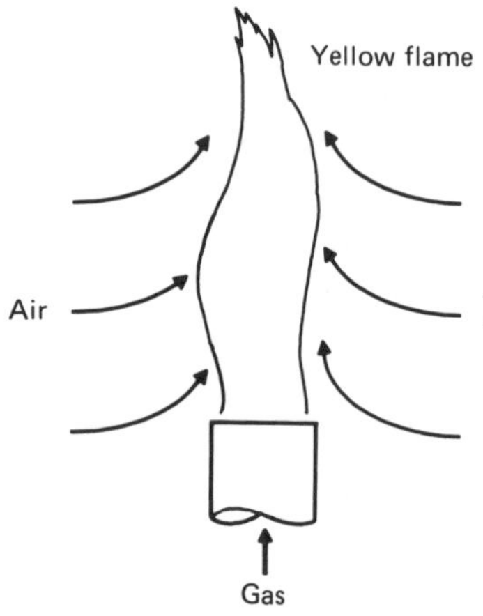

FIGURE 2-1 _____

Gas burner used for lighting purposes

have a common source. Both oil and natural gas are considered to be the fossil remains of plant and animal materials.

Natural gas has been used for lighting and heating purposes for thousands of years, but only within the last century has its use been common. The ancient Chinese used natural gas from shallow wells to heat seawater for the reclaiming of salt. The original development of a natural gas well for commercial production in the United States occurred in the early 1800s in New York State. Inhabitants of the area of Fredonia, New York, were intrigued by a spring from which gas flowed that could be ignited. A young gunsmith saw the possibilities of the gas, dug a shallow well to recover it, and collected and used it for lighting and heating. From this beginning, the first commercial gas company in the United States was organized.

The first common use of natural gas was for lighting purposes. Gas was burned in a simple "fishtail" burner, and the natural incandescence from the flame provided light. Figure 2–1 depicts a typical gas burner used for lighting purposes. The fishtail burner is very simple. Because air for **combustion** does not mix with the gas except in the flame, carbon particles in the fuel burn slowly and glow while they burn. A yellow flame results that produces more light than heat.

Until the development of the electric light in the late 1800s, gas lighting was the principal use for natural gas. As electric light was perfected, the gas industry looked for other uses for its product; hence the development of gas-burning heating units.

The value of natural gas as a heating fuel was recognized early, but its widespread use had to await the development of distribution piping systems. As the distribution systems spread out from the wellheads where the gas was abundant, furnace manufacturers developed gas-burning equipment that would utilize the fuel. Natural gas quickly became the most popular of the heating fuels. One of the reasons is that gas is convenient; it does not have to be delivered by truck or stored on the premises. This factor and its relative abundance make natural gas a good fuel for commercial and industrial use, as well as for domestic heating.

Natural gas to be used for commercial purposes is either recovered at the wellhead of oil wells or produced at wells drilled for the specific purpose of collecting it. In the past, the presence of natural gas with oil was considered a nuisance, and the gas was burned off or wasted at the wellhead. Now natural gas has become valuable in its own right, and it is recovered and used.

Major gas fields are found in many different parts of the world, including the United States. The gas from these fields is distributed by large pipelines owned by major distribution companies. Through the use of interconnected systems and local distribution systems, the gas is available in almost all areas of the United States where the population is great enough to create a market. The development of the distribution system has been the most important feature in making natural gas the most commonly used fuel. Its ready availability and convenience make it popular.

Natural gas is generally distributed by a local utility company. This company is responsible for the installation and maintenance of the pipelines, as well as ensuring a constant supply of the gas. Mains, or large gas lines, bring the gas to the property line of the consumer. The utility brings a service line from the main to the gas meter adjacent to the building where the gas will be used. The installation and maintenance of the piping within the building is the responsibility of the building owner.

Pressure in the gas main is relatively high to help distribute the gas. A **pressure regulator** at the meter

**Combustion**    The process of burning; the release of heat energy through the process of oxidation

**Pressure regulator**    A device to regulate the amount of pressure in a system

**Water gauge (wg)**    A manometer or device for measuring pressure in a duct or piping system; or the pressure measured by a manometer in inches of water

**Manometer**    A device for measuring low pressure in a liquid or air

**Resistance***

**Pressure drop***

**Gasify**    To change a liquid to a gas

reduces that pressure for distribution in the building. Regulators are also used on each appliance to reduce the pressure at the burner. The pressure in the lines is measured in inches, **water gauge** (in. wg). This measurement refers to the height, in inches, of a column of water that the pressure would support. In. wg is measured by an instrument called a **manometer.**

When a gas distribution system is designed for a particular building, the amount of gas needed for each heating unit has to be considered. The gas is measured in cubic feet per hour (cu ft/hr) and is determined by the heating capacity of each unit. Pipe is sized to deliver the cu ft per hour, but the flow rate also depends on the amount of **resistance** to the flow. A certain amount of friction always exists between the gas and the pipe walls and fittings. The accumulated value of this friction is the resistance. The longer the pipe, the greater the resistance. This resistance causes a difference in pressure between the two ends of the piping system. The difference is called **pressure drop.** The pipe is sized so the required amount of gas flows with a minimum amount of pressure drop.

A pipe sizing chart is used to size the pipe properly considering the cu ft of gas required and the pressure drop. The pressure of the gas at the unit is critical.

Natural gas is colorless and odorless. The chemical composition varies with its source, but methane ($CH_4O$) is always a major constituent. Since methane is made up of hydrogen and carbon, the fuel is called a hydrocarbon, as are all of the other common fuels. Most natural gas contains some ethane ($C_2H_6$) and a small amount of nitrogen. The heating value of natural gas averages about 1000 Btus per cu ft of gas. In some cases, the value may be higher, and in some it may be lower.

**Liquefied Petroleum Gas.**    LPGs are propane and butane, or a mixture of the two. Propane and butane are by-product gases of the petroleum industry. Because they are in a liquid state, they are usually used in areas where pipelines are not run or as a reserve source of fuel supply.

LPGs have been used as fuels since the early days of petroleum production. Both propane and butane are

*Term defined in text.

found in the gaseous state, but they will liquefy under moderate pressure. They are liquified and stored and transported in tanks under pressure. The LPGs can then be made available to areas where gas lines do not run. Both propane and butane are used widely as fuels. Butane is also valuable as a source of chemicals for many other processes.

LPGs are transported from the source of supply by tank trucks or train cars especially built for that purpose. At the local distributor level, the LPGs are transferred to smaller tank trucks that carry the liquid fuel to customers. Each time the fuels are transferred from one tank to another, they are pumped in the liquid state so the pressure is maintained. The consumer must also keep the fuel in a tank under pressure. The fuel is not allowed to **gasify** until it enters the burner of the appliance that will burn it. In the appliance, LPG burns in the same way as natural gas.

Both propane and butane are made up of hydrogen and carbon, as are other fuels. Butane has a chemical formula of $C_4H_1O$, with traces of other gases. Propane has a formula of $C_3H_8$. The heat content of the LPGs is considerably higher than that of natural gas. Butane produces 3200 to 3260 Btu per cu ft of gas burned. Propane produces approximately 2500 Btu per cu ft. Table 2–1 provides comparative values.

# Fuel Oil

Fuel oil is a liquid fuel derived from petroleum. Petroleum, or crude oil, comes from the well as a complex mixture of chemicals, mostly hydrocarbons. Refineries separate the mixture into various fuels, such as gasoline, kerosene, and light and heavy fuel oils. The fuel oils themselves are separated into different grades, which are numbered. The most commonly used fuel oil for small heating units is No. 2. The heavier oils, called residual oils, are heavy, tarlike oils and are normally used in stationary power plants and ships.

Oil has been used as a heating and lighting fuel for thousands of years. The original source was from seeps, places where the petroleum came to the surface of the earth naturally. Here the oil formed in pools and could be dipped up in any container. The original oil burners were open containers of oil. The oil burned either on the surface or on a wick that was placed in the oil.

**Combustion air***

**Distillation**    The process of boiling, or changing a liquid to a gas, and then condensing it back to a liquid

The first commercial oil burners were pot burners, open burners in which the oil burned on the surface and **combustion air** mixed with the flame above the point of combustion. Combustion air is air that provides oxygen for the combustion process. This type of burner is effective and is still used in very small heating units, but regulating the amount of fuel burned is not easy, and the burner is not very efficient.

In the mid-1800s, power oil burners were developed that mixed combustion air with the oil and introduced the oil-air mixture into a combustion chamber under pressure. This innovation made it possible to control the burning rate and also produced a much more efficient flame for the combustion of oil. Now, most oil-burning heating units use some type of power burner, and manufacturers are continually making improvements to increase the combustion efficiency of burners. Figure 2–2 shows a modern power oil burner commonly used in residential or small commercial heating units.

Crude oil, the source of fuel oil, is found in pools and layers that lie trapped in the earth's crust. Wells are drilled in the area of the pools, and pumps are used to raise the oil to the surface. The crude oil is then transported to refineries. Most of the oil is transported in pipes, but some is carried by barge, tank car, or tank truck.

All liquid fuels are products, or by-products, of the refining process. The way the petroleum is refined has more to do with the fuel quality than does the source of the crude oil. Refining consists of separating and recombining the hydrocarbons of the crude oil into specialized products like gasoline, kerosene, and fuel oil. The basic process of the refinery is **distillation.** The hydrocarbons are separated into different fractions or groups, having approximately the same range of boiling points. The lighter fractions, called distillates, are represented by gasoline, kerosene, and the lighter fuel oils. The remainder, or residual products, are heavy fuel oils, asphalt, and tars. The light fuel oils are generally used for domestic and light commercial heating. The heavier oils are used for commercial and industrial heating.

In the modern refining process, distillation is only the beginning. Heat and pressure are used to change the hydrocarbon structure of the petroleum so that the

FIGURE 2-2 _____

Power oil burner commonly used in residential or small commercial furnaces crude oil will yield more of the lighter products and proportionately less of the heavier ones. A diagram showing some of the products of refining is found in Figure 2–3.

Fuel oil is stored in large tanks at the refinery until it is taken by tank trucks or tank cars to the distributing companies. For local use, dealers deliver the fuel oil in trucks to customers. The oil is stored in tanks on the site where it will be used. These tanks may be located either above or below ground. The oil is pumped from the tanks to the heating units to be fired. Lighter oils, No. 1 and No. 2, can be pumped without any processing as long as they are stored where the temperature does not fall below 0°F. When heavier fuel oils are used, heaters have to be installed in the tank and around the oil lines to ensure an adequate flow of oil to the burner.

Fuel oil is made up mainly of carbon and hydrogen, with approximately 85% carbon and 15% hydrogen. Fuel oil also contains traces of oxygen, nitrogen, and sulfur. No. 2 fuel oil, the most common grade for residential or light commercial heating, produces approximately 140,000 Btu of heat per gal burned. When No. 2 fuel oil is used in a properly adjusted burner, the combustion efficiency is 80%, or 112,000 Btu per gal; that is, the usable heat value will be 112,000 Btus per gallon of oil burned. Table 2–2 gives some characteristics and the heating value for the various grades of fuel oil.

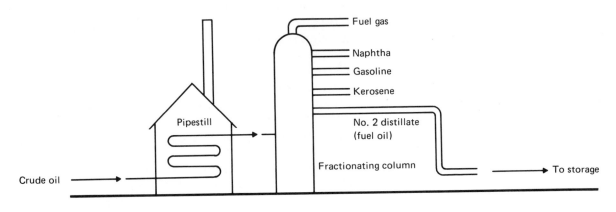

FIGURE 2-3

Some of the products of the refining process. Crude oil is separated into lighter and more volatile products.

TABLE 2–2

Typical Gravity and Heating Values of Standard Grades of Fuel Oil

| Grade no. | Gravity, API | Weight, lb per gallon | Heating value, Btu per gallon |
|---|---|---|---|
| 1 | 38–45 | 6.95–6.675 | 137,000–132,900 |
| 2 | 30–38 | 7.296–6.960 | 141,800–137,000 |
| 4 | 20–28 | 7.787–7.396 | 148,100–143,100 |
| 5L | 17–22 | 7.94–7.686 | 150,000–146,800 |
| 5H | 14–18 | 8.08–7.89 | 152,000–149,400 |
| 6 | 8–15 | 8.448–8.053 | 155,900–151,300 |

Source: Reprinted with permission from American Society of Heating, Refrigerating, and Air-Conditioning Engineers, Inc., *ASHRAE Handbook, 1977 Fundamentals* (New York, 1977)

# Coal

It is not known when coal was first used as a fuel, but remains found indicate that coal has been used for at least 4000 years. The Chinese were using coal some 3000 years ago, and the use of coal as a fuel is mentioned in the Old Testament. Only in the last 200 years has it been used extensively, however.

The English, the first to see the commercial value of coal, used it widely during the 1700s to power the Industrial Revolution. And although coal was recognized in the United States by some of the earliest explorers and settlers, not until the mid-1700s was it developed for commercial use. Mining began in western Pennsylvania in 1759. Important beds of coal were discovered in eastern Pennsylvania in 1791. As the United States industrialized, the use of coal as a fuel became important. By the mid-1800s, the growth of the railroads created a direct market for coal and also enabled the distribution of coal to other places where industry could use it.

From the late 1800s to about 1950, coal was used extensively in industry and as a fuel for heating residences and commercial buildings. With the development of distribution systems for natural gas and electricity, however, the use of coal for residential heating has been almost completely abandoned. Coal continues to be important as a source of heat for commercial and industrial use. And currently, as other sources of energy become scarcer and more expensive, the use of coal is increasing in the United States.

Coal is a fairly abundant mineral in the United States and is mined commercially in many places. It is mined both underground and in open-pit mines. The mining process in the underground mines is relatively expensive and often dangerous. Consequently, the open-pit concept is used whenever the coal bed is near enough to the surface of the earth to make mining feasible. Large quantities of coal can be obtained at lower cost by the open-pit method.

Coal is often shipped directly from the mines to the point of use by direct rail lines or even by conveyors. Such is the case in many of the large generating plants, which have been built in recent years. After the initial cost of the distribution system is absorbed, this type of transportation is less expensive.

One of the greatest problems in the use of coal for fuel is that it burns "dirty." Because coal is a solid fuel, burning it generally leaves undesirable by-products. Although coal is a hydrocarbon and thus similar to natural gas and fuel oil, coal has a much higher percentage of carbon and contains other minerals. When coal is burned, in the solid form, high efficiency is difficult to achieve. Soot, clinkers (hard particles of ash), and other undesirable by-products are produced in the combustion process. These by-products must be removed, a process that increases the cost of using the fuel. In the past, when coal was used commonly as a fuel, the contamination from the smoke and soot was considered a necessary evil. But with the increased recognition of the necessity of maintaining a clean environment, preventing pollution is now considered important. Coal-burning heating equipment can be installed so that the coal burns reasonably clean, but the special equipment and processing of the fuel before burning add to the cost of the process.

The Btu content of coal varies greatly depending upon the exact chemical content of the coal. Carbon makes up the most important constituent of coal, and hydrogen and other chemicals are also found in it. Coal is broken into different grades depending on the chemical composition and the hardness. The "soft" bituminous coals found in the western part of the United States have different burning qualities and Btu content per pound than do the "hard" anthracite coals of the East. The burning properties and the amount of pollution created by the combustion of coal are determined by the hardness and the chemical composition. See Table 2–3 for the chemical makeup and heat value for the various standard grades of coal.

## 2.3   OTHER SOURCES OF HEAT

Fuel-burning equipment traditionally has been the most common source of heat. Other sources are practical and often used too. Electric heat has become as common as gas and oil heat in most parts of the United States. Solar heat and geothermal heat are both being used more commonly.

Electricity is a form of energy rather than a fuel, but when electricity is used for heating, it is often included in descriptions as a fuel. Since about 1950, the use of electricity as a source of heat has increased rapidly. Today electricity is used in new buildings as often as the older forms of fuel. In areas served by large elec-

**TABLE 2–3**

Chemical Composition and Approximate Heating Value of Standard Grades of Coal

| Rank | Btu per Lb | | Constituents, Percent | | | | | |
| | Moist, Mineral-matter-free[a] | Moist, as Received | Oxygen | Hydrogen | Carbon | Nitrogen | Sulfur | Ash |
|---|---|---|---|---|---|---|---|---|
| Anthracite...................... | 14,600 | 12,910 | 5.0 | 2.9 | 80.0 | 0.9 | 0.7 | 10.5 |
| Semi-Anthracite............... | 15,200 | 13,770 | 5.0 | 3.9 | 80.4 | 1.1 | 1.1 | 8.5 |
| Low-Volatile Bituminous....... | 15,350 | 14,340 | 5.0 | 4.7 | 81.7 | 1.4 | 1.2 | 6.0 |
| Medium-Volatile Bituminous.... | 15,200 | 13,840 | 5.0 | 5.0 | 79.0 | 1.4 | 1.5 | 8.1 |
| High-Volatile Bituminous $A$.... | 14,500 | 13,090 | 9.2 | 5.3 | 73.2 | 1.5 | 2.0 | 8.8 |
| High-Volatile Bituminous $B$.... | 13,500 | 12,130 | 13.8 | 5.5 | 68.0 | 1.4 | 2.1 | 9.2 |
| High-Volatile Bituminous $C$..... | 12,000 | 10,750 | 21.0 | 5.8 | 60.6 | 1.1 | 2.1 | 9.4 |
| Sub Bituminous $B$............. | 10,250 | 9,150 | 29.5 | 6.2 | 52.5 | 1.0 | 1.0 | 9.8 |
| Sub Bituminous $C$............. | 9,000 | 8,940 | 35.8 | 6.5 | 46.7 | 0.8 | 0.6 | 9.6 |
| Lignite....................... | 7,500 | 6,900 | 44.0 | 6.9 | 40.1 | 0.7 | 1.0 | 7.3 |

[a] (Btu as received) $\times$ 100 $\div$ (100 $-$ 1.1 Ash)

Source: Reprinted with permission from American Society of Heating, Refrigerating and Air-Conditioning Engineers, Inc., *ASHRAE Handbook, 1977 Fundamentals* (New York, 1977)

trical utilities, calculating the heat loss of buildings in **kilowatt per hour** (kWh), a term related to electric energy, rather than in Btu/h, a term related to heat energy, is now common practice.

Currently there is a great deal of interest in, and much development work being done with, solar energy, geothermal heat, improved heat pumps, reclaimed heat from buildings, and other sources of heat. These sources are being used experimentally, and in some cases practically, for heating. It can be expected that new **technology** related to these sources will be developed, and others found, that may change many of our traditional ways of heating.

## Electrical Energy

The use of electricity as a source of energy for heating has been increasing since the use of electricity became common. With power lines being carried into virtually all populated areas since the early 1900s, the use of electricity for heating has made a great deal of sense. In many areas where the power was reasonably inexpensive because of generation by water power, electricity could be used for heating purposes at less cost than more traditional fuels.

The earliest experimenters recognized that heat could be produced by connecting a wire to an electric battery. Whenever electric current passes through a conductor that has some resistance to the flow of the electricity, heat is produced by that resistance. When the electric light was being developed in the 1880s, the same principle was used to produce light.

Practical electric heating units have been used since the early 1900s. The first were small units that simply heated by radiation off the hot wires. During the 1920s and 1930s, systems were developed in which the wires were embedded in ceilings, floors, or walls of buildings. The surfaces became radiant panels. These systems were not particularly successful because most buildings at that time were not insulated the way they should have been for that kind of heating system.

During the 1950s, electric furnaces similar to gas or oil furnaces were developed, and electric heat could then be used in circulated air systems. The development of electrical furnaces made it possible to provide

**Kilowatt per hour (kWh)**    An electrical term related to the amount of work or heat produced by one kilowatt in one hour's time

**Technology**    The science of the industrial arts

**Electromagnetic generation**    The generation of electricity by the means of moving a conductor through a magnetic field

all of the elements of a comfortable heating system while using the electricity as a source of heat. About the same time, electric baseboard heating and electric wall panel heaters were also developed, and their use became fairly common.

With the use of more and better insulation in buildings, electric radiation systems, with the heat in the ceilings, have become quite common. In this system, wires are embedded in the ceiling finish material, and the entire ceiling surface is heated by the wires, providing a source of radiant heat for the room below. Adequate thermal insulation must be provided above the ceiling. Although this type of heating does not provide some of the requirements of a "total comfort" system, such as circulation of air or filtration, it does have the advantage of being relatively inexpensive to install. Of the buildings that are being heated with electricity today, approximately 60% are heated by radiant systems and 40% by circulated air systems.

Electricity can be generated by several different methods. The only practical method developed so far for producing commercial quantities of electrical power is by using large dynamos for generation. This method is called **electromagnetic generation.** Electromagnetic power generation is the conversion of one kind of energy (mechanical) to another (electrical). The mechanical power to turn the dynamos for the generation of electricity is derived from waterpower, steam power, or any other reasonable and economical power source. Waterpower is usually used for the generation of electricity where such power is available, generally in the western United States. The only drawback is that mechanical energy is needed to turn the rotors on the dynamos, and the cost and energy used to produce the mechanical power can be considerable. Also some power is lost in the conversion because of inefficiencies in the system.

Dams and reservoirs are found along most large rivers in the western United States. Although the water is used for irrigation, the main reason for the construction of the dams is to provide electric power generation. Water-generated power is relatively inexpensive. Where waterpower is not available, dynamos are often driven by steam turbines. The steam is produced by burning gas, coal, or oil to heat the boilers. In some areas, steam from natural geothermal sources is used for generating electricity.

The electric utility companies use interconnected distribution systems to carry the electricity from the point of generation to almost any other point in the United States. Areas so remote they do not have electric power lines are rare. Because of this extensive distribution system, electricity is a convenient energy for heating purposes.

The utility company's distribution system brings the power to the point of use; the owner provides the internal distribution system. Service lines bring the electricity from the power lines to the building. See Figure 2–4. The service lines then go through the meter at the building and to the main service panel where the disconnect and fuses, or circuit breakers, are located. A service panel with circuit breakers and branch circuit connections is shown in Figure 2–5.

At the main service panel, the power is distributed to subpanels, which serve the various branches in the building where the individual loads, such as heating units, are located. The heating system is normally on a separate branch out of a subpanel that is fused and provided with a disconnect for the heating circuit only. Since the electrical power is distributed in the main transmission lines at a high voltage, the power is always transformed down to a lower voltage for distribution to local areas.

The method of transforming determines the exact

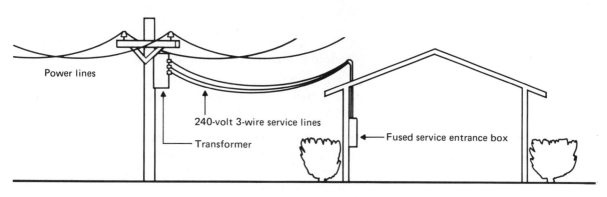

FIGURE 2-4

Service lines bringing in electricity from power lines into a building

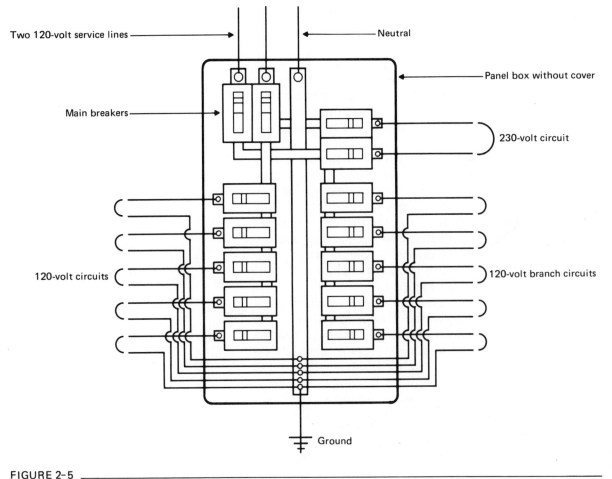

Two 120-volt service lines

Neutral

Panel box without cover

Main breakers

230-volt circuit

120-volt circuits

120-volt branch circuits

Ground

FIGURE 2-5
Service panel used to provide branch circuits, disconnects, and overcurrent protection

voltage and phase that will be provided to a particular building. The most common voltages are:

1. 460 or 480 volts (V), 3 phase,
2. 208, 230, or 240 V, 3 phase,
3. 120 or 240 V, 1 phase.

For heating purposes, voltages of at least 208 or 240 V, either 1 or 3 phase, are required.

When electricity is used for heating purposes, the energy conversion is 100% efficient since there is no loss for flue gases or combustion efficiency. The rated output of the heating equipment is 100% of the electrical energy input. Each kilowatt of electricity produces 3413 Btu/h of heat.

## Alternative Ways to Heat

Because our supply of traditional fuels is limited and because we are rapidly depleting them, a greatly accelerated search for new sources of fuel and energy is now going on. Scientists are looking for ways to tap some of the sources of energy that are obvious in nature but not always easy to use. Among these sources are solar energy, wind power, wave power, and geothermal heat. Not all of these alternatives have been developed to the point where their use is practical, but some are being used to some extent. All of them, plus others, will certainly be developed over time.

**Solar Heat.**    Probably humans have been interested in solar heat since they first started reasoning. It is obvious that there is a great deal of heat energy in the rays that come to the earth from the sun. Although we have used this energy in some ways since before recorded history, only in the last 100 years has research and technology made it possible for us to use much of that available.

There are basically two ways to heat directly with solar energy. The first is called passive. In a passive system, the building itself, or parts of the building, are warmed by the sun's rays, and the interior of the building is then warmed. The other method, an active system, has collector cells that collect the energy from the sun's rays in a medium such as water or air, and the medium is then used to heat the building spaces.

The principles of a passive solar system have been used since people started facing their buildings toward the south to catch the sunlight in the winter. Active systems are a newer technology and are still under development.

A complete description of the different methods of heating with solar energy will not be attempted in this text, but certainly developments of solar systems will continue and the utilization of solar energy for heating purposes will become more common.

Heat pump*

Coefficient of performance (COP)*

**Geothermal Heat.**    The term geothermal refers to the heat found within the earth. Scientists accept the fact that the center of the earth is molten rock. Evidence to support this belief is seen in various places where the heat affects the surface in such manifestations as boiling springs, geysers, and, more dramatically, in volcanoes. Under some circumstances, this heat can be tapped and utilized as a source of energy. Either the hot water that comes to the surface naturally can be used, or wells can be drilled into the earth and pumps used to bring the hot water to the surface. In many places in the world, natural hot water and steam are used as a source of heat energy for both domestic and commercial heating. But the technology is fairly new, and much work is now going on to develop it more fully.

**Heat Pumps.**    A **heat pump** is a mechanical device that moves heat from one place to another. The pump does not actually produce heat but will move the heat from a place where it is not needed to one where it is. Heat pumps make it possible for us to pick up heat from the ground, a water source, or the air and move it into a building that we want to heat. Much less energy is needed to move the heat from one place to another than to produce heat by combustion or by direct use of electricity. For this reason, the heat pump is efficient and economical.

A heat pump is a refrigeration unit. The heat is picked up from the heat source, or heat sink, by a refrigerant. The pressure and temperature of the refrigerant are raised, and then the heat is given up at the place where it is needed. The actual commercial development of the heat pump as a heating device goes back to the early 1900s, but the use of it as a common heating unit dates back only to about 1950.

Most heat pumps are driven by electrical motors. But because of the principle upon which the pumps operate, the amount of heat that they put out is much more than that produced by the same amount of electricity converted directly into heat through resistance heaters. A **coefficient of performance** (COP) of 3.5 to 1 is common. The COP is the ratio of the amount of energy used to the amount of energy made available as heat.

**Wood Heating.**    As a result of the energy and fuel shortage and consequent high prices, many people are using wood as a fuel to heat their homes. This use is not really new, since wood is probably the oldest fuel used by people, but it is new to most people in this generation.

Wood is generally less expensive to purchase than other fuels, but in most cases, the equipment used for burning it is so inefficient that much energy is wasted. Wood is usually burned in a stove or space heater that heats one area of a building. It can be burned in a central furnace if the furnace is designed for that purpose. Wood should be burned in an air-tight unit designed to give maximum efficiency.

The actual heat value of wood as a fuel varies greatly depending upon the type of wood used, its density, and how dry it is, but the efficiency of the heating unit really determines heat output. Table A–1 in the Appendix gives the weight and heat value for some typical woods used for fuel in the United States.

## 2.4  SUMMARY

Traditionally the most common way of producing heat for comfort has been to use a fuel that can be burned. In the burning or combustion process, heat energy is released from the fuel. The most common of fuels, listed in the order of their development and use, is as follows: wood, coal, fuel oil, and natural gas. In recent years, electricity, which is not a fuel but a form of energy, has become important as a source of heat.

The use of the various fuels has progressed from the simple to the more complex as society's needs, and the technology to use the fuels, has changed. Wood, the first fuel used, could be burned in an open fire, but the fuels used today require complex and sophisticated equipment. The efficiency of the equipment is much better than in the past.

At the present time, we are seeing the development of equipment and methods of using our traditional fuels more efficiently and also the development of new sources of energy for the production of heat. Heating technologists today should understand the traditional fuels and their use and also keep aware of new developments that are made in the heating industry.

## 2.5  QUESTIONS

1. Provide a definition for *fuel*.

2. The four most common fuels in use today are: (a) by-product gases, (b) coal, (c) natural gas, and (d) fuel oil. List them in the order of their importance as fuels.

3. Name an important source of heat that is technically not a fuel but an energy and is often used today.

4. Match the fuel or energy in the first column with the Btu values in the second column.

   1. Natural gas        a. 3200 Btu
   2. Fuel oil           b. 2500 Btu
   3. Butane             c. 3413 Btu
   4. Propane            d. 140,000 Btu
   5. Electricity        e. 1000 Btu

5. Electromagnetic generation of electricity is always produced by water turning a generator. True or false?

6. The two main methods of collecting solar energy for heating purposes are called _____ and _____ systems.

7. A heat pump is a _____ system, used to move heat from a place where it is not needed to a place where it is wanted.

8. Wood makes a good fuel, but most of the equipment used to burn it is not very efficient. True or false?

9. Heating technologists today should not be interested in new developments because all of the ways to produce heat have already been developed. True or false?

## 3.1 INTRODUCTION

During combustion, chemicals in the fuel combine with oxygen from the air. In the process, new chemical compounds are formed, and heat is given off. We are interested in the heat, but analysis of the process is also important. This chapter covers the amount of heat that is given off during combustion, the products that are formed during combustion, and the requirements that make the combustion process safe.

## 3.2 COMBUSTION PROCESS

Combustion, the process of burning, is the release of heat energy from fuel through the process of oxidation. Combustion takes place when chemicals in the fuel combine with oxygen. The combustion process is the same process as the rusting of metal, but rusting is generally slow, and the heat produced is not noticeable. Burning a fuel is a relatively rapid process with a considerable amount of fuel and oxygen involved, so the amount of heat produced is greater.

In order for combustion to take place, three elements must be present during the process. They are:

1. Fuel,
2. Oxygen,
3. Heat.

See Figure 3–1.

The **chemical elements** in the fuel represent a form of energy. Energy can be neither destroyed nor created, but it can be changed into other forms. This principle is the law of conservation of energy. In the combustion process, this energy is changed into heat and light, which are also forms of energy.

## Fuels and Combustion

Fuels are composed of hydrogen atoms and carbon atoms in different combinations according to the particular fuel being considered. When a fuel goes through the combustion process, the atoms of carbon and hydrogen in the fuel combine with oxygen atoms from the air to form new elements, and heat is given off in

# CHAPTER
# 3
# Combustion

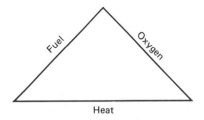

FIGURE 3-1 _____

Combustion triangle representing the three necessary elements for combustion to take place

**Chemical elements**    Basic chemical components of a fuel

**Kindling temperature**    The temperature at which a fuel and oxygen mix will begin to burn

the process. Methane ($CH_4$), which is most common in gaseous fuels, has a chemical composition of four hydrogen atoms to one carbon. When methane combines with oxygen at a high enough temperature (called the **kindling temperature**) to start a reaction, the carbon atom combines with two atoms of oxygen to form carbon dioxide ($CO_2$), and two atoms of hydrogen combine with one atom of oxygen to form water ($H_2O$). Figure 3–2 shows how the chemicals in methane and oxygen combine in the combustion process and the products that result.

Fuels with a chemical composition different from methane produce slightly different results in their products of combustion, but since hydrogen and carbon are the main chemical components of all fuels, the results are similar.

Heating technicians do not need a complete understanding of the chemical processes in combustion, but they need a knowledge of the basic process, since flue gas analysis, which measures the chemicals in the products of combustion, is one method they use to measure the efficiency of the process.

## Oxygen for Combustion

The oxygen for the combustion process is obtained from air that is mixed with the fuel, or combines with the fuel in the flame, during the combustion process. Normal air, as we know it in our atmosphere, is made

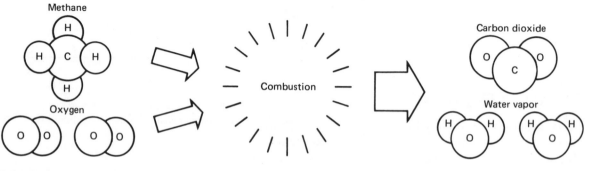

FIGURE 3-2

Combustion of methane and oxygen. In the combustion process, the elements in the fuel combine with oxygen to produce heat and other combinations of the same elements.

up of 76% nitrogen and 23% oxygen. There are traces of other elements in the air, but they do not affect the combustion. The nitrogen passes through the combustion process unchanged, and only the oxygen combines with the other elements in the fuel. The diagram in Figure 3–3 shows how the nitrogen passes through the process.

## Heat for Combustion

In a typical burner, the fuel and air are mixed in the proper proportions to form a combustible mix. The final requirement for combustion to take place is that a kindling temperature be reached. The chemical reaction of the oxygen in the air and the hydrocarbons in the fuel will take place only if the temperature is high enough to start the process. To start combustion, a spark or pilot flame is used. After the flame is established, the heat from the fire itself is great enough to keep the process going. If at any time the temperature of the flame is reduced below the kindling temperature, combustion ceases and the flame goes out. The ignition

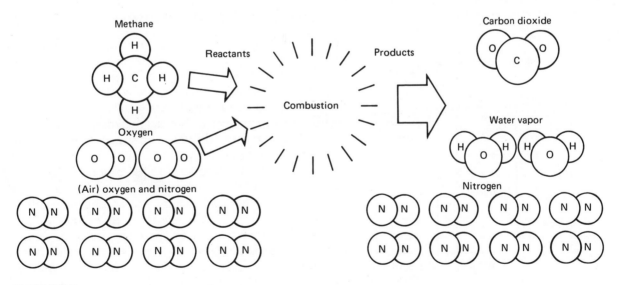

FIGURE 3-3 _____

In the combustion process air is used to supply oxygen. Air also contains nitrogen, but this passes through the process unchanged. The products of combustion contain the same amount of nitrogen as the air used in the process.

**Products of combustion**    The chemical elements and compounds that are formed in the combustion process

**Complete combustion\***

**Incomplete combustion\***

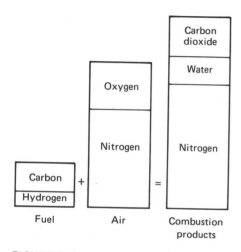

FIGURE 3-4 _____

Products of complete combustion, none of which are harmful to people

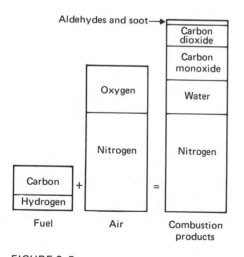

FIGURE 3-5 _____

Products of incomplete combustion, some of which are harmful to people

devices that generate the heat for kindling the fire are discussed in Chapters 7, 8, and 11.

## 3.3  PRODUCTS OF COMBUSTION

Whenever a fuel is burned completely, the **products of combustion** are harmless. But if not enough air is present during the combustion process, some of the resulting products will be harmful.

## Complete Combustion

When sufficient air is present during the combustion process, the chemical reaction is such that the carbon and the hydrogen in the fuel combine completely with the oxygen in the air. This process is called **complete combustion.** The new compounds, or combinations of elements formed, are products that are harmless to humans and the atmosphere. The main products are water ($H_2O$) and carbon dioxide ($CO_2$). Figure 3–4 shows these products of complete combustion in relation to the fuel and air.

## Incomplete Combustion

When not enough air is present during the combustion process to react with the hydrogen and carbon completely, products that are harmful to people are formed. This process is called **incomplete combustion.** During the process, some of the oxygen and the hydrogen still combine to form water, but as the remaining oxygen and carbon combine, there is a shortage of oxygen. The resulting compound formed is carbon monoxide (CO) instead of carbon dioxide ($CO_2$). Carbon dioxide is a harmless gas; carbon monoxide is a deadly poisonous gas. Other products of incomplete combustion are aldehydes and soot. Aldehydes are toxic chemical compounds that are irritating to the eyes and nose. Soot is unburned carbon. Figure 3–5 shows the products of incomplete combustion in relation to the fuel and air.

Incomplete combustion also occurs when a flame

\*Term defined in text.

is cooled below its ignition temperature—for example, when the flame **impinges,** or strikes, a cool metal surface. Unburned particles of carbon in the flame stick to the metal surface as soot. Anytime soot is observed in a flame, incomplete combustion is taking place. Figure 3–6 illustrates how soot is formed on a pan bottom. Incomplete combustion can occur inside a furnace unit.

## 3.4  COMBUSTION AIR

Burners in heating equipment are designed to mix air and fuel in the proper proportions for complete combustion. The designer of the heating system must make sure that a sufficient amount of air can get to the equipment for combustion; the heating technician who sets up the unit must adjust the burner for the proper amount of air.

### Ignition Limits

Combustion can take place only when the **air-fuel mix** is "right." A rather precise amount of oxygen per part of fuel is needed to form the combustion mix. Figure 3–7 shows that for natural gas, the mix has to be between 4% and 14% gas, and the rest air, or combustion will not take place. Less than 4% gas is too lean a mixture, and more than 14% gas is too rich. Each fuel has its critical mix proportions. Combustion will not take place unless the mix is right for the particular fuel being used.

### Air Quantities Required

Each fuel requires a certain proportion of air to the amount of fuel being burned. If not enough air is available, dangerous products of combustion are formed. Figure 3–8 shows the amount of air required for proper combustion for natural gas.

Since the air-fuel mix is so critical, extra air must be provided to the burners. This air is called **excess air.** A system should always be designed to provide at least 50% excess air, or half again as much air as is actually needed for complete combustion.

**Impinge***

**Air-fuel mix**    The mixture of air, to provide oxygen, with a fuel to make a combustible combination

**Excess air***

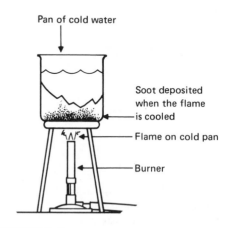

Pan of cold water

Soot deposited when the flame is cooled

Flame on cold pan

Burner

FIGURE 3-6 _____
Incomplete combustion. Soot, or unburned particles of carbon, is deposited on pan bottom when flame is cooled by pan.

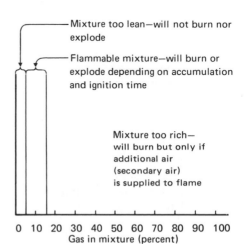

Mixture too lean—will not burn nor explode

Flammable mixture—will burn or explode depending on accumulation and ignition time

Mixture too rich— will burn but only if additional air (secondary air) is supplied to flame

0  10  20  30  40  50  60  70  80  90  100
Gas in mixture (percent)

FIGURE 3-7 _____
Percentage requirements of air-fuel mix for combustion of natural gas

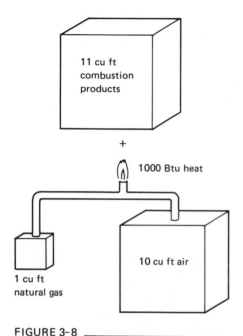

FIGURE 3-8 _____

Combustion air requirement for complete combustion of natural gas

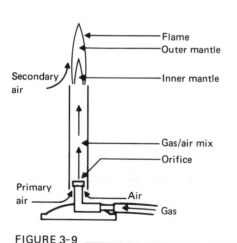

FIGURE 3-9 _____

Cutaway view of a Bunsen burner

## 3.5   FLAME TYPES

Two types of flame are created by the combustion processes: blue and yellow. Each type is characteristic of the way the air is mixed with the fuel in the firing process. The burners used in the heating unit provide the air and fuel mixing, so ultimately the flame type is related to the type of burner used and to the type of fuel burned.

With some burners and fuels, a large proportion of the combustion air can be mixed with the fuel before the fuel is burned, and the blue flame results. An example is a burner that mixes air with **gaseous fuels,** such as natural or manufactured gases. In a burner for **liquid fuel,** such as fuel oil, or solid fuel, such as coal, the air cannot be mixed with the fuel before burning. In this case, the flame generated is yellow.

## Blue Flame

If the air and the fuel are premixed—that is, before being introduced at the burner ports—the flame produced will be blue. This flame is very clean. This flame is also called a Bunsen flame, after the burners used in laboratory work. In Figure 3–9, a cutaway view of a Bunsen burner shows the fuel being introduced through an orifice, or opening, in the burner body, or tube, and air is also introduced at this point. The fuel and air then mix as they travel up the burner body. When they are fired at the burner port, they produce a clean-burning blue flame. The blue flame is produced by the rapid burning of the hydrogen and the carbon in the fuel.

Gaseous fuels are easily mixed with air before the combustion takes place. The flames they produce are usually blue.

## Yellow Flame

If the air cannot be mixed with the fuel before the fuel reaches the burner head, or if insufficient air is mixed with the fuel before the fuel is fired, the flame will be yellow. The yellow in the flame is caused by the carbon particles that are not completely burned. They exist inside the mantle of the flame and become incandescent from the heat. They do finally burn, but not until they mix with air that is drawn in from the sides of the flame.

Yellow, or incandescent, flame burners were used in nineteenth-century gaslights for lighting purposes. Because of the incandescence, the yellow flame burner gives off a great deal of light but less heat. The light is an unwanted form of energy for our purposes. See Figure 3–10 for a typical yellow flame burner.

Liquid fuels will not burn in the liquid state. They must be vaporized, and then the vapors mixed with air before they will burn. Figure 3–11 shows a simple pot burner with oil vapor and air burning. Fuel oil, a liquid, is vaporized and mixed with air as it leaves the burner face. An oil burner face is shown in Figure 3–12.

Because of the difficulty of mixing the vaporized fuel and the air, a yellow flame is normally produced with an oil burner. Oil burners have been developed that work almost as efficiently as blue flame burners, but unless the fuel and air can be mixed thoroughly before ignition, it is difficult to produce as efficient a flame with an oil burner as it is with a gas fuel burner.

Solid fuels, such as coke, wood, and coal, can be burned only as vapors. The heat from the process must vaporize the fuel, and then the vapors must mix with air for combustion to take place. The type of flame produced depends on the amount of air that can be mixed with the vapors. The amount of air is determined by the design of the burner and the burner system. Solid fuel burners are not covered in detail in this text.

**Gaseous fuels**   Fuel that is in the gaseous state when introduced into the burner

**Liquid fuel**   Fuels that are introduced into the burner in the liquid state and vaporize only as they are burned

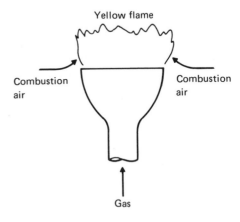

FIGURE 3-10 _____
Yellow flame burner, in which the combustion air mixes with the fuel in the flame itself

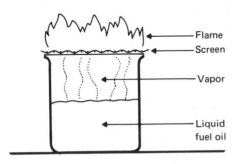

FIGURE 3-11 _____
Simple pot burner with oil vapor and air burning. Vapor burns after air is mixed above screen and vaporization is caused by heat from flame.

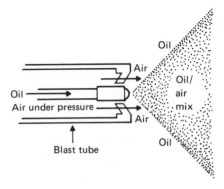

FIGURE 3-12 _____
Diagram of an oil burner face showing how the oil is vaporized and then mixed with air to make a combustible mix

## 3.6  COMBUSTION EFFICIENCY

The elements in the air and the fuel always combine in exact proportions in the combustion process. The elements in the combustion air react with the elements in the fuel so that the elements in the flue gas, while they have combined into different compounds, still equal in volume the elements in the air and the fuel. The bar graph in Figure 3–13 illustrates this principle.

By analyzing the elements in the fuel and those in the air, a technician can predict what elements will appear in the flue gas. By measuring the actual elements in the flue gas, the efficiency of the process can be determined.

In a combustion process operating at maximum efficiency, the oxygen in the air, approximately 21%, combines with the carbon and hydrogen in the fuel and forms carbon dioxide and water vapor. The nitrogen in the air goes through the process unchanged. The products of combustion have carbon dioxide and nitrogen in a fixed proportion depending upon the amount of carbon and hydrogen in the fuel and the amount of combustion air used.

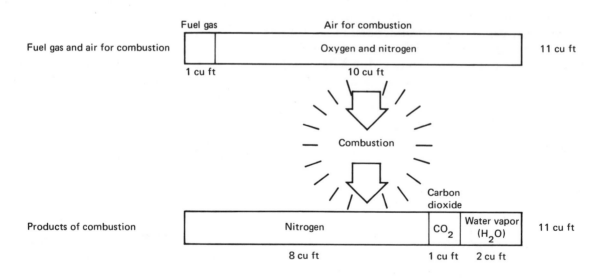

FIGURE 3-13 _____

Graph showing how the products of combustion always have the same volume as the fuel and the air used in the combustion process

For natural gas or fuel oil, the percentage of carbon dioxide is around 10%. The maximum possible would be 12%. Table 3–1 shows the results of an analysis for a natural gas combustion process.

# Flue Gas Analysis

Flue gas analysis is used to check the operating efficiency of a burner in a furnace or a boiler. The results indicate if the equipment is in good condition and/or if it needs adjustment or replacement. The description here will be for the procedure to use on a residential or small commercial unit.

Three tests are required for making a simple analysis: (1) a $CO_2$ analysis, (2) a smoke spot test, and (3) a stack temperature reading. The tests are made by sampling the flue gas with the appropriate instruments and by measuring the temperature rise through the furnace.

# Instruments Used for Analysis

The $CO_2$ analyzer is a device that gives the $CO_2$ content of the flue gas in percentages. The analyzer is a tubular instrument containing a liquid that will absorb $CO_2$. See Figure 3–14.

To use the analyzer, the technician draws flue gas out of the vent where the gas leaves the heating unit. The flue gas mixes with the liquid in the analyzer. As the gas mixes with the liquid, the change in volume of the liquid indicates the amount of $CO_2$ present. The analyzer is calibrated to give the $CO_2$ reading in percentages.

The smoke spot tester shows whether combustion is complete by indicating the presence of smoke in the flue gas. Smoke is an indication that all of the carbon in the fuel is not burned in the combustion process. The tester is a vacuum pump with a filter paper holder located where the air that is pumped out of the flue passes through the paper. Figure 3–15 shows a typical smoke spot tester.

To use a smoke spot tester, the technician inserts a flexible tube connected to the end of the tester pump into the flue, and a sample of flue gas is drawn out by the pump. The sample is drawn through a piece of filter

TABLE 3–1

Flue Gas Analysis for Natural Gas Combustion Process, by Volume

| Element | Percent of total volume |
|---------|------------------------|
| $CO_2$ | 8.86 |
| $O_2$ | 10.70 |
| CO | 0.00 |
| $N_2$ | 80.44 |
| Total | 100.00% |

FIGURE 3-14

$CO_2$ analyzer being used by a technician for a combustion test

FIGURE 3-15

Smoke spot tester

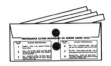

FIGURE 3-16 _____

Filter paper and smoke spot comparison chart

paper by the pump. Stain on the paper indicates the presence of smoke in the flue gas. The stain on the paper is compared with a set of test stains to indicate the amount of smoke. Figure 3–16 shows a sample of the filter paper and the comparison chart used for grading the sample.

The stack temperature is measured with a thermometer that reads high temperatures. The stack temperature is then compared with the room temperature to show the temperature rise through the unit.

## Procedure

During a flue gas analysis, the burner is first adjusted so it is operating correctly. The burner is run for ten minutes so it will stabilize, and then the $CO_2$ sample is taken from the flue below the point at which any air enters—that is, below the draft diverter or the barometric damper. A smoke spot test is made at the same time from the same place. The smoke spot test should indicate only a trace of smoke, and the $CO_2$ reading should be between 8% and 10%. If the smoke is more than a trace, then more combustion air should be admitted at the burner. The $CO_2$ should be rechecked to make sure it does not go below 8%. If the smoke cannot be kept to a trace, the burner should be checked to see that all of the parts are operating correctly, and the furnace should be checked for air leaks into the combustion chamber.

To complete the analysis, the technician should measure the flue gas temperature. The procedure is to insert a high-temperature thermometer in the flue at the same place where the smoke spot and $CO_2$ reading were taken. See Figure 3–17 for an illustration of a typical flue gas thermometer.

The flue gas temperature should be compared to the room temperature. The difference between the two, or flue gas temperature minus room temperature, is the net stack temperature.

## Percentage Efficiency

The slide rule or graph supplied with the $CO_2$ analyzer is used to find the operating efficiency of the unit after the tests are made. Such a slide rule is illustrated in Figure 3–18.

FIGURE 3-17 _____

Stack temperature thermometer being used on a stack for checking gas temperature

The slide rule is made up of a body and interchangeable tongues. Each tongue is calibrated for a specific fuel. Make sure the tongue used is the one calibrated for the fuel used in the unit tested. The tongue is set so the net stack temperature, or temperature rise through the unit, appears in the window on the upper right. Find the $CO_2$ percentage, which is read on the analyzer in the window in the right center of the slide rule, and the combustion efficiency in percentage will be shown on the rule opposite the $CO_2$ figure.

The combustion efficiency should be 75% or more for a unit in reasonably good condition. If the unit cannot be adjusted to give at least 75% efficiency, it should be reconditioned or replaced.

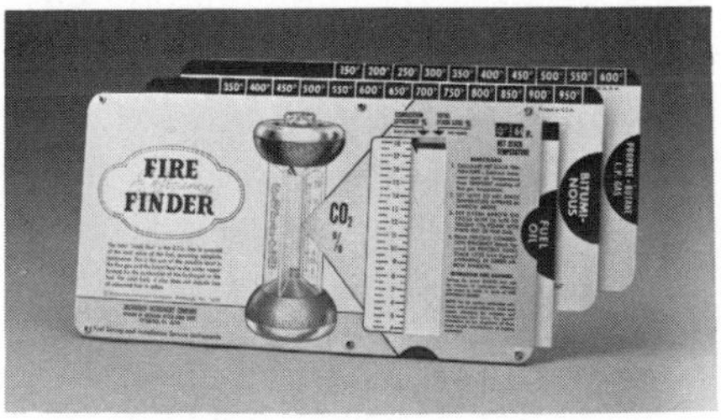

FIGURE 3-18
Combustion efficiency slide rule showing the various tongues used for different fuels

## 3.7  SUMMARY

The controlled production of heat is necessary to maintain comfortable conditions in homes and business buildings. It is also important in the production of materials and equipment. The control of heat requires understanding the combustion process.

Fuel, oxygen, and a high enough temperature to start a reaction are required for the combustion process. The use of different fuels and the amount of fuel used will determine how much heat is produced in a given time. The method used to introduce the air, which provides the oxygen, into the process determines the type of flame produced. The application of the heat to start the process is a function of the burner.

Under proper firing conditions, the combustion process is safe and will not produce dangerous products. If insufficient air is furnished to the flame, however, by-products can be produced that are extremely hazardous to people and animals. A designer must always be sure that a burner has not only enough air for proper combustion but excess air, to prevent the development of a dangerous situation.

## 3.8  QUESTIONS

1. Combustion is the release of _____ energy from fuel during the process of _____ .

2. Name the three elements that are necessary for combustion to take place.

3. The combustion process is the combining of _____ and _____ in the fuel, with _____, and heat and light is given off in the process.

4. The temperature required to start the reaction between the chemicals in the fuel, and oxygen in the air is called the _____ _____ .

5. When the chemicals in a fuel combine completely with the oxygen in the combustion air, the products of combustion are harmless. True or false?

6. Incomplete combustion generates products that are harmful. True or false?

7. All mixtures of air and fuel are combustible, regardless of the percentage of air to fuel. True or false?

8. The extra air furnished to a burner as combustion air is called _____ _____, and at least _____% should be provided.

9. When the fuel is mixed with the combustion air before it is introduced at the burner ports, a _____ flame is produced by the combustion process.

10. When the combustion air combines with the fuel during the actual burning process, as when solid or liquid fuels are used, the flame produced will be _____ .

11. Fuels will burn only after they have been vaporized. True or false?

12. Combustion efficiency can be tested by measuring the amount of _____, the quantity of _____ in the flue gas, and the _____ of the gas.

13. When the smoke spot test reveals just a trace of smoke, the $CO_2$ reading should be between _____ and _____ $CO_2$ for a properly adjusted burner.

14. If a furnace cannot be adjusted to give approximately _____% efficiency in burning, it should be rebuilt or replaced.

## 4.1 INTRODUCTION

In a typical heating application, it is often necessary to generate heat at one point in a building, while heating rooms or spaces that are remote from that point. For example, the furnace in a large building is often located in an equipment room in the basement of the building but this furnace is used to heat the entire building. Because of the difference in location, some **medium** must be used to carry the heat to the spaces where the heat is needed. Air, water, and steam are the most common mediums used for this purpose. Each of these mediums is discussed in this chapter, with emphasis on air and water.

## 4.2 HEAT EXCHANGE

A part of the furnace or boiler called a **heat exchanger** is used to extract heat from the fire and products of combustion and to heat the air or water that is then used to heat the building. The actual design of heat exchangers is covered later in this text, but a brief description is given here. A heat exchanger is any device that extracts heat from one medium and transfers it to another. Figure 4–1 illustrates the type of heat exchanger used in hot air furnaces.

In the case of a hot air furnace, the exchange is made between the hot products of combustion on the inside of the exchanger and the air that passes over the outside. In a hot water or steam boiler, the exchange is made between the hot products of combustion on the one side of the boiler tubes and the water that is on the other side of the tube walls.

### Rate of Heat Exchange

The formula in Chapter 1 relating to the amount of heat involved when a temperature difference occurs in any given material will help explain what happens when heat is transferred through the walls of a heat exchanger. In that formula,

$$\text{Btu} = ms(T_1 - T_2)$$

*Term defined in text.

## CHAPTER

# 4

# Heating Mediums

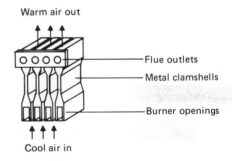

Warm air out

Flue outlets

Metal clamshells

Burner openings

Cool air in

FIGURE 4-1
Warm air heat exchanger

**Medium**   A gas or liquid used to carry heat from one place to another in a heating system

**Heat exchanger***

But when heat passes through any material, because of a temperature difference between two mediums on each side of the material, the formula must be revised to account for the movement of the mediums, the thin film of air or water on each side of the material, and the fact that the heat is removed on one side as fast as it is made available on the other. Therefore, it is necessary to use a factor called a **heat transfer factor,** or **conduction factor,** in the calculations. A heat transfer factor, normally called a U factor, is one that has been calculated by considering the variables. The heat transfer factor is actually a number that defines the amount of heat that will pass through a given material in one hour per square foot of material, per 1°F temperature difference on each side of the material. The formula for heat transfer by conduction is:

$$Btu/h = A \times U \times (T_1 - T_2)$$

where:

$Btu/h$ = amount of heat transferred in Btu per hour

$A$ = area of material to be considered

$U$ = conduction factor for given material

$(T_1 - T_2)$ = temperature difference between the two sides of the material

In practical applications, technicians do not need to calculate the heat output for the heat exchanger used. The manufacturer of the equipment furnishes the Btu/h output for the heat exchanger as part of the catalog data for the heating unit.

## 4.3   AIR AS THE HEATING MEDIUM

In a warm air system, the furnace has a burner in which combustion takes place, and a heat exchanger where most of the heat is extracted from the flames and the products of combustion. Figure 4–2 provides a cutaway view of a furnace showing the heat exchanger in place.

The heat that is extracted by the heat exchanger from the flue gases is added to the air distributed through the building for heating. In the usual design of exchangers, there is only a thin metal separation between the hot products of combustion and the circulating air, so the rate of exchange is very high; 80% is the usual efficiency of a warm air exchanger.

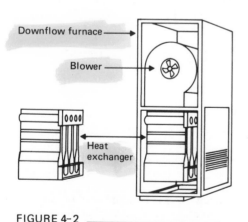

Downflow furnace

Blower

Heat exchanger

**FIGURE 4-2**

A warm air heat exchanger and a cutaway view of a downflow furnace with the heat exchanger in place

The air being used as the heating medium is blown directly into the room where the heat is wanted or is distributed through a duct system to the rooms to be heated. The circulation of the air is accomplished by a blower, or air mover. Hence the term forced air is used to describe this type of system.

Provision must be made for return air in the system to get back to the unit, to ensure complete circulation. As the warm air circulates through the rooms requiring heat, the air gives up its heat to the room. The air that is now at normal room temperature is recirculated back through the furnace, to start the cycle over.

## Temperature Rise of the Air

**Temperature rise** through a furnace is the difference between the temperature of the air entering the furnace and the air leaving. The term thus indicates the amount of heat the air has picked up as it passed through the heat exchanger. Temperature rise is a function of the amount of heat available, the efficiency of the exchange, and the amount of air going through the exchanger.

Each type of furnace is designed to work best at some particular temperature rise. Combustion-type furnaces should have from 80°F to 100°F temperature rise, electrical furnaces less than 50°F, and heat pumps only about 20°F. It is important to check the temperature rise through a furnace since a high temperature rise can indicate that the heat exchanger may be too hot and can be damaged. A low temperature rise indicates that too much air may be circulating for comfort.

Temperature rise can be determined by measuring the entering and leaving temperatures with a thermometer and then subtracting the entering temperature from the leaving, as in the following formula:

$$TR = T_L - T_E$$

where:

$TR$ = temperature rise

$T_L$ = temperature leaving

$T_E$ = temperature entering

This direct approach is a good one to use on an existing system when access to the unit is easy.

When a distribution system is designed, the quan-

Heat transfer factor (conduction factor)*

Temperature rise (TR)*

**Sensible reading**    A temperature that can be sensed by a human being or measured by a standard thermometer

**Hydronic**    A heating system that uses a liquid as the primary heating medium

**Terminal device**\*

tity of air to circulate is often decided upon for a building of a given size, independently of the size of furnace used. In this case, the temperature rise, through the furnace to be used, should be checked mathematically to make sure the air quantity is correct for the heat exchanger. This check can be done by formula. The formula used for figuring temperature rise of air when it is heated is:

$$TR = \frac{Btu/h}{cfm \times 1.08}$$

where:

$TR$ = temperature rise

cfm = quantity of air in cu ft per minute

1.08 = constant, based on specific heat of air × specific weight of air × 60 min in 1 hr

This formula is for air at standard conditions. *Standard conditions* refers to temperatures and relative humidity condition considered standard. In most heating-only applications, it is not necessary to be concerned with moisture changes in the air as the temperature changes, so the temperatures in the formula are for dry bulb or **sensible readings** only.

To use the formula, take the rated heating output of the furnace in Btu/h and divide that by the quantity of air in cfm going through it, times the constant, 1.08. The result is the temperature rise of the air as the air passes through the exchanger.

## EXAMPLE 4–1

If a furnace is rated at 64,000 Btu/h output, and the blower is adjusted to deliver 960 cfm of air, what is the temperature rise of the air through the furnace?

**Solution**

$$TR = \frac{Btu/h}{cfm \times 1.08}$$

$$= \frac{64,000 \text{ Btu/h}}{960 \text{ cfm} \times 1.08}$$

$$= 61.7°F$$

The answer is 61.7°F. The temperature of the air leaving the furnace will be 61.7°F higher than the temperature of the air entering the furnace.

## 4.4  WATER AS THE HEATING MEDIUM

In a hot water, or **hydronic,** heating system, the circulating medium that carries the heat to the point of use is water. The burner is located under a boiler, and water is heated by the flames and products of combustion. A picture of a small gas-fired boiler is shown in Figure 4–3. A number of different types of boilers are used, but they all have the same basic purpose: heating water.

The hot water is circulated through a piping system to the various rooms or zones in which heat is needed. The water is directed through some kind of a **terminal device,** such as a radiator or a coil. The heat from the water is used to warm air, which in turn is used to heat the space. The system uses a supply and a return pipe, as well as pumps, to circulate the water. Water is an excellent medium to carry heat because of its high specific heat capacity. But since the ultimate purpose is to heat air, hydronic systems are most practical when the heat is to be utilized at some distance from the point of generation.

To understand why a hydronic system transfers heat so well, we must look at the specific heat, or the amount of heat the water will hold per unit. Water is used as the basis of comparison for the specific heat of all substances, and the specific heat of water is considered 1.0. Thus, 1 Btu of heat is required to raise the temperature of 1 lb of water 1°F. The value 1.0 is relatively high compared to that of other materials listed. Table 4–1 shows the specific heat of some common materials, including water. Thus water is a good medium for carrying heat, such as in a hydronic heating system. Where the specific heat of water is 1.0, the specific heat of air is 0.24, which shows that water will carry approximately four times as much heat as air.

## Temperature Rise of Water

It is sometimes necessary to calculate the temperature rise of the water through a boiler, which is only a heat exchanger, when designing a hydronic system. The amount of heat carried by the water is a function of the temperature of the water, and the quantity of water in gallons per minute (gpm), that is circulating.

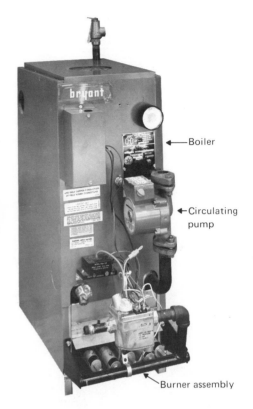

FIGURE 4–3

Typical small boiler showing the circulating pump and burner location

TABLE 4–1

Specific Heat of Some Common Materials

| Substance | Specific heat (Btu/lb/°F) |
| --- | --- |
| Air | 0.24 |
| Aluminum | 0.22 |
| Copper | 0.093 |
| Glass | 0.21 |
| Ice | 0.50 |
| Iron (steel) | 0.115 |
| Silver | 0.056 |
| Steam | 0.48 |
| Water | 1.00 |

Since water is used as the standard for defining specific heat, and we know that 1 Btu is required to raise the temperature of 1 lb of water 1°F, a formula can be derived using these data to find the temperature rise of water when water is heated.

The usual way of measuring the rate of flow of water in a system is by gpm, and since specific heat is calculated per pound of water, the formula has to convert pounds of water to gallons and hours to minutes. The formula is:

$$TR = \frac{Btu/h}{gpm \times 500}$$

where:

$TR$ = temperature rise of the water flowing through the boiler

Btu/h = total heat output rating for the boiler

gpm = quantity of water passing through the boiler in gpm

500 = a constant derived by multiplying the specific heat of water × the minutes in an hour × the weight of water in pounds per gallon

To figure the temperature rise of water through a given boiler, divide the rated Btu/h capacity of the boiler in Btu/h by the amount of water flowing through it in gpm, times 500. The answer will be the temperature rise in degrees Fahrenheit of the water.

## EXAMPLE 4–2

If a boiler with a rated output capacity of 450,000 Btu/h is installed on a system in which water is circulated at a rate of 36 gpm, what will the temperature rise of the water be?

**Solution**

$$TR = \frac{Btu/h}{gpm \times 500}$$

$$= \frac{450,000 \ Btu/h}{36° \times 500}$$

$$= 25°F$$

The temperature of the water leaving the boiler will be 25°F higher than the entering temperature.

# 4.5  STEAM AS A HEATING MEDIUM

**Condensate**    A liquid obtained by condensation of a gas or vapor

When steam is used as a medium to carry heat, the system is similar to that described for hot water. The main difference is found in the boiler and the piping for distribution and return of **condensate.** Since the heat from the burner is used to raise the temperature of water high enough to convert it to steam and since steam has so much more volume than the liquid, the system operates at pressures above atmospheric. The steam, under pressure, circulates to the terminal devices in the system, where heat is extracted from the steam by those devices. Heat extraction condenses the steam into water, and so the return system conducts water back to the boiler, to start the cycle over.

Steam is seldom used as a medium in small or midsized heating jobs, and since the quantity of heat in a given quantity of steam varies with the pressure of the steam, the calculations for the temperature rise in a steam boiler become too cumbersome for the scope of this text. Steam tables giving the Btu content of steam per pound of steam at various pressures and temperatures are normally used when designing a steam system and should be referred to by the designer of any such system.

# 4.6  SUMMARY

The practical use of heat produced by the combustion process requires the distribution of that heat throughout the building or the spaces to be heated. Since the products of combustion themselves may be harmful to people in the building, the heat must be extracted from the products of combustion and distributed through the building by using some medium. The most common mediums used for this purpose are air, water, and steam.

The heat is extracted from the products of combustion by a heat exchanger. The flue gases and heating mediums are completely separated from each other in the heat exchanger. The rate of heat exchange between the flue gas and the medium can be calculated accurately, and the amount of heat carried by the medium can be predicted. The calculations enable designers to determine exactly how much air, water, or steam must be distributed to the different areas of a building to provide the required heat to those areas.

Each medium has particular characteristics that makes its use more practical for some applications than others. Since air is the medium finally used for heating the spaces occupied by people, air should be used as the primary medium whenever possible. If heat has to be transferred long distances, water or steam should be the primary medium and air should be the secondary medium. The best medium to use in each individual case is determined by the designer after analyzing all of the heating requirements of the building.

## 4.7  QUESTIONS

1. Since heat is often generated at one point in a building but distributed throughout the building, a heating _____ is used to carry the heat.

2. The most common mediums used for space heating application are _____, _____, and _____.

3. The part of a furnace or boiler where heat is extracted from the fire or products of combustion is called the _____ _____.

4. Heat passes through the walls of a heat exchanger because of the _____ _____ existing on either side of the walls of the heat exchanger.

5. A conduction factor, or U factor, gives Btu/h per 1 sq ft, per degree difference between the two sides of the material. True or false?

6. When air is used as a heating medium, all of the heat produced by the burning fuel is transferred to the air used for heating the building. True or false?

7. If a warm air furnace has an output rating of 60,000 Btu/h, and the air is circulating through the furnace at a rate of 780 cfm, what will the TR of the air be?

8. A hydronic system is one that uses a liquid medium to carry heat. True or false?

9. Hydronic systems are usually used when the _____ _____ is fairly long.

10. What would the TR be of the water going through a boiler at a rate of 27.7 gpm, if the heating output of the boiler was 240,000 Btu/h?

11. Steam heating systems are often used in heating small buildings. True or false?

# CHAPTER
# 5

# Warm Air Systems and Heating Equipment

## 5.1 INTRODUCTION

The two major types of systems used for heating buildings are the warm air system and the hydronic system. This chapter covers the warm air system, and Chapter 10 covers the hydronic system.

In a warm air system, air is heated in a furnace and circulated through a building to maintain a desired temperature in the building. Many different types of furnaces are used in this system. These furnaces are categorized by fuels, heating capacity, building use, physical arrangement, and specific application. This chapter describes the total system and the general types of furnaces.

## 5.2 WARM AIR HEATING SYSTEM

A warm air heating system is one in which the heating medium is air. This air is heated as it goes through a furnace. The heated air is then circulated to various areas in a building, where heat is needed, through a system of ducts. The air is distributed into rooms through supply registers and is then returned to the furnace through return air grilles and return air ductwork. The air is circulated by a blower that is part of the furnace or system.

The temperature of the air in the building is controlled by a **thermostat** located in the building. A thermostat senses the temperature of the air and "brings on" the furnace if the temperature drops. When the temperature of the air rises, the furnace is "turned off."

Warm air systems are most practical in buildings where the furnace can be located relatively near the areas needing heat. Warm air ducts are usually fairly large and require a considerable amount of space in the building frame.

## 5.3 GENERAL TYPES OF FURNACES

Three main types of furnaces are available. The first is related to the fuel or energy used to produce the heat, the second is related to the type of job that the unit will be used on, and the third is related to physical characteristics of the unit itself. Each of these is discussed in detail in the following sections.

## Furnaces for Different Fuels

Furnaces burn gaseous fuels, liquid fuels, and solid fuels; others use electric energy to produce heat. Each of these types of furnaces has some features in common, but each is also different from the others in some ways. Gas furnace burners and heat exchangers are different from those used for oil. Electric furnaces use resistance heaters instead of burners and heat exchangers. Details concerning gas, oil, and electric furnaces will be given in the following chapters, but since furnaces that burn solid fuels are not common in smaller sizes, they will be covered only in a general way.

## Residential and Commercial Furnaces

In the second category, units by size, units used for residences and smaller commercial jobs are generally smaller in heating capacity and also are more compact in physical size than are units built for commercial or industrial applications. The control systems are less sophisticated on residential units than on commercial units. Some options, such as multiple fuel use, are not available on residential units but are common on commercial units. Commercial units are built with more concern about utility and less about appearance, and they may include accessories important to commercial jobs.

## Physical Characteristics

Since nearly every application is different, heating units must be built to fit different types of installations. This concern brings us to the third category for furnaces: physical configuration. Furnaces in this category are identified by the direction the air passes through the unit. In this respect, furnaces are built to operate

**Thermostat***

---

*Term defined in text.

**Upflow furnace**    A furnace in which the air to be heated enters through the bottom or one side and is discharged through the top

**Downflow furnace**    A furnace in which the air to be heated enters at the top and is discharged down through the bottom

**Horizontal furnace**    A furnace in which the airflow is horizontal (Air to be heated enters one end and is discharged out the other.)

as **upflow, downflow,** or **horizontal furnaces.** These three configurations enable designers to find units that will fit almost any job.

**Upflow Furnaces.**    A typical upflow furnace has provisions for the return air to enter near the bottom of the unit. The air will come through the bottom of the cabinet or on one side. The air is drawn through the filters, passes through the **heat exchanger,** and goes out at the top of the unit. Figure 5–1 shows a typical upflow gas-fired furnace.

When upflow furnaces are used, they are placed below the ductwork. The air can go up from the furnace and straight into the ducts. An example is a system used in a residence with a basement, where the ductwork is located on the ceiling of the basement.

**Downflow Furnaces.**    In a downflow furnace, the air enters the unit at the top, passes through the filters, is pulled through the blower, and is then blown down through the heat exchanger. The air is discharged vertically out of the bottom of the unit. Figure 5–2 shows a typical downflow gas-fired furnace.

Downflow furnaces are most practical when the ductwork is located below the unit. The air can then go straight down from the unit into the ductwork. This type of system is often used when the floor is a concrete slab-on-grade and the ductwork is located under the slab. One of the advantages of using a downflow furnace in this type of installation is that the furnace can be located in a closet on the main floor of the building; therefore a special equipment room does not have to be provided under the floor.

**Horizontal Furnaces.**    A horizontal furnace is one that has the air pass through it horizontally. That is, the air enters horizontally on one end, passes through the furnace, and is discharged horizontally on the other end. The air enters the furnace, passes through the filter, is drawn through the blower, goes across the heat exchanger, and is discharged. This type of furnace is available in both left- and right-hand discharge, and the

FIGURE 5–1 _____

Typical upflow gas-fired furnace

unit has to be ordered accordingly. Figure 5–3 shows a typical horizontal gas-fired furnace.

Horizontal furnaces are most often used when the ductwork is on the same level as the unit and when the height of the space the furnace is to be installed in is limited. A good example is the use in a crawl space or in the attic of a building. In warmer climates where supply registers can be located overhead, a horizontal furnace is often used in the attic, and the ductwork is run in the attic. The supply registers are then dropped down through the ceiling into the spaces to be conditioned. In colder climates where the registers should be located on the floor, the horizontal furnace is placed in the crawl space for the best heating. The ducts run under the floor with the **boots** for the registers going up through the floor.

## 5.4  UNITS FOR SPECIAL APPLICATIONS

In addition to the usual type of heating units where the heat source, filters, blowers, and other parts are all contained in one cabinet, other types of units are used for special applications. These types include duct heaters to be installed directly in the duct of a system, unit heaters that are self-contained and are not connected to any duct system, and radiant heaters that provide heat to a particular location but do not heat the air in the space.

### Duct Heaters

A **duct heater** is a heating unit that is normally installed in some part of a duct system remote from the blower unit. A duct heater contains a burner or heating elements, a heat exchanger, and the controls for the burner. A cabinet encloses these parts. Duct connections on each end connect to a duct system. The duct heater is used for heating air, just like any other furnace. Air must go across the heat exchanger to prevent burn-out of the exchanger. The air supply comes from a blower located in a part of the system remote from

**Heat exchanger**    A device that is used to transfer heat from the flue gases in a furnace to the air used to heat a building

**Boots**    A sheet metal box that connects a register or grille to the ductwork

**Duct heater**    A heating unit that is installed directly in an air duct (The air is circulated by a remote blower.)

FIGURE 5-2 _____

Typical downflow gas-fired furnace

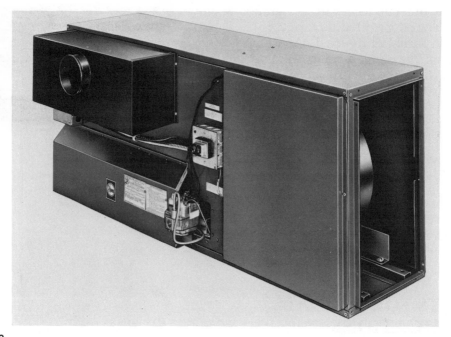

FIGURE 5-3 _____

Typical horizontal gas-fired furnace

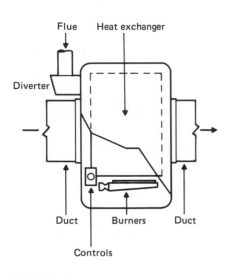

FIGURE 5-4 _____

Drawing of duct furnace with cutaway
showing burners in relation to heat
exchanger

its own location. The illustration in Figure 5–4 shows
the arrangement of parts in a duct heater.

Duct heaters are manufactured to burn or use any
of the common fuels or energy available to any heating
unit. Duct heaters are often used to provide heat to one
part of a building or one zone in a building rather than
to the entire building from a central system. These
heaters are often used in branch ducts and can be controlled individually to provide **zone control.**

For each of the fuels or energy used, the burners,
exchangers, elements, and other heat-producing components are similar to those found in self-contained
units.

Controls for duct heaters vary slightly from those
found in central station units in that the air mover is
remote from the location of the heater, and so the
**blower controls** are not located near the blower. If a
duct heater is used as a zone heater rather than as the
primary heating unit in the system, then it is common
to provide **flow switches** to "prove," or provide evi-

dence of, the flow of air in the duct before the duct heater burner can come on. This device is used to eliminate the danger of fire or damage to the unit or building because of high temperature in the unit if no air is moving across the heat exchanger.

## Unit Heaters

Another type of heating unit, the **unit heater,** is used when close control of the temperature in a space is not required. A unit heater is self-contained, with the burners or heating elements, heat exchanger, controls, and an air mover all in one cabinet. This type of heater is not intended for use with a duct system. The air mover draws air in through the back of the unit, blows it through the heat exchanger, and is discharged out the front through a louver on the face. Figure 5–5 shows the parts of a unit heater.

Unit heaters are used for heating large open spaces, such as warehouse or factory bays. Such heaters are usually suspended from the ceiling. The louver on the front of the unit directs the air down to the occupied space. Unit heaters can be fired with gas or oil, or electricity, steam, or hot water may be used as the source of heat. Because the unit heater is not used with a duct system and does not have to move air against the friction in the ducts, a propeller fan is used as an air mover. The propeller fan can move the air efficiently in the free-flow application.

The components in the unit heater, such as burners, heat exchanger, and controls, are similar to those found in self-contained furnaces for central installation on duct systems.

## Infrared Radiant Heaters

A third type of heating unit is often used for specialized applications. It is the infrared radiant heater. This type of heater emits **infrared rays** that do not heat the air but do heat any objects or surfaces that they strike. The typical infrared heater has a heat source, such as a gas burner, that heats ceramic elements to a very high temperature, and the rays are then given off by the hot element. Figure 5–6 shows an infrared heater.

In a typical application, infrared heaters are sus-

**Zone control**    Temperature control in a building by individual zones as opposed to the entire building as a whole

**Blower control**    A device used in a heating unit to turn the blower on when the burner comes on

**Flow switch**    A device that senses whether air is moving in a duct

**Unit heater***

**Infrared rays**    Invisible rays that have a longer wavelength than visible light but can be used to transfer energy from one place to another

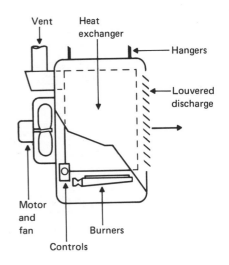

FIGURE 5-5

Cutaway drawing of unit heater

FIGURE 5-6 _____

Infrared heater with radiant surface in back of reflector cabinet

pended from the ceiling, and the rays are directed downward. The rays are changed into heat energy only when they strike the floor, people, or other objects in the line of sight of the heater. The air around the objects is then heated to some extent by convection as the air circulates across the warm objects.

Infrared heaters are practical as a source of heat in any building where it is not important that the air in the space be heated but where it is important to provide heat to the floor area or to people on the floor level.

## 5.5  SUMMARY

The purpose of any heating system is to provide a comfortably, warm atmosphere in a building. The system achieves this end by controlling the temperature of the air in the building by means of a warm air furnace.

Several types of warm air heating units are available. A unit should be selected only after the designer considers the fuel to use, physical requirements, and the airflow patterns required. In addition to the typical self-contained furnaces available, there are specialized units that may be better for particular applications.

# 5.6  QUESTIONS

1. The two major types of heating systems used for heating buildings are _____ _____ and hydronic.

2. In a warm air system, the temperature of the air in the building is controlled by a _____ thermostat that turns the unit off and on.

3. The three main categories of warm air furnaces are:
   a. Related to _____ or _____ used to produce heat.
   b. Related to _____.
   c. Related to _____ characteristics of the unit itself.

4. All furnaces are alike, regardless of the type of fuel burned. True or false?

5. Control systems are usually less sophisticated on _____ heating units than they are on _____ units.

6. The three basic physical configurations of furnaces, in respect to airflow, are _____, _____, and _____.

7. Match the terms in the first column with the correct description in the second column.

   1. Upflow furnace
   2. Downflow furnace
   3. Horizontal furnace

   a. Available in left or right hand
   b. Ductwork located below the unit
   c. Return air enters on the bottom or sides

8. A duct heater always has an air mover located in its cabinet. True or false?

9. When using a duct heater, it is necessary to provide a _____ _____ to prove airflow through the unit on a call for heat.

10. Unit heaters are always connected to ductwork. True or false?

11. Unit heaters have their own air mover. True or false?

12. Infrared heaters warm the air in a space. True or false?

13. Infrared rays are a form of energy. This energy is changed into _____ energy when the rays strike an object.

# CHAPTER
# 6
# Warm Air Furnaces

## 6.1 INTRODUCTION

Warm air furnaces are the most commonly used types of heating units for small and medium-sized buildings. These furnaces are manufactured in various styles and for use with different fuels. Regardless of the style or fuel used, each type of furnace has some components that are similar: cabinets, blowers, filters, burners, and heat exchangers.

Cabinets, blowers, and filters are covered in detail in this chapter, while burners and heat exchangers are covered in a more general way. Burners and heat exchangers are explained in more detail in the chapters on each system.

## 6.2 WARM AIR FURNACE

A warm air furnace is the unit that provides the warm air for use in a warm air heating system. There are many types of furnaces. Each type is different from the others in some respects, but all of them have certain parts that are nearly the same. These similar parts are described in this chapter.

A unit is made up of five major components:

1. Cabinet,
2. Blower,
3. Filter,
4. Burners,
5. Heat exchanger.

In addition, each type of furnace has a set of controls that operate it. Figure 6–1 shows a furnace with these parts identified.

### Cabinet

Most furnaces have a sheet metal cabinet that holds the other components and forms an enclosure to direct the air through the blower and across the heat exchanger. Figure 6–2 shows a furnace cabinet.

The cabinet is made of sheet metal, painted to protect the metal from corrosion. In smaller units, the cabinet is rigid enough to act as a support for the other components. Larger furnaces usually require a frame

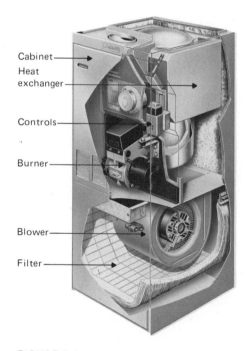

Cabinet
Heat exchanger
Controls
Burner
Blower
Filter

FIGURE 6-1 _____
Cutaway view of upflow oil-fired furnace showing major components

to support these parts. Much consideration is given to the appearance of the cabinet for residential furnaces, but units for commercial or industrial applications are designed for utility rather than appearance.

The cabinet is built to enclose the components on all sides except the front. Openings are provided where the return and supply air duct connections are made. If the unit has options regarding the location of the return air, such as an upflow furnace usually has, then knock-outs are provided where the openings can be cut. Knock-outs are indentations in the sheet metal that show where openings can be made.

The front of the cabinet is usually a removable door. This door covers a vestibule that encloses the burner and the controls. The heat exchanger is behind another panel in the back of the vestibule, and the burner, or burners, extends through this panel into the space below the heat exchanger. The front panel is slotted or louvered so combustion air can get into the burner vestibule.

The filters are located in a separate compartment with the blower. This compartment has an access door that can be removed for changing filters or for servicing the blower.

## Blower

Most heating units for use in small- to medium-sized buildings are designed as package units. The blower, filter, burners, and controls are in one cabinet. The blower, or air mover, used in these units is almost always a centrifugal blower. Such blowers are designed to move the air efficiently in the duct system and with relatively low power consumption. The blowers can also be adjusted to deliver different air quantities.

Medium- to high-pressure air distribution systems are used in some larger buildings. The higher pressure enables the distribution of large quantities of air through smaller ducts. In these systems, axial-flow fans are often used. The axial-flow fan can develop higher pressure in the duct system than the centrifugal fan, but it uses more power in the process.

**Centrifugal Blower.**    The most commonly used blower for a furnace is a centrifugal blower. This type of blower

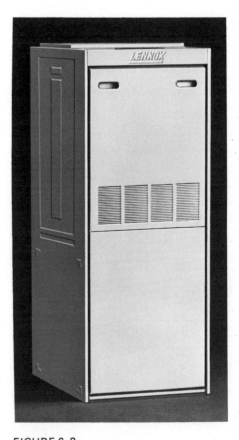

FIGURE 6-2
Modern furnace cabinet

**Impingement**    Dirt or dust particles in the air that hit or strike the filter media

FIGURE 6-3 _____

Wheel used in centrifugal blower showing the vanes connected to the rim

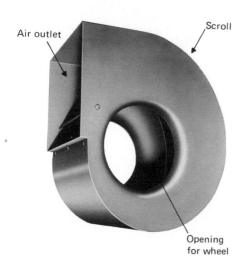

FIGURE 6-4 _____

Centrifugal blower scroll

uses a wheel that looks like a paddle wheel with small blades to move the air. These vanes, as they are called, are located around the outside of the wheel. A centrifugal blower wheel is shown in Figure 6–3.

The wheel rotates on a shaft parallel with the vanes and at right angles to the direction of air flow. Air is pulled in through the middle of the wheel and is thrown out through the vanes as they spin. The air is discharged through an outlet at one point on a sheet metal scroll surrounding the wheel. Figure 6–4 shows a blower scroll without the wheel.

Many smaller furnaces in residential applications use a direct-drive blower-motor combination, as shown in Figure 6–5. In a direct-drive blower, the wheel hub is connected directly on to the shaft of the motor, and the wheel speed is the same as the motor speed. Normally a multispeed motor is used on a direct-drive blower application, so some control can be maintained on the air output, which is directly related to wheel speed.

On some furnaces and most larger units, the blower is belt driven. In this application, the motor is located adjacent to the blower, and the drive is through pulleys and a belt. Figure 6–6 shows a furnace with a belt-drive blower. Such an arrangement makes it possible to change pulley sizes to provide blower speed adjustment for controlling air output. Most belt-driven systems have an adjustable pulley on the motor. This pulley allows for changing the blower speed, which changes the air quantity through a set range. The motor is mounted in such a way that belt tension can be adjusted as the pulley size is changed.

**Axial-Flow Blower.**    In applications where air has to be provided at medium to high pressure, an axial-flow blower is used. This blower has a wheel that looks like a turbine wheel. The blades are at right angles to the shaft on which it turns and are inclined so they move the air like a propeller does. The shaft itself is parallel to the direction of the airflow. The wheel is enclosed within a cabinet, which is normally part of the ductwork. An axial-flow blower is shown in Figure 6–7.

The motor that drives the axial-flow blower may be within the cabinet or adjacent to it. An axial-flow blower is usually found on duct systems on larger jobs.

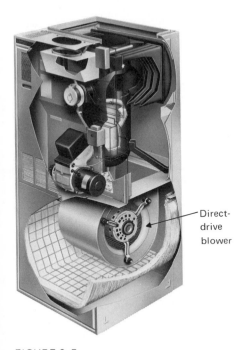

FIGURE 6-5 _____
Cutaway view of upflow oil furnace with
direct-drive blower

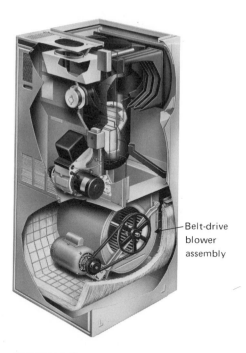

FIGURE 6-6 _____
Belt-drive blower in modern furnace

# Filters

One of the advantages of a circulated air system is that all of the air in the building can be constantly filtered and cleaned. Air filters in the heating unit or in the return air duct system ahead of the unit perform this operation. In some cases, filters may be placed in the return air grilles where the air is taken out of each room. As the air in the building passes through the return air system, the air is cleaned as it goes through the filters.

Of the many different types of filters, the most common ones fall into one of two major categories. The first is mechanical filtration. It is represented by mesh-type filters that clean the air by **impingement** or by straining. The second category is electronic filtration. In this type, unwanted particles of dust and dirt are removed from the air by electronic means.

FIGURE 6-7 _____
Axial-flow blower used for medium- and
high-pressure distribution systems

**Filter Efficiency.**    The most important variations found in filters relate to the cleaning efficiency of the filter. Standard disposable filters furnished with most residential furnaces are not very efficient when compared to filters used for commercial applications. The mesh of standard filters is efficient in filtering out large particles of dust and dirt but not small particles.

To rate filter efficiency, it is necessary to define particle size as related to the dust and dirt that must be removed from the air. The table shown in Figure 6–8 shows the approximate particle sizes of some common materials.

The particle sizes in Figure 6–8 are given in **microns.** One micron equals 0.001 millimeter (mm), or 1/25,400 of an inch. A typical sand particle is approximately 100 microns across; a particle of water in airborne fog is 10 microns in diameter. An optical microscope can be used to magnify particles down to

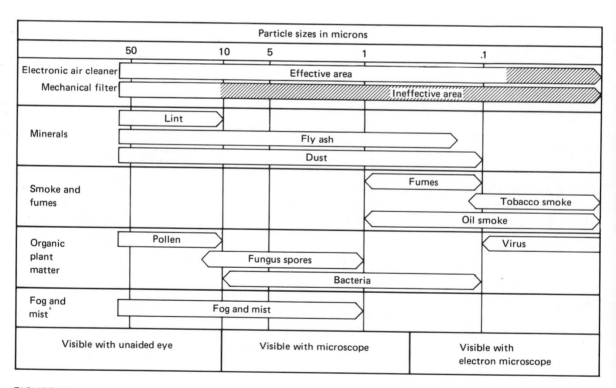

FIGURE 6-8 _____

Size of some airborne particles in microns

0.1 micron in size, but an electron microscope is required for smaller particles.

Mesh-type filters are relatively efficient in removing particles from the air down to somewhere between 1 and 10 microns, and an electronic filter will remove particles as small as 0.03 micron. The actual efficiency of filters is measured in the laboratory by using dust made up of particles of a known size and then measuring the percentage of the dust removed from the air under a given set of conditions.

**Mechanical Filters.**    Two different principles of filtration are involved in mechanical filters: impingement and straining. In a filter that uses the impingement principle, the dust and dirt are collected on the filter media by sticking, or impinging, on a sticky surface. The surface is made sticky by coating the media with oil. In the straining filter, the fibers of the media are close enough together so the particles are simply caught between the strands. Usually the impingement filter is more efficient.

Many different materials are used for media in mechanical filters, including fiberglass, plastic foam, shredded paper, and metal mesh. The media is usually formed into a pad or blanket and is used in that form. It may be enclosed in a cardboard frame and the frame then used in the filtering unit. If the pad is used alone, it is normally installed in frames that are part of the unit.

The performance of a mechanical filter in removing contaminants from the air is dependent upon the air going through the filter at a slow enough velocity and in such a path that it has to make rapid changes in direction. As the air goes through the filter, the velocity of the contaminants will cause them to continue in a straight line, even as the air changes direction, and they will impinge on the filter material and stick there. The efficiency of a filter can be increased by increasing the depth, thus making it necessary for the air to go through more changes of direction and consequently making it more likely that the dirt and dust in the air will impinge on the media.

Mechanical filters of the straining or impingement type can be used to filter particles from the air from about 100 microns down to 1 micron in size. By making

**Micron**    A unit of measurement; 0.001 millimeter, or 1/25,400 of an inch

FIGURE 6-9 _____
Deep bed filter used for medium-high
efficiency in mechanical filtration

the filters thicker and using material that is denser, the efficiency can be improved to where particles as small as 0.01 micron can be removed. For particles smaller than this, deep bed filters of special design are required. A deep bed filter is thicker and has more depth of filtering material. A deep bed filter for medium-high filtering efficiency is shown in Figure 6–9.

To select the proper filter, the designer must first decide on the percentage efficiency desired. Then he or she will select a type and size of filter from a manufacturer's catalog that will provide that efficiency with the amount of air that has to be filtered. The size of the filter must be checked to ensure that it will fit the unit; and the **pressure drop** through the filters must be checked against the allowable pressure drop for the system so the blower can deliver the proper quantity of air.

**Electronic Filters.**    Electronic filters are used in both residential and commercial applications when a high degree of filtration efficiency is desired. Figure 6–10 shows a typical electronic air cleaner as used for residential applications.

The electronic filter does not use filter media except as a prefilter. The actual cleaning is accomplished by passing the air from the system through a grid of wires that have been electrically charged with high-voltage direct current electricity. This prefiltering **ionizes** the dust particles in the air. The air, with the dust particles, then passes through a series of plates that are also electrically charged, and the ionized dust is attracted to and adheres to the plates by electrostatic attraction. Not only is the general efficiency of the electronic filter high, but the air cleaner will remove extremely small particles from the air. Figure 6–11 provides information on the general efficiency of the electronic filter.

An electronic filter is sized by using manufacturers' data sheets as to its efficiency. In the sheets, the efficiency is shown for various cfm through the filter. The designer selects the filter that will give the desired efficiency at the quantity of air required for the job.

Special applications, such as rooms that require high-efficiency filtration, normally use filters that have great density and large area, and, in many cases, they are very deep. The resistance to airflow is high in such

FIGURE 6-10 _____
Electronic air cleaner used for residential
or small commercial applications

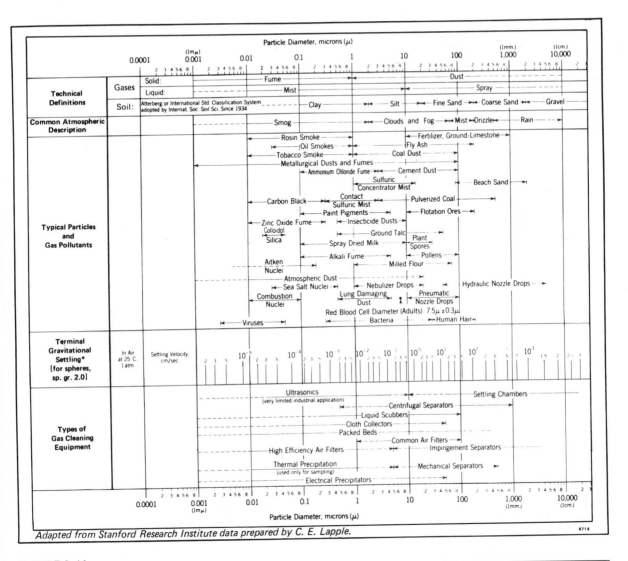

Adapted from Stanford Research Institute data prepared by C. E. Lapple.

FIGURE 6-11 _____

Size and characteristics of some airborne particles and filter efficiency related to them

systems and is allowed for in the design of the total air system. A bag filter used for high efficiency cleaning is shown in Figure 6–12.

In some cases, filters are used for odor removal or control. Special chemical filters, such as charcoal filter beds, are used for this purpose.

**Pressure drop**   The difference between the air pressure on either side of a filter or other restriction in a duct system

**Ionize**   Create an electrostatic charge on particles by displacing some of the free electrons on the atoms of the particles

FIGURE 6-12 _____

Bag filter used for high efficiency and long life in mechanical filtration system

# Burners

Burners are the next major component to be discussed. A general description is given here to help readers understand the function of the burner in a unit, but details concerning specific burners are given in the chapters for each fuel.

The three main functions of any burner, regardless of the fuel used, are to:

1. Meter the flow of fuel and air for combustion,
2. Provide proper mixing of fuel and air for combustion,
3. Supply the fuel-air mix to a burner face where the mix can be ignited.

The first function, metering, is performed by a part called an orifice in a gas furnace. The orifices are located at the primary air opening end of the burner tubes, where the gas enters the tube. In an oil burner, the metering device is called the nozzle. The nozzle is at the end of the oil line, at the burner face. A typical oil burner is shown in Figure 6–13. The burner face is inside the furnace and is not visible in the picture.

The second function, fuel and air mixing, occurs in the burner tube in a gas furnace and at the burner face in an oil furnace.

The third function, supply of the fuel-air mix, takes place at the burner ports in a gas furnace and at the burner face in an oil furnace.

Although burners for different fuels vary in appearance and in the parts used to achieve the function, the result achieved is the same: safe and controlled combustion.

# Heat Exchangers

The heat exchanger in a warm air furnace is the component that extracts heat from the flames and products of combustion and transfers that heat to the air that is used as the heating medium.

In a warm air system, it is necessary to separate the products of combustion from the air used to carry the heat. Products of combustion may contain harmful or dangerous gases that cannot be safely introduced into a habitable space. The heat exchanger is the part of the

FIGURE 6-13 _____

Oil burner in residential oil-fired furnace

unit where this separation takes place. Figure 6–14
shows a heat exchanger that is used in a gas-fired, warm
air furnace. The model shown has a curved design that
increases the efficiency and ensures quiet operation.
The furnace blower blows the air to be heated up between the sections of the exchanger where it is heated.
The products of combustion go through each section
and are completely separated from the air on the outside.

Gas furnace heat exchangers are normally made of
cast iron or sheet steel. Most of them currently used
are made of steel. The steel is formed into sections, and
the sections are joined together to make the exchanger.
Oil furnace heat exchangers are normally round, drum-
shaped sections. The burner flame fills the lower part,
or fire pot, and the upper part serves as the exchanger.
The air from the blower passes over the outside of the
fire pot and exchanger, and the flame and flue gas is
contained inside.

Although there are many types and styles of heat
exchangers, each one performs the same functions in
the unit: they extract heat from the flames and products
of combustion, and they separate the products of combustion from the air used to heat the building.

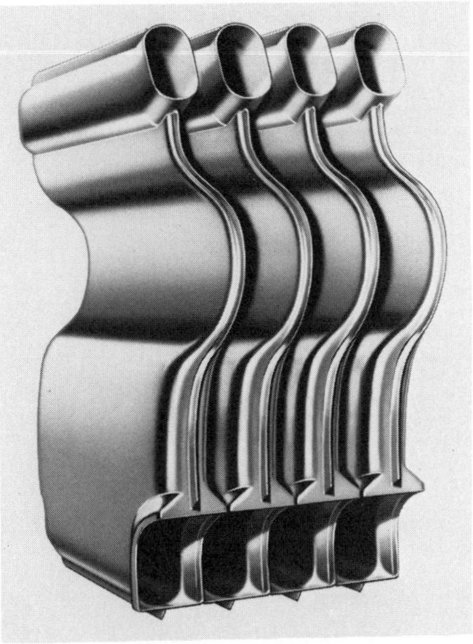

FIGURE 6-14
Heat exchanger used in warm air furnace

## 6.3  SUMMARY

Proper control of the combustion process for the production of heat is necessary. This control is provided
by burning fuel in a heating unit that produces the heat
at a specified rate and also provides a medium to distribute the heat. Air and water are the most common
mediums.

Warm air heating equipment is made up of a cabinet, filter, blower, burner, heat exchanger, and necessary controls. The performance of each part is
important. The way each interacts with the others
makes the safe and controlled production of heat possible.

## 6.4  QUESTIONS

1. There are two basic types of heating units. One that uses
   air as a medium is called a _____ _____ furnace.

2. List the six major components of a warm air furnace.

3. A furnace cabinet provides two important functions. One is to hold the components together. What is the other?

4. A centrifugal blower will move air against low to medium resistance in a duct system with a minimum use of energy. True or false?

5. Axial flow fans are used on high pressure air systems. True or false?

6. Direct-drive centrifugal blowers are often driven by _____ _____ motors.

7. Use of an adjustable motor pulley on a belt-driven blower makes it possible to regulate the _____ output of the blower.

8. Name the two basic types of filtration systems used for circulated air systems.

9. Particle size is important in rating filter _____.

10. The two principles of filtration used in mechanical filters are _____ and _____.

11. Electronic filters are more efficient than mechanical filters in filtering out large particles. True or false?

12. Name the three main functions of any burner.

13. A heat exchanger is used to extract _____ from flames and products of combustion and to transfer it to the _____ medium.

## 7.1 INTRODUCTION

Natural gas is one of the most commonly used fuels. It follows, then, that natural gas burning equipment is common. This chapter covers the different types of natural gas furnaces. The chapter contains a general description of the parts used in gas-fired furnaces, such as the burner and heat exchanger, and the typical applications for the different types of furnaces. Combustion safety controls are also covered.

While natural gas is the most commonly used fuel gas, other gases are also used for heating. Most important among these are the liquified petroleum gases. Only minor modifications have to be made to a unit to change from one gas to the other. The descriptions of the furnaces in this chapter can apply to all gas-fired furnaces.

## 7.2 MAJOR PARTS

A gas-fired furnace has six major parts:

1. Cabinet or frame,
2. Filter,
3. Blower,
4. Burners,
5. Heat exchangers,
6. Controls.

The first three of these were covered in detail in Chapters 5 and 6, and the controls will be covered in a later chapter. This chapter deals with burners, heat exchangers, and safety controls.

## Burners

A furnace burner is the part of the unit where combustion takes place. The burner is designed to control combustion so heat is produced safely and at a fixed rate. Two basic types of burners are used in gas-fired heating equipment: atmospheric and power. Atmospheric burners generally are used in smaller units that are installed inside buildings. Power burners are used on larger units and especially those installed outside a building, such as in rooftop units.

CHAPTER

# 7

# Gas-Fired Furnaces

**Primary air**    Air that is mixed with fuel gas before firing

**Secondary air**    Air that helps support combustion by mixing with the fuel gas in the actual burning process

**Orifice***

**Burner port(s)***

**Entrained air**    Air that is drawn into the stream of gas coming out of the orifice

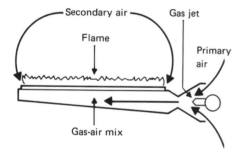

FIGURE 7-1 _____

Drawing of atmospheric burner showing how the gas and air mix and are carried to the burner ports where combustion occurs

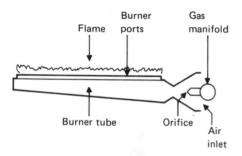

FIGURE 7-2 _____

Parts of atmospheric burner

**Atmospheric Burners.**    Atmospheric burners, as the name implies, rely on atmospheric pressure to supply air for combustion. Figure 7–1 shows an atmospheric burner and indicates where air enters the burner tube at the front and also where the air is drawn into the flame. The air that mixes with the gas in the tube is called **primary air;** the air that is drawn in at the flame is **secondary air.**

The mixture of gas and air must be in the proper proportions for combustion to be achieved. The design of the burner is such that this mixture is achieved by natural draft, from the air surrounding the unit.

The description given here is of an individual burner. A furnace actually has a number of these burners fastened together into an assembly called a burner tray. Each section of the heat exchanger has one burner. The number of burners and sections of the exchanger determines the heating capacity of the unit.

The construction of an atmospheric burner is simple and practical. See Figure 7–2 for an illustration of a burner. The burner is a tube, open at one end, where air and gas enter. The gas is brought to the open end of the burner tube through a pipe called a manifold and flows into the tube through a fitting called an **orifice.** A hole in the orifice is drilled to the proper size to meter the flow of the gas. The tube has a series of **burner ports** or openings along the top where the gas-air mix, now a combustible mixture, leaves the tube and is ignited.

When the gas enters the burner tube out of the orifice, it enters at a high enough velocity to create a low-pressure area around it. Air is **entrained** with the gas because it is pulled into this low-pressure area. The air that is entrained is the primary combustion air. Normally, there is a shutter on the open end of the tube, around the orifice. The shutter can be adjusted to regulate the amount of air drawn in so that the proper mix of gas and air is achieved. Figure 7–3 illustrates the air adjustment shutter located on the burner tube face.

The gas-air mix flows down the burner tube and out of the burner ports on the top of the tube. This mix

*Term defined in text.

is ignited by a pilot light or spark as it leaves the ports. Although the gas and air in the tube may be in the right proportion to allow ignition, additional air for combustion is always necessary. This additional air flows in through the front of the burner vestibule and mixes with the gas at the flame during the actual combustion process. This air is the secondary air. A properly designed burner will achieve smooth, quiet, and efficient combustion.

The gas pressure for an atmospheric burner should be set at 3-1/2 in. wg pressure so that it will function properly. Also, the primary air has to be adjusted correctly, and there must be sufficient secondary air for complete combustion.

The flow of gas to the burner is controlled by a gas valve similar to the one shown in Figure 7–4. Many types of valves are used on heating equipment, but the basic operation of most of them is quite similar. The valve, located in the gas line ahead of the manifold, is an electric-solenoid-operated valve. When a solenoid, which is part of the valve assembly, is energized, the valve opens, and gas flows through it. When the solenoid is not energized, the valve is closed, and no gas flows.

Some gas valves have built-in gas pressure regulators, and some contain automatic pilot safety valves. If there is no pressure regulator in the valve, there will be one in the gas line ahead of it. There will also be a pilot safety valve in the line if it is not part of the valve assembly.

The function of the pressure regulator is to regulate the line gas pressure so that it is suitable for the orifices used in the burner. The pilot safety valve provides safety shutoff if the pilot is not lit on a call for heat.

The gas pressure is regulated to the 3-1/2 in. wg at the pressure regulator by an adjusting screw under a protective cap on the stem of the regulator. The cap is removed and the screw turned to adjust the pressure. This action is done while the pressure in the gas line downstream from the regulator is checked with a manometer or a gauge.

After the pressure is set, the primary air is adjusted by turning the primary air shutter on the front of each burner. The adjustment is right when the flame burns evenly along the surface of the ports. The flame should

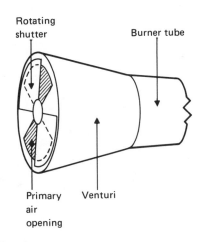

FIGURE 7–3

Primary air adjustment shutter on atmospheric burner

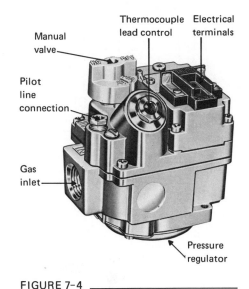

FIGURE 7–4

Typical gas valve

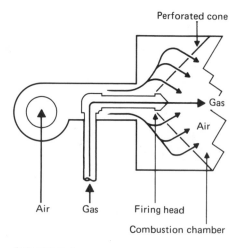

FIGURE 7-5 _____

Combustion air and gas mixing at the firing head in a gas power burner

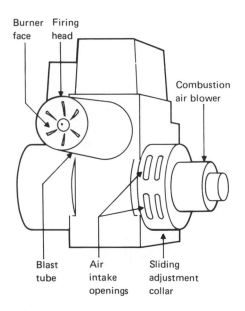

FIGURE 7-6 _____

One type of combustion air adjustment used on a power burner

be blue with just a touch of yellow in the tips. If there is too little primary air, the flame will be lazy. It will burn very softly and have a great deal of yellow in it. If there is too much primary air, the flame will be hard. The velocity of the gas-air mix through the ports can be great enough to cause the flame to burn up above the burner, causing a noisy flame.

If not enough secondary air is available at the burner location, the flame will appear very soft and lazy and will often burn out through the front of the burner ports. In some cases, the flames can roll out enough to burn the wiring and the controls in the **burner vestibule.** Usually, insufficient secondary air will occur when the furnace room is partitioned off and provisions are not made to bring air into it from other parts of the building or from outside.

**Power Gas Burners.**    A power gas burner is one in which combustion air is provided by a blower instead of by atmospheric pressure. Gas is introduced through a firing head, which is a nozzle, and air from the combustion blower is introduced around the head. The diagram in Figure 7–5 illustrates the construction of a gas power burner.

A typical gas power burner has an orifice in the gas line leading to the burner head. This orifice is sized to control the flow rate of the gas and thus regulate the heat output of the unit. The burner face is a perforated, diverting nozzle. In some burners, it is round, and in others it may be square or rectangular. The gas flows out of the nozzle and mixes with the combustion air. This gas-air mix is ignited by an ignition device, such as a spark plug. Since the proper amount of air has to be mixed with the gas for good combustion, there is an adjustment on the combustion air intake, at the blower. The drawing in Figure 7–6 shows one type of air adjustment used on power burners.

The complete power burner assembly has a gas valve, a pressure regulator, and ignition safety controls. The valve and the pressure regulator are similar to those used with an atmospheric burner, but a standing pilot is usually not used with a power burner so a pilot safety is not needed. Instead, a flame-surveillance type of safety is used. In this type of safety, a **light-sensitive cell** or **flame-rod** type of sensor monitors the fire, both on start-up and continually during a call for heat.

The light-sensitive type of safety is similar to that used on an oil-fired burner. The flame-rod safety has an electric conductor located so that it is in the flame when the burner is operating. When the fire is burning, the flame itself and the conductor form an electric circuit. This circuit is part of a safety circuit, and the flow of electric current through it is monitored by an electronic control device. If on a call for heat, the flame is not present within a safe time, the control device will shut the unit down on safety. The control device will also shut the unit off "on safety" if the fire is extinguished during a call.

A power burner can be designed as a two-stage burner, with a low-fire and a high-fire stage. A two-stage burner has one-half the heat output at low stage as it does at high stage. This difference allows a designer to select a unit that will more nearly match the heat load on a building at two different load conditions. When a two-stage burner is selected, two gas valves are used; one controls gas flow for low-stage fire, and both are used for high fire. When a unit is two staged, the combustion air flow rate has to be regulated for each stage. There must be less air for low stage and more for high. This regulation is accomplished by using a motorized damper on the combustion air intake. When the thermostat calls for one-stage fire, the damper is opened the proper amount, and on a call for two stage, the damper opens more.

Power burners are usually used in larger furnaces or in units that are located outside. When a heating unit is installed outside a building, the wind can affect the air pressure around the unit. In such a case, an atmospheric burner is hard to control. Because a power burner is not dependent upon atmospheric pressure for its operation, a power burner can be used outside as well as inside.

## Heat Exchangers

The heat exchanger is one of the most important components of the heating unit. Its purpose is to allow the exchange of heat from the products of combustion to the air.

The design of the exchanger varies somewhat depending on the burner used. Two basic types of heat exchangers are used in warm air heating equipment:

**Burner vestibule**    The space in the front of the furnace where the burner and controls are mounted

**Light-sensitive cell**    Any one of several types of photoelectric cells used for flame sensing

**Flame rod**    An electric conductor used as a flame-sensing device

the clamshell and the drum type. The clamshell exchanger is normally used with atmospheric burners, and the drum type is used with power burners.

**Materials Used in Heat Exchangers.**    Most of the heat exchangers manufactured today are made of sheet steel. Cast iron was used in the past, but because of the greater mass of material in cast iron, heating of the sections was slower, and consequently cooling of the metal was also slower on an off cycle. This slowness was an advantage in furnaces without blowers because of the desirability to have a constant, steady source of heat to move the air. In today's forced-air systems rapid transfer of heat is more desirable, so lighter and thinner metal is used in the exchangers.

So that the heat exchanger has a long life expectancy, special steels or special coatings often are used in its construction. The special materials make the exchanger more resistant to corrosion caused by acids produced by the products of combustion or to burnouts from the flame itself.

Aluminized steel is often used for the construction of exchangers. This steel has been coated with aluminum in a process that bonds a thin coat of aluminum to the outer surface of the steel. This process makes the steel more resistant to corrosion.

Glass- or ceramic-coated steel is also used for exchangers. The coating in this case is a thin layer of glass or ceramic material that is melted onto the steel exchanger. The glass or ceramic makes an impervious coating on the steel that will resist corrosion as long as the coating is intact.

Stainless steel is another material used for heat exchangers when extra durability is desired. Stainless steel is resistant to most of the types of corrosion that may occur with heat exchangers. It is an excellent material for heat exchanger construction.

All of the special coatings or materials used for heat exchangers are more expensive than plain steel and add to the cost of the original unit. In applications where corrosion may be a special problem, however, the cost difference may easily be offset by replacement costs. In some cases, manufacturers give a better warranty on the exchanger made out of special materials than they do on the plain steel one.

**Clamshell Exchanger.**   The clamshell exchanger is the most common type of heat exchanger. It is used in smaller gas-fired furnaces almost exclusively. Drawings of a typical heat exchanger and a downflow furnace with the heat exchanger installed are shown in Figure 7–7.

The clamshell exchanger consists of multiple sections, or "clams." Each is formed of two pieces of steel welded together in such a way that they form a tight chamber inside. The burners are then inserted in these sections, at the bottom, as shown in Figure 7–8.

A number of sections are joined together to make up an exchanger. In most furnaces, each section is designed for an input of approximately 25,000 Btu/h of heat. The capacity of a furnace can be estimated by multiplying the number of sections in the exchanger by 25,000. The result is the approximate Btu/h input capacity of the furnace.

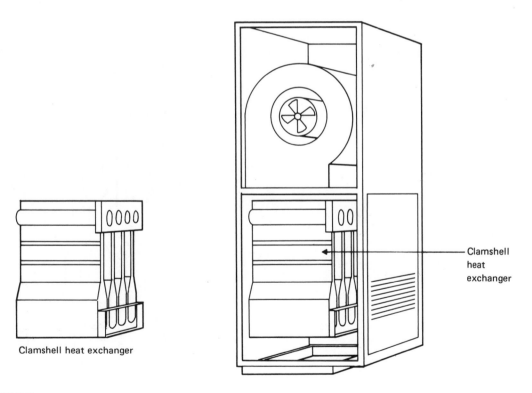

Clamshell heat exchanger

Clamshell heat exchanger

FIGURE 7–7

Clamshell heat exchanger and cutaway of downflow furnace with exchanger installed

**Spot weld**    A method of welding two pieces of thin metal together by clamping them between two metal electrodes and passing an electric current through the metals between the electrodes

**Metal fatigue**    Breaking of metal caused by continued bending and flexing

**Fire pot***

**Refractory**    A material that will not melt at the operating temperature of the burner; a device used to separate the heat exchanger surface from the flames in a furnace

When the burners are fired, the flames and products of combustion rise through the inside of the clamshell sections, and the flue gases are exhausted at the top of each section through an opening that leads to the flue. The air that is being heated passes up through the openings between the clamshell sections. Figure 7–9 shows a diagram with the airflow around the outside of the sections of a heat exchanger.

As the burner cycles on and off, the various sections of the heat exchanger heat and cool. In this process, the metal expands and contracts, so provisions must be made in the construction of the clamshells to keep the metal from distorting. The usual arrangement consists of internal ties and braces or **spot welds** to prevent this distortion. These ties make the clamshell somewhat rigid and can create problems. If the metal is restrained too rigidly, it will bend at the points of at-

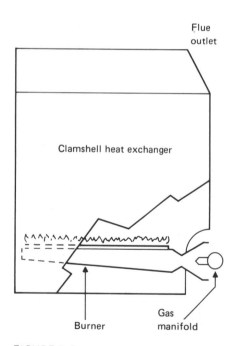

FIGURE 7–8
Clamshell heat exchanger with cutaway showing burner installed

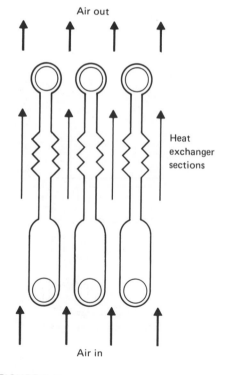

FIGURE 7–9
Airflow in clamshell heat exchanger

tachment and will break due to the continuous movement. Also, if the attachments are too rigid, the heat exchanger will be noisy because of stresses created at the points of connection. The noises are popping or cracking sounds heard when the exchanger is heating up or cooling down. Manufacturers have to provide for some movement of the heat exchanger in the unit to prevent this noise or breaking due to **metal fatigue,** or both.

**Drum-Type Exchanger.**     A second type of heat exchanger used is the drum type. The drum type of heat exchanger is normally used with a power burner. The first is contained in the lower part of the heat exchanger, which is called the **fire pot.** The fire pot is made of heavy gauge metal and is lined with **refractory** materials. The fire pot is actually part of the exchanger. Figure 7–10 shows a typical design for a drum-type exchanger. The flames from the burner are contained in the fire pot, and the hot products of combustion rise out of the pot into a larger section at the top. The hot products then flow through a connection to the flue pipe. The air to be heated flows around the outside of the fire pot and the upper section, where it picks up heat from the hot surfaces. The hot air then is used for heating the building.

Some drum-type heat exchangers have an additional section where the products of combustion flow through secondary tubes after passing through the main exchanger. In such an operation, more hot metal surface is exposed to the air. This type of exchanger is called a primary-secondary, or multipass, exchanger. A sectional drawing showing the construction of a multipass exchanger appears in Figure 7–11.

**Flue Connections.**     The basic function of the heat exchanger is to keep the products of combustion separate from the air being heated. Thus, the heat exchanger must be connected to a vent or a flue to exhaust the flue gases. Openings are made in the top of the clamshells or drum and then connected to the flue. In an atmospheric burner, the flue openings from the clamshells are connected to a device called a draft diverter. The draft diverter connects to the flue. On a drum-type exchanger, the connection is directly to the

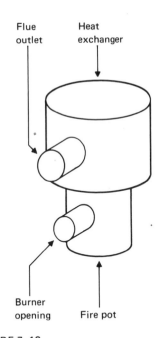

FIGURE 7–10 _____
Drum-type heat exchanger

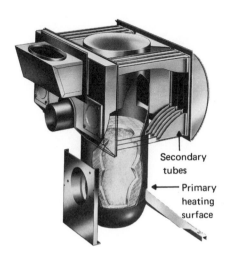

FIGURE 7–11 _____
Cutaway view of multipass heat exchanger

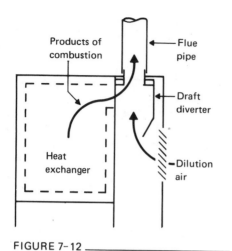

FIGURE 7-12 _____

Flue gas leaving heat exchanger on a
gas furnace

flue. Figure 7–12 shows the path that the flue gas takes as it leaves a clamshell exchanger and goes into the flue or vent.

The draft diverter on an atmospheric burner heat exchanger is used to allow **dilution air** to mix with the products of combustion and to prevent the natural draft effect from pulling the flames out of the firebox.

# Heating Capacity

Gas-fired furnaces are available in many different sizes and heating capacities. Manufacturers make different sizes available so designers can better match the heat loss for any given job. To provide different heat capacities, the manufacturer has to consider three factors:

1. Heat value of the gas to be used,
2. Flow rate of the gas into the unit,
3. Efficiency of the heat exchanger.

Each of these will be covered in the sections that follow.

**Heat Value of Gas.**    The heat value of the gas is determined by the type of gas to be fired. Natural gas, the most common of the gases used, has a nominal value of 1000 Btu per cubic foot. For every cubic foot of gas that the furnace uses, 1000 Btu of heat will be produced. A 100,000 Btu/h furnace burns 100 cu ft of gas per hour.

Liquid petroleum gases (LPG) come in the liquid state and are rated in Btu per gallon. The heat value of these gases per gallon varies as the chemical composition varies. The values are shown best by the tables found in Figure 2–1 in Chapter 2.

**Gas Flow Rate.**    The flow rate of the gas into the burner is controlled by the pressure of the gas in the line and by the size of the hole drilled in the orifice at each burner head. The gas pressure is controlled by a pressure regulator. Units are commonly designed to produce a specific amount of heat per burner. This means that each orifice is sized to allow the quantity of gas to flow per hour that will produce that amount of heat exactly. The hole in the orifice is sized according to the size shown on an orifice sizing chart. One is found in Table A–2 in the Appendix. This chart shows the orifice size required for specific flow rates at various gas pressures.

**Efficiency.**    The third factor is the efficiency of the burner and the heat exchanger combination used in the furnace. The efficiency rating of a unit gives the amount of heat available for heating a building as a percentage of that produced in the combustion process. The normal efficiency factor is 80%. Some heat has to be allowed to go up the flue to provide draft. About 20% of the total produced goes up the flue, so 80% is left for heating. The amount of heat produced in the combustion process is called the **input** of the furnace, and the amount left over after deration is called the **output.**

**Unit Sizes.**    Gas furnaces are manufactured in many different physical sizes and heating capacities in each of the styles available. The physical sizes are determined by the capacity of each unit. The capacities for residential furnaces start at about 75,000 Btu/h. The capacities increase in increments of approximately 25,000 Btu/h, up to 300,000 Btu/h. Much larger units are made for commercial or industrial applications. Figure 7–13 is a page from a manufacturer's catalog showing some of the sizes available in one type of furnace.

# Combustion Safety Controls

To ensure that the fuel-air mix is fired as soon as it is introduced to a burner, combustion safety controls are used. Because the mix is explosive, it must be fired as soon as it is introduced. Any delay in firing could cause a buildup of the mixture, and if an accumulation was fired, the result could be a disastrous explosion. The furnace is designed to burn fuel at a steady, sustained rate but not to burn large quantities at one time.

Combustion safety controls monitor the combustion process to make sure that firing takes place within a reasonable time after gas starts to flow. If firing does not take place, the control shuts the gas valve; or if the flame goes out during a firing period, the valve will be shut off.

The two most common types of combustion safety

**Dilution air**    Air that is allowed to mix with the products of combustion after they leave the heat exchanger

**Input**    Heat, in Btu/h, produced by a furnace as related to fuel use

**Output**    Actual heat, in Btu/h, available for heating after a furnace has been derated for efficiency loss

PP/PRODUCT PRESENTATION — HEATING

# RESIDENTIAL TYPE SINGLE ZONE HEATING

Up-flo, down-flo and horizontal models are available in gas, oil or electric heat. Most are designed to accept air conditioning. A large selection of evaporator coils, outdoor condensing units, electronic air cleaners and humidifiers can be added for economical all-season total comfort.

All models are equipped with either resiliently mounted belt drive or multispeed beltless blowers. A durable, enamel finish is baked-on for lasting good looks. Matching return air cabinets have been designed to compliment up-flo installations.

**Gas furnaces** feature the exclusive Lennox DURACURVE ® heat exchanger. It's revolutionary design reduces metal fatigue, ticking and resonance. An optional DURAGLASS II ® coating helps fight corrosion for added heat exchanger life. Quiet continuous-slot burners provide complete combustion. They end irritating popping and flashback.

**Electric furnaces** contain strong nichrome heating elements with higher temperature capabilities and longer life. They are placed directly in the airstream to give instantaneous heat transfer — no waste. Completely automatic controls efficiently sequence elements. Double protection is provided by circuit breakers and individual element temperature limit controls.

*Conservator* **Electric ignition and optional vent damper for highest gas heat efficiency. Also on G12 "E" models.**

**GAS UP-FLO MODELS**

| Model Number * | Input Btu/h | Tons of Add-on Cooling | Flue Size (in.) | Net Weight (lbs.) | Dimensions (in.) | | |
|---|---|---|---|---|---|---|---|
| | | | | | H | W | D |
| G12D2-40 | 40,000 | 1½ or 2 | 4 | 143 | 49 | 16¼ | 26⅛ |
| G12D2E-40 | 40,000 | 1½ or 2 | 4 | 143 | 49 | 16¼ | 26⅛ |
| G12D2-55 | 55,000 | 1½ or 2 | 4 | 143 | 49 | 16¼ | 26⅛ |
| G12D2E-55 | 55,000 | 1½ or 2 | 4 | 143 | 49 | 16¼ | 26⅛ |
| G12D2-82 | 82,000 | 1½ or 2 | 5 | 158 | 49 | 16¼ | 26⅛ |
| G12D2E-82 | 82,000 | 1½ or 2 | 5 | 158 | 49 | 16¼ | 26⅛ |
| ▲ G11Q3E-82V Conservator | 82,000 | 2, 2½ or 3 | 4 | 167 | 49 | 16¼ | 26⅛ |
| G12Q3-82 | 82,000 | 2, 2½ or 3 | 4 | 166 | 49 | 16¼ | 26⅛ |
| G81-85-110 | 85,000 / 110,000 | 2½ thru 5 | 5 | 275 | 53 | 26 | 26⅛ |
| G12D-110 | 110,000 | ---- | 5 | 185 | 49 | 21¼ | 26⅛ |
| ▲ G11Q3E-110 | 110,000 | 2 thru 3 | 5 | 194 | 49 | 21¼ | 26⅛ |
| G12Q3-110 | 110,000 | 2 thru 3 | 5 | 192 | 49 | 21¼ | 26⅛ |
| G12Q3E-110 | 110,000 | 2 thru 3 | 5 | 192 | 49 | 21¼ | 26⅛ |
| G12Q4-110 | 110,000 | 3½ or 4 | 5 | 200 | 49 | 21¼ | 26⅛ |
| G12Q4E-110 | 110,000 | 3½ or 4 | 5 | 200 | 49 | 21¼ | 26⅛ |
| ▲ G11Q3E-110V Conservator | 110,000 | 2½ and 3 | 5 oval | 194 | 49 | 21¼ | 26⅛ |
| ▲ G11E-110V Conservator | 110,000 | 2, 2½ or 3 | 5 oval | 193 | 49 | 21¼ | 26⅛ |
| G12Q3-137 | 137,000 | 2½ and 3 | 6 | 236 | 53 | 26¼ | 26⅛ |
| G12Q3E-137 | 137,000 | 2½ and 3 | 6 | 236 | 53 | 26¼ | 26⅛ |
| G11Q4E-137 | 137,000 | 3½ or 4 | 6 | 252 | 53 | 26¼ | 26⅛ |
| G12Q4-137 | 137,000 | 3½ or 4 | 6 | 252 | 53 | 26¼ | 28⅛ |
| G12Q4E-137 | 137,000 | 3½ or 4 | 6 | 252 | 53 | 26¼ | 26⅛ |
| G12Q5-137 | 137,000 | 3½ thru 5 | 6 | 262 | 53 | 26¼ | 26⅛ |
| G12Q5E-137 | 137,000 | 3½ thru 5 | 6 | 262 | 53 | 26¼ | 26⅛ |
| G11Q3E-137V Conservator | 137,000 | 2½ or 3 | 6 oval | 239 | 53 | 26¼ | 26⅛ |
| ▲ G11E-137V Conservator | 137,000 | 3, 3½, 4 or 5 | 6 oval | 249 | 53 | 26¼ | 26⅛ |
| G12Q5-165 | 165,000 | 3½ thru 5 | 6 | 278 | 53 | 31¼ | 26⅛ |
| G12Q5E-165 | 165,000 | 3½ thru 5 | 6 | 278 | 53 | 31¼ | 26⅛ |
| ▲ G11Q5E-165V Conservator | 165,000 | 4 or 5 | 6 oval | 301 | 53 | 31¼ | 26⅛ |
| ▲ G11E-165V Conservator | 165,000 | 3, 3½, 4 or 5 | 6 oval | 278 | 53 | 31¼ | 26⅛ |
| G81-165 | 165,000 | 6 | 6 | 520 | 58 | 38 | 28 |
| G11E-200V | 200,000 | 3½, 4 or 5 | 7 | 291 | 53 | 31¼ | 26⅛ |
| G81-220 | 220,000 | 7½ or 11 | (2) 5 | 675 | 58 | 51½ | 28 |

▲ Also available in 50 Hz models.

FIGURE 7–13

Page from manufacturer's catalog showing some of the sizes of units available

controls used on furnaces are (1) pilot safety and (2) flame surveillance.

**Pilot Safety Controls.**    Until recently, most gas-fired furnaces using atmospheric burners utilized a standing pilot, or a pilot flame that was constantly burning, to fire the gas-air mix on a call for heat. In such systems, it is necessary to provide some means of monitoring the pilot flame to make sure it is burning so the gas will be ignited when the gas valve opens. The device that is used most commonly to do this monitoring is called a **thermocouple.** It is used with a safety valve. The thermocouple is located so the pilot flame heats the end of the device when the flame is lit. Figure 7–14 shows some of the types of thermocouples available from one manufacturer.

When a thermocouple is heated, it generates a low-voltage electrical signal. This signal is conducted to a magnetic coil in a shutoff valve located in the gas line ahead of the control valve. In some cases, this shutoff valve is part of the main gas valve itself. A typical gas valve with the safety shutoff coil and valve is shown in Figure 7–15.

When the electric coil in the shutoff device is energized by the electric current from the thermocouple, the gas can flow through the valve. If the pilot flame is out, no current is generated, and the shutoff valve will not open. Consequently, gas will not flow into the burners.

There are other less commonly used types of sensing devices for flame monitoring. Among them are warp switches and remote bulb temperature-pressure devices. In a warp switch, heat from the flame is converted directly into mechanical action, which is used to control a valve. In the remote bulb type of control, the heat causes pressure in a tube, and the pressure is used to generate a mechanical motion, which controls a valve.

**Electric Ignition Safety.**    Standing pilot ignition systems may be wasteful of gas since the pilot flame is always on. Instead, electric ignition can be used on gas furnaces to save gas. This type of ignition system ignites the gas-air mix with an electrical spark. The ignition system may be part of the original package or

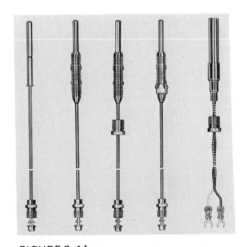

FIGURE 7-14 _____
Types of thermocouples available from one manufacturer for pilot safety application

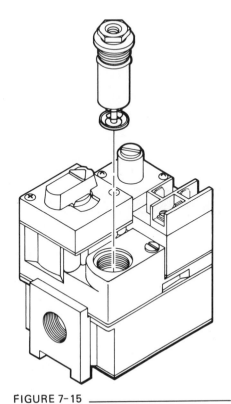

FIGURE 7-15 _____
Gas valve with pilot safety built in

**Thermocouple**    A device that when heated produces a low-voltage electric signal

**Thermistor**    A solid-state device in which the resistance to an electric current varies as the temperature varies

used as a conversion on a conventional gas furnace. On the original ignition system, the spark is used to fire the gas-air mix directly on a call for heat. If the system is used on a furnace with a standing pilot, the spark is used to fire the pilot each time the thermostat calls for heat.

When electric ignition is used, a flame-proving method of ignition safety is used. This method is usually one of three types: (1) a bimetal warp switch, (2) a thermocouple generating a low-voltage electric signal, or (3) a solid-state system using a **thermistor** to monitor a low voltage signal. In each case, ignition safety is provided.

**Power Burner Combustion Safety.**    When a power burner is used, the flame must be monitored both on start-up and during the "on" cycle. On a call for heat, a combustible fuel-air mixture enters the combustion chamber. If ignition does not take place, the accumulation of this mix is highly dangerous. Ignition of a large quantity of fuel and air would cause an explosion.

The most common method of monitoring ignition when a power burner is used is by a type of cell that "sees" the flame when the burner is on and that controls an electric signal generated by a control center. The electric signal, in turn, is used to control the main valve or operating relay for the burner.

The cell that is used to scan the flame is an electronic device in which the resistance to electrical flow varies with the amount of light the cell absorbs. Figure 7–16 is a photograph of a light cell showing the mounting and connecting wires. The more light that strikes the cell, the lower the resistance to the flow of electricity; the less light that strikes the cell, the higher the resistance. Therefore, it is possible to monitor a flame by sending a low-voltage electrical signal through the cell and metering the amount of current flowing. The higher the resistance, the lower the current; the lower the resistance, the more current that will flow. The low-voltage signal is generated by a control center, and the current monitoring device is located there also. The electric circuit that controls the gas valve is also routed through the control center, so the operation of the gas valve is dependent upon the signal received

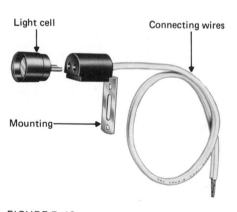

**FIGURE 7–16** _____
Light cell, mounting bracket, and connecting wires used on a combustion safety device

through the sensor. Figure 7–17 shows a control center used with a light-sensing type of sensor.

## 7.3  SUMMARY

All gas-fired furnaces, regardless of the particular gas fuel used, have many similarities. The fuel must be mixed with the air in the right proportions to burn efficiently, and provisions must be made to control the combustion process. Also, the furnace has to be built so that heat from the combustion process can be extracted from the flames and products of combustion and used to heat air.

The two main components of the furnace that produce the heat, and extract it, are the burners and the heat exchanger.

Two types of burners are used. The first is the atmospheric burner. This burner gets all of the combustion air from the atmosphere by natural draft. The other type of burner is the power type. This burner uses a blower to provide the combustion air. Each type of burner has special characteristics that make it more practical than the other to use in some circumstances.

Also, two main types of heat exchangers are used for gas-fired furnaces: the clamshell and drum-type exchangers. The clamshell exchanger is normally used with the atmospheric burner and has separate sections for each burner to fit into. The drum-type exchanger has a fire pot where the flame is produced by the burner and a separate section where most of the heat is transferred to the air. This type of exchanger is most often used with a power burner.

Some type of ignition safety control must be provided on each furnace. When the air-gas mix is introduced into a furnace through the burner, the mixture is explosive. The mixture must be fired at a continuous rate, or it can create a highly dangerous condition. Since various means are used to fire the gas-air mix, there are different types of flame proving, or combustion safety, devices used to make sure the ignition takes place. Among these are warp switches, thermocouples, and flame surveillance. Each device is used on various types of heating units. The choice of which to use depends

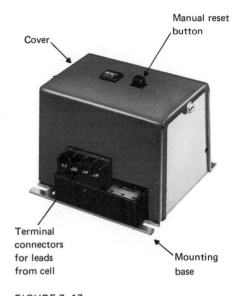

Cover

Manual reset button

Terminal connectors for leads from cell

Mounting base

FIGURE 7–17

Control center with light-sensing type of sensor

upon the type of burner used and the application the furnace is designed for.

## 7.4 QUESTIONS

1. There are great differences in gas-fired heating units that burn different gas fuels. True or false?

2. Two main types of gas burners are used. They are the _____ and the _____ burner.

3. Combustion air for the atmospheric burner is provided by _____ _____.

4. Match the term in the first column with the proper description in the second column.

   1. Primary air      a. Mixes in the flame
   2. Secondary air    b. Mixes with the flue gas
   3. Entrained air     c. Mixes with the fuel
   4. Dilution air      d. Drawn into the fuel stream

5. Atmospheric burners work well in heating units that are installed on rooftops because they are outside. True or false?

6. The combustion air provided for firing in a power burner provides all of the air needed for good combustion of the fuel. True or false?

7. The two main types of heat exchangers used for gas-fired heating units are _____ and _____.

8. Special steels, or coatings, are used in the construction of heat exchangers because heat exchangers made of them are more _____ to _____ than are mild steel exchangers.

9. A clamshell heat exchanger is made up of several sections, or "clams," attached to each other. The heating capacity of the heat exchanger is partially determined by the number of clams used. True or false?

10. A clamshell heat exchanger is always a quietly operating exchanger, regardless of how it is built. True or false?

11. In a drum-type heat exchanger, the flames are contained in a _____.

12. Combustion safety controls are used only on furnaces in commercial buildings. True or false?

13. The two main types of combustion safety control systems used on gas-fired furnaces are _____ and _____.

## 8.1 INTRODUCTION

Oil-burning heating equipment is common in residential, industrial, and commercial applications. Although many of the components of an oil furnace are similar to those used in gas furnaces, there is a great difference in the oil burners and the heat exchangers.

Since the burner in an oil furnace is of a special type, the heat exchanger is special also. This chapter describes the heat exchangers most often used in oil-burning equipment and explains the function of the exchanger in relation to the burner. The chapter also examines the types of combustion safety controls used to ensure proper firing of the oil-air mix in the oil burner.

## CHAPTER

# 8

# Oil-Fired Furnaces

## 8.2 MAJOR PARTS

Oil-fired heating units are manufactured in many different styles and varieties—such as upflow, downflow, and horizontal. Figure 8–1 shows these major configurations.

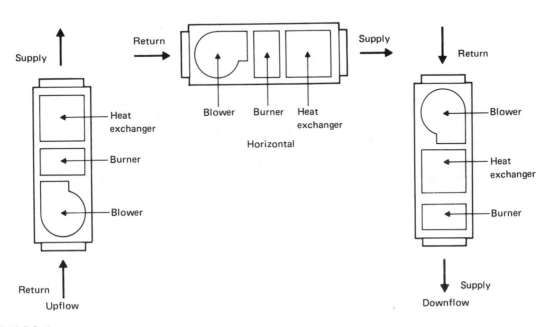

FIGURE 8-1

Three normal configurations of furnaces

FIGURE 8-2 _____

Typical oil burner suitable for a residential furnace

Oil-fired furnaces do differ from gas furnaces in some important respects. The basic difference is found in the way the fuel is fired. An oil burner is basically a power burner. That is, it does not get its combustion air strictly by atmospheric pressure; rather, the air is supplied by a blower. Since the oil is in a liquid state but can be burned only in the vapor state, one of the main functions of the burner is to vaporize the oil and then mix it with air for burning. As the firing methods vary between gas and oil, so do the burners, heat exchangers, and safety controls.

## Oil Burners

The majority of the burners used in oil furnaces are power burners. Some small units, and especially room heaters, use **pot-type burners,** but these are seldom used in furnaces and will not be covered in this text. A photograph of a typical oil burner used in an oil-fired furnace is shown in Figure 8–2.

Oil is in a liquid state when it is introduced at the burner head. Since oil will not mix with air for combustion when it is in this state, the oil must be broken down into small droplets at the burner head so it can be mixed with air. This process is called **atomization.** Atomization is one of the functions of the burner. Oil is delivered to the **burner nozzle** at a fairly high pressure from a pump that is also part of the burner assembly. Air is delivered to the firing head from a combustion air blower. As the oil passes through the nozzle, it is broken up into very small droplets, mixes with the combustion air, and is then sprayed in a cone-shaped pattern into the fire pot. Figure 8–3 shows how the oil is atomized and mixed with air at the burner face.

The oil burner assembly is made up of various parts to perform the functions just described. These parts are: an oil pump, combustion air blower, gun assembly, blast tube, burner face, and ignition assembly. Each of these parts, in turn, is made up of several more parts.

The **oil pump** is a rotary pump that brings oil from the storage tank and delivers it to the nozzle at the burner head at a pressure of 100 lb per sq in. (psi). It is motor driven, normally by the same motor used to drive the combustion blower. The pump is self-lubricating and adjustable for outlet pressure. There is an

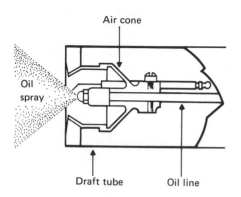

FIGURE 8-3 _____

Atomized oil and air mixing at the firing head of an oil burner for proper combustion adjustment screw on it and a port for a pressure gauge.

The **combustion air blower** is a small centrifugal blower, usually driven by the same motor that operates the pump. The blower provides air to the blast tube assembly so the air is available at the burner face. Because the proportion of air to oil vapor is critical, the combustion air blower can be adjusted to regulate the amount of air the blower puts out.

The **gun assembly** includes the oil pipe that runs from the oil pump to the nozzle at the burner face. The pipe goes through the center of the blast tube. The nozzle actually fits on the end of this pipe. The support bracket for the gun assembly is adjustable so that the nozzle can be positioned correctly in the burner face.

The large tube that carries the air from the combustion blower to the burner face is called the **blast tube.** The gun assembly fits inside it. The tube goes through an opening in the fire pot wall and is the connection between the oil burner outside the unit and the burner face inside.

The perforated plate on the inside end of the blast tube is the **burner face.** Combustion air passes through the perforations, and the nozzle is located in a hole in the center. Ignition **electrodes** are located just above and in front of the nozzle. The oil-air mix is ignited as it enters the fire pot.

The **ignition assembly** includes a transformer for raising the voltage of the electricity, leads from the transformer to the electrodes, and the ignition electrodes themselves. On a call for heat, the oil pump, the combustion air blower, and the ignition system are all energized at the same time. As the oil leaves the nozzle and is mixed with the air from the perforated face, a spark is generated between the two ignition electrodes, and the oil-air mix is ignited. The spark occurs adjacent to the cone of oil-air but not in it, as shown in Figure 8–4.

The flame that is produced by the oil burner is a concentrated flame; it appears in one general locality. Consequently, the flame can be contained in a fire pot where the hot lining material helps produce the high temperature needed for efficient combustion.

*Term defined in text.

**Pot-type burner**    A burner for liquid fuels in which the fuel flows into a vessel where it is vaporized by heat and burned in the same vessel

### Atomization*

**Burner nozzle**    A steel nozzle through which fuel oil is introduced at the burner face (The oil is partially vaporized, and the rate of flow is metered.)

### Oil pump*

### Combustion air blower*

**Gun assembly**    The oil line, burner nozzle, and ignition electrodes (These parts are mounted inside the blast tube and connect the oil pump to the burner face.)

### Blast tube*

### Burner face*

**Electrodes**    Steel rods that conduct electricity from an ignition transformer and produce a spark for ignition of the oil-air mix

### Ignition assembly*

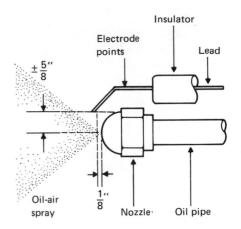

FIGURE 8-4 _____

Ignition of the oil-air mix in an oil burner

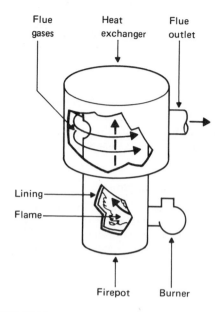

## FIGURE 8-5

Wrap-around or drum-type heat exchanger

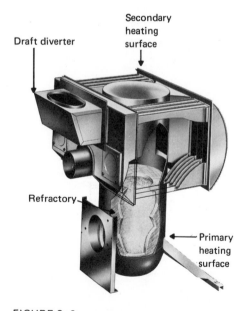

## FIGURE 8-6

Cutaway drawing of oil heat exchanger showing the secondary heating section in relation to the primary section

# Heat Exchangers

Oil-fired heat exchangers used with power burners are wrap-around, or drum type, as shown in Figure 8–5. Since a power burner produces a single flame, the heat exchanger used with it is designed to perform best with this flame pattern.

In an oil-fired heat exchanger, the combustion process takes place in a fire pot. The fire pot is the lower part of the exchanger and is lined with ceramic or fire-clay material. This material reduces the possibility of the flame's burning out that part of the heat exchanger and also produces an area of intense heat, which helps the combustion process.

After firing, the hot products of combustion rise out of the fire pot area and go into the heat exchanger proper. This part of the exchanger is called the **drum.** Here the metal is heated by the hot products of combustion from the inside, while the air blown over the outside surface is heated. This hot air is used to heat the building. After passing through the drum, the products of combustion go into a breaching, or vent, connection, which is connected in turn to the flue or chimney.

Some drum-type heat exchangers have a secondary section where the hot flue gases pass through additional tubes after leaving the drum, or primary section. See Figure 8–6. A heat exchanger with a secondary section is called a two-pass exchanger. The main advantage of this type of exchanger is that more heat can be extracted from the flue gases with a smaller exchanger.

# Heating Capacity

The heating capacity of an oil furnace, rated in Btu/h, is the function of the heat value of the oil burned, the rate of firing the oil, and the efficiency of the unit. Combustion and heat exchanger efficiency are figured together and are determined by combustion efficiency tests that include the performance of both.

**Unit Operation.**    Oil, being a liquid fuel, is measured in gallons, and the flow rate of oil into a burner is rated in gallons per hour (gph). The oil is pumped to the burner at a pressure of 100 psi and introduced at the firing head through a nozzle that controls the flow rate and breaks the oil down into small droplets to facilitate its mixing with air. The nozzle is sized to allow the oil to flow at a given rate—usually from about 0.65 gph to 1.50 gph for residential and smaller commercial units. Larger nozzles are used for large commercial and industrial units.

**Heat Value.**    Fuel oil comes in different grades, designated by numbers from 1 to 6. The smaller numbers relate to oil that has a lower viscosity, as well as other characteristics. See Table 8–1 for a table of specifications for the various grades of fuel oil.

Each of the different grades of oil has a different heat value, or Btu per gallon rating. The approximate value of the different grades is shown in Table 8–2.

If the combustion is complete in the firing process, which means that all of the fuel and air are combined in the process, then the combustion efficiency is 100%; if any of the hydrogen or carbon in the oil is not burned in the process, efficiency is something less than 100%. The burner and the fire pot are designed to work together. If the burner gets out of adjustment, then the balance between it and the fire pot is upset, and the efficiency is affected. Because an oil burner is a complex device, keeping it properly adjusted is not easy. The amount of combustion air used in the firing process also affects the efficiency of the burner. It is necessary to provide between 20% and 50% more air to the burner than that needed for combustion, as a safety precaution. The excess air passes through the combustion process unchanged and goes through the heat exchanger with the products of combustion. The presence of this excess air reduces the efficiency of the exchanger. The best efficiency that can be expected from an older model oil-fired unit is 80%. Manufacturers are continually improving the burners and heat exchangers, so higher efficiencies may be possible on newer models.

**Unit Sizes.**    No. 2 burner oil—that used for fuel in smaller furnaces—is nominally rated at 140,000 Btu/h per gallon. The smallest residential furnace is usually designed to use a 0.65 gph nozzle, and so the input of this furnace would be:

140,000 Btu/gal × 0.65 gph = 91,000 Btu/h

Drum*

TABLE 8–1

## Description and Specification of Common Grades of Heating Oil

| Grade of Fuel Oil[b] | Flash Point F | Pour Point F | Water and Sediment % by Vol. | Carbon Residue on 10% Residuum % | Ash % by Wt. | Distillation Temperatures F | | | Saybolt Viscosity Seconds | | | | Kinematic Viscosity Centistokes | | | |
|---|---|---|---|---|---|---|---|---|---|---|---|---|---|---|---|---|
| | | | | | | 10% Point | 90% Point | End Point | Universal at 100 F | | Furol at 122 F | | at 100 F | | at 122 F | |
| | Min. | Max. | Max. | Max. | Max. | Max. | Max. | Max. | Max. | Min. | Max. | Min. | Max. | Min. | Max. | Min. |
| 1. A distillate oil intended for vaporizing pot-type burners and other burners requiring this grade[d] API gravity 35 (min.) | 100 or Legal | 0 | Trace | 0.15 | | 420 | | 625 | | | | | 2.2 | 1.4 | | |
| 2. A distillate oil for general purpose domestic heating for use in burners not requiring No. 1 API gravity 26 (min.) | 100 or Legal | 20[c] | 0.10 | 0.35 | | [e] | 675 | | 40 | | | | (4.3) | | | |
| 4. An oil for burner installations not equipped with preheating facilities | 130 or Legal | 20 | 0.50 | | 0.10 | | | | 125 | 45 | | | (26.4) | 5.8 | | |
| 5. A residual type oil for burner installations equipped with preheating facilities | 130 or Legal | | 1.00 | | 0.10 | | | | | | 150 | 40 | | | 32.1 | (81) |
| 6. An oil for use in burners equipped with preheaters permitting a high viscosity fuel | 150 or Legal | | 2.00[f] | | | | | | | | 300 | 45 | | | (638) | (92) |
| PS 300 | 150 or Legal | | 1.0 | | | | | | | | 40 | 25 | | | | |
| PS 400 | 150 | | 2.0[f] | | | | | | | | | 60 | | | | |

[a] Recognizing the necessity for low sulfur fuel oils used in connection with heat-treatment, non-ferrous metal, glass and ceramic furnaces and other special uses, a sulfur requirement may be specified as 0.5% for No. 1, 1.0% for No. 2, and no limit for Nos. 4, 5 and 6.

Other sulfur limits may be specified only by mutual agreement between the buyer and seller.

[b] It is the intent of these classifications that failure to meet any requirement of a given grade does not automatically place an oil in the next lower grade unless in fact it meets all requirements of the lower grade.

[c] Lower or higher pour points may be specified whenever required by conditions of storage or use. However, these specifications shall not require a pour point lower than 0 F under any condition.

[d] No. 1 oil shall pass test for corrosion made in accordance with paragraph 15 *ASTM Specifications for Fuel Oils*, D 396 – 48.

[e] The 10 percent point may be specified at 440 F maximum for use in other than atomizing burners.

[f] The amount of water by distillation, plus the sediment by extraction, shall not exceed 2.00 percent. The amount of sediment by extraction, shall not exceed 0.50 percent. A deduction in quantity shall be made for all water and sediment in excess of 1.0 percent.

PS 300 and PS 400 are Pacific Specification numbers. Balance of table from *Commercial Standard* CS 12-48 and *ASTM Materials Specification for Fuel Oils* D 396 – 48.

PS 300 is a residual oil for use without preheating in furnaces and burners requiring a low viscosity fuel, and commonly described as light fuel oil, domestic fuel oil, or low viscosity fuel oil.

PS 400 is a residual oil for use in furnaces and burners equipped with preheaters permitting a high viscosity fuel, and commonly described as industrial fuel oil, heavy fuel oil, or high viscosity fuel oil.

Source: Reprinted with permission from American Society of Heating, Refrigerating, and Air-Conditioning Engineers, Inc., *ASHRAE Handbook, 1977 Fundamentals* (New York, 1977).

TABLE 8–2 _____

Typical Gravity and Heating Value of Standard Grades of Fuel Oil

| Grade No. | Gravity API | Weight lb/gal | Heating Value Btu/gal |
|-----------|-------------|---------------|-----------------------|
| 1  | 38–45 | 6.950–6.675 | 137,000–132,900 |
| 2  | 30–38 | 7.296–6.960 | 141,800–137,000 |
| 4  | 20–28 | 7.787–7.396 | 148,100–143,100 |
| 5L | 17–22 | 7.940–7.686 | 150,000–146,800 |
| 5H | 14–18 | 8.080–7.890 | 152,000–149,400 |
| 6  | 8–15  | 8.448–8.053 | 155,900–151,300 |

Source: Reprinted with permission from American Society of Heating, Refrigerating and Air-Conditioning Engineers, Inc., *ASHRAE Handbook, 1977 Fundamentals* (New York, 1977).

This furnace would be rated as a 91,000 Btu/h unit, a measure of input. The actual output of the unit would be 80% of the input, or:

$$91,000 \times 0.80 = 72,800$$

The actual amount of heat available for heating a building is 72,800 Btu/h.

Residential oil furnaces are usually available in capacities ranging from 75,000 Btu/h input to 300,000 Btu/h. Commercial and industrial models are available in much larger sizes.

## Combustion Safety Controls

The combustion safety controls on an oil furnace require proving of the flame, or the burner will not stay on. Proving means that the flame must be established on a call for heat, or the burner will be shut down.

**Stack Switch Safety.**    On older models of oil-fired furnaces, flame proving is usually accomplished by a thermostat-type device placed in the flue outlet of the furnace. This device is called a **stack switch.** Figure 8–7 illustrates a stack switch installed in a flue.

A stack switch has a bimetal sensing element that goes in the flue. Changes in temperature of the flue gases cause the bimetal element to expand or contract.

**Stack switch***

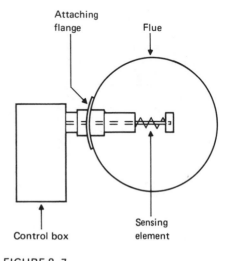

FIGURE 8-7 _____

Stack switch installed in a furnace flue

**Cad cell***

**Manual reset***

A mechanical connecting link from the bimetal opens or closes an electric switch in a control box that controls the burner.

Some stack switches are self-contained with a control box mounted on the flue with the switch. Others have the control box external to the actual switch, and the two are connected electrically. On a furnace using a stack switch as a combustion safety control, the temperature rise of the flue gases is monitored by the switch on a call for heat. If the temperature of the flue gases does not go up to a predetermined temperature within a certain time after the call, then the control shuts the unit down on safety lock-out.

**Cad Cell Safety Control.**    In most oil-fired furnaces manufactured since about 1973, the safety device is a flame scanning unit. The most common one is a **cad cell** unit. The unit uses a cell with a chemical, cadmium sulfide, on the sensing surface. This chemical is sensitive to light. The cell is placed in a position in the burner so it can "see" into the fire pot. Electrical leads from the cell to a control center provide an electrical circuit between the two. Figure 8–8 shows a cad cell located in the blast tube of an oil burner. On a call for heat, if the cad cell does not "see" fire, it will lock out the burner through the control center.

The cad cell control reacts more quickly than the stack switch to fire outage, since the light-sensitive cell and the electronic controls act almost instantaneously. A stack switch works with a bimetal switch, which is much slower. The cad cell system primary control locks out the burner through a series of relays and controls, just as the primary control does on a stack switch. Figure 8–9 shows a primary control used with a cad cell; it also shows the wiring diagram of the control.

**Reset.**    One important feature of any of the combustion safety systems is the requirement that the safety have a **manual reset** in case of a lockout. In this case, the control must be physically reset. The primary control or stack switch control has a reset button or lever. This feature is included so the unit can be checked for the cause of failure in case of a lock-out. A safety control should never be reset continuously on a failure to

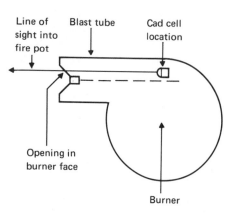

FIGURE 8-8 _____

Cad cell control sensor installed in the blast tube

fire. Each time the control is reset, a quantity of oil is introduced into the fire pot. If the unit is not firing, a buildup of oil can occur and a dangerous situation created.

## 8.3  SUMMARY

Although all of the furnaces commonly used to heat residential or smaller commercial buildings look alike, there are some major differences in the burners, heat exchangers, and the combustion safety controls used for each of the different fuels.

Oil-fired furnaces basically use a power type burner rather than an atmospheric type; have a drum-type heat exchanger; and the combustion safety controls sense main fire ignition rather than prove a pilot light.

In an oil burner, the oil is partially vaporized as it leaves the nozzle in the burner face, and combustion air is mixed with it there. Ignition is caused by an electric spark at the burner face. The power burner produces a large flame that burns in a fire pot, which is part of the heat exchanger.

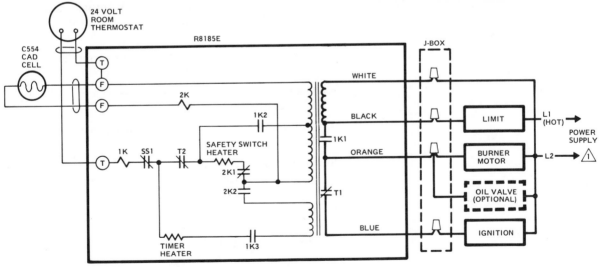

**FIGURE 8-9** ————————————————————————————————————

Control center used with cad cell for combustion safety and wiring diagram of the center

The products of combustion then pass up through the heat exchanger proper, which is located above the fire pot, and air passing around the outside of the exchanger is heated by the exchanger.

The combustion safety controls are one of two types. A stack switch senses temperature rise in the flue on ignition; a cad cell control actually sees the flame. The stack switch is generally found on older units, and the cad cell safety is used on more of the newer models. The combustion safety controls are used to prevent flooding of the fire pot with oil before ignition, which would create a highly dangerous condition.

## 8.4  QUESTIONS

1. The basic difference between oil- and gas-fired furnaces is the method used to _____ the _____.

2. In a typical oil burner, the oil is supplied at the nozzle at: (a) 40 psi, (b) 80 psi, (c) 100 psi, (d) none of these.

3. In an oil burner, the combustion air is mixed with the oil in the oil pump. True or false?

4. In an oil-fired furnace, combustion actually takes place in the: (a) draft tube, (b) air cone, (c) burner assembly, (d) fire pot.

5. The fire pot is lined with ceramic material that gets very hot during a firing cycle. True or false?

6. Most oil-fired unit heat exchangers are of the wrap-around, or _____, type.

7. The lower part of a drum type heat exchanger is called the _____ _____.

8. Heat exchangers that have a second section for the hot gases to pass through are called primary-secondary, or _____ _____ exchangers.

9. A stack switch combustion proving device actually senses the temperature of the flue gas as a method of proving combustion. True or false?

10. A cad cell combustion proving device has to be located in a unit in such a way that the cell can "see" the flame. True or false?

11. Both the stack switch and the cad cell control the burner through a control box called a _____ control.

12. Combustion safety controls should always have _____ reset in case of lock-out.

# CHAPTER
# 9
# Electric Furnaces

FIGURE 9-1 _____

Typical electric furnace used for residential or light commercial buildings

## 9.1  INTRODUCTION

The major difference between a warm air electric furnace and a combustion furnace is that the electric furnace does not have a burner. Instead, it heats the air with electric resistance coils. Otherwise, the unit is similar to a combustion unit, although the controls are usually more complex than those on a gas or an oil furnace. This chapter gives a general description of the electric furnace and a specific description of the heating elements and controls.

## 9.2  MAJOR PARTS

The general appearance of an electric furnace is similar to that of a gas or an oil furnace. No fire is needed in the operation of an electric furnace, so it does not require combustion air or a flue to get rid of the products of combustion. The cabinet of an electric furnace, however, is almost the same as any other furnace. Electric furnaces are available in upflow, downflow, and horizontal air flow, just as the gas and oil units are. They are also available in a variety of types, such as self-contained furnaces, unit heaters, and duct heaters. Figure 9-1 shows a typical electric furnace.

Many of the parts found on an electric furnace are the same as those in combustion furnaces: the filters, the blowers, and some of the controls. Figure 9-2 presents a cutaway view of the parts of a typical electric blower coil unit.

Parts that are the same as those in the other types of furnaces are discussed in other chapters and will not be repeated here. Instead, the chapter focuses on the parts that are unique to electric furnaces: the heating elements and the controls.

### Heating Elements

Electric heating units do not have burners and heat exchangers like combustion-type furnaces. They pro-

*Term defined in text.

duce heat through **electric resistance coils.** These resistance coils are generally one of two kinds: (1) nichrome wire or (2) Calrod.

**Nichrome Elements.    Nichrome elements** are made with resistance wire of a nickel and chromium alloy. This combination of metals provides a wire with the proper resistance to electrical flow and with the tensile strength when heated to make a good heater. Coils of the wire are suspended in a metal frame so the air can pass readily through them. The wire is insulated from the frame to prevent an electrical short circuit. Figure 9–3 shows such an element.

The nichrome wire has a high resistance to the flow of electric current. When the wire is energized by an electric power source, the wire becomes very hot. The air that is to be heated is blown across the hot elements and thus is heated itself. The heated air is then used for heating the building.

**Calrod Elements.    Calrod** is the brand name for elements that are constructed with a resistance wire inside a protective tube. The wire itself is surrounded by an insulating material. Figure 9–4 contains an illustration of the construction of a Calrod unit.

**Electric resistance coils**    Wire coils made of material with a high resistance to electric current flow (When current passes through the wire, the wire becomes hot.)

**Nichrome element**\*

**Calrod**\*

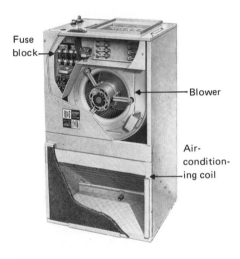

**FIGURE 9-2** _____
Electric heating blower coil unit showing the internal parts

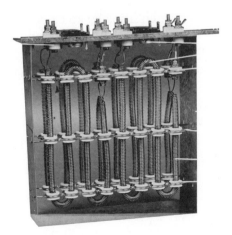

**FIGURE 9-3** _____
Electric heating element of the exposed-wire type

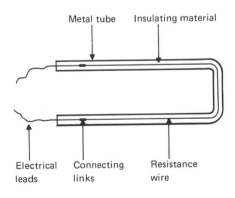

**FIGURE 9-4** _____
Drawing of a Calrod-type electric/resistance heater showing the construction details

**Watt (W)**    An electrical term used to describe electrical power (Technically, it is the amount of electrical power being used when 1 volt of electrical potential moves 1 amp of current; 1 watt of electrical power is equivalent to 3.416 Btu/h of heat energy.)

**Kilowatt**\*

**Primary limit**\*

**Secondary limit**\*

When electric current passes through the resistance wire, the wire gets hot. The insulating material and the outer metal tube get hot also. The air passing over the outer tube is then heated.

The main difference between the nichrome unit and the Calrod unit is that the air is not directly exposed to the hot resistance wire in the Calrod unit.

# Heating Capacity

The heating capacity of individual furnaces is determined by the number of heating elements used in each furnace. The more heating elements, the greater the heating capacity. Manufacturers give the heating capacity for the units in Btu/h or **watts** (W), and the output is always given for a specific voltage and phase of power used. Voltage is the electrical force, and phase is related to the alternations per second. Btu/h is a term related to heat energy, and watts is related to electrical energy. When electrical energy is converted to heat energy, 3.416 Btu/h is produced for each watt of electrical energy converted. Since 1 W is a relatively small amount of energy, the term **kilowatt** (kW), which is 1000 W, is often used in calculations. Watts can be converted to Btu/h by multiplying W times 3.416, and kilowatts can be converted by multiplying kW times 3416.

These relationships can be clarified by an example. The heat output of an electric heating unit is found by using the following formula:

$$Btu/h = 3.416 \times W$$

when:

Btu/h = heat output in Btu per hour
3.416 = Btu/h per W of electric power used
W = electric power used by the furnace

The wattage rating of the furnace is provided in the manufacturer's catalog.

## EXAMPLE 9–1

Determine the heating capacity of a furnace rated at 15 kW by the manufacturer at 240 V single-phase power (240/1/60).

**Solution**

Multiply 15 kW times 1000 to get 15,000 W. Then multiply the 15,000 W times 3.416:

$$\text{Btu/h} = 3.416 \times \text{W}$$
$$= 3.416 \times 15,000$$
$$= 51,240$$

Therefore, 51,240 Btu/h is the rated capacity of the furnace.

All electric heating elements are 100 percent efficient. All of the electrical energy input to the element is converted to heat energy. When designers are selecting an electric furnace, they should be sure that the voltage shown in the manufacturer's literature, for a particular unit, is the same as that available on the job.

Manufacturers make a large number of different capacity furnaces by using heating elements of varying kW rating or various numbers of elements. The smaller units for residential or small commercial use start with 2-1/2 kW elements and increase in approximately 2-1/2 kW increments to about 30 kW. This range provides a wide selection of capacities, ranging from 8540 Btu/h to 102,480 Btu/h. Commercial and industrial units are available in much larger sizes.

## Safety Controls

Combustion safety controls are not needed on electric heating units since these units do not have a combustion system. Nevertheless, they still need safety controls to shut off power in case of a malfunction. **Primary limits** are used to protect against high temperature in case of a blower failure. **Secondary limits** are used to protect against high amperage in the operating circuit due to an electrical short circuit. Figure 9–5 shows the limits found on an electric furnace and a diagram of how they are wired into the circuit.

The primary limit is a thermostatic device with a bimetal switch in it. If the temperature of the air around the bimetal gets too high, the switch will open. The switch is wired into the control circuit so that if it opens, the electric element is deenergized. This limit

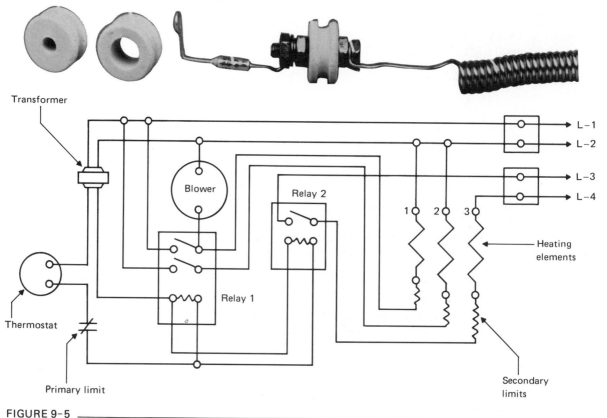

**FIGURE 9-5**

Electric heating element showing the limits and a wiring diagram showing how the limits are wired into the furnace circuits

is mounted on the heating element in such a way that the limit senses the temperature of the air going through the element. In case of a blower failure or any other malfunction that decreases the quantity of air flowing, the temperature of the air rises, and the limit switch opens.

The secondary limit is a fusible link. This limit functions in the same way as a fuse does in any other electric circuit. If an overcurrent condition is sensed, the limit overheats and melts, breaking the circuit. The overcurrent condition is generally caused by a short circuit. The fusible link cannot be reset. If it opens by melting, the element must be replaced after the cause of the short circuit is found.

## 9.3  SUMMARY

Electric furnaces heat air by passing it through hot elements that have been heated by electric power. The furnaces also use electricity for operating their blowers and controls. The units are similar in appearance to furnaces that use a combustion process for heating the air, although they do not require a flue.

Electric furnaces are available in upflow, downflow, and horizontal air flow and in a variety of types.

The heat-producing devices in electric furnaces are electric resistance heating elements. Two different types are used: bare nichrome wire elements and Calrod elements. In both types of elements, the heat is produced when a current of electricity flows through a wire with high resistance to current flow, and the wire gets hot because of that resistance.

The heating capacity of individual furnaces is determined by the number of heating elements used in each furnace. Normally the elements are rated in watts (W), or kilowatts (kW). These terms are related to the amount of electric power the unit will use to produce heat. One W of electric power will produce 3.416 Btu/h of heat, and one kW will produce 3416 Btu/h.

Combustion safety controls are not needed on electric furnaces, but limit controls are required to shut off the power in case of malfunction. Two such limits are normally used on each set of elements in a furnace: the primary limit and the secondary limit. The primary limit is a fusible link. This link is installed in the line voltage circuit to the element. In case the circuit becomes overloaded or starts to draw too much current, the link melts and the circuit is opened. The secondary limit is a thermostatic device, which opens the control circuit if the temperature of the air around the elements goes too high. This condition would occur if the blower went off or anything else stopped the flow of air through the elements.

## 9.4  QUESTIONS

1. Electric furnaces do not require any combustion to produce the heat. True or false?

2. Electric furnaces are available in many types, as are gas and oil furnaces, such as upflow, _____, and _____.

3. Heat is produced in electrical resistance elements because wire gets hot when an electric current can flow easily through it. True or false?

4. Nichrome wire elements are exposed directly to the _____ that is being heated by the furnace.

5. Calrod elements are heating units that are enclosed in a metal tube full of electric insulating material. True or false?

6. Select the proper value from the list to complete this sentence: 1 kW of electric power will produce _____ of heat energy.

   a. 3.416     b. 3416     c. 1000     d. None of these

7. Electric furnaces, like gas and oil furnaces, have to be derated 20% for combustion efficiency. True or false?

8. A 12 kW furnace will produce which of the following Btu/h, at 240/1/60?

   a. 51,240     b. 5124     c. 40,992     d. None of these

9. Two types of safety controls are used to prevent over-heating of the elements in an electric furnace. One is a thermostatic type used to measure air temperature; the other is a _____ _____ type used to sense electric _____ overload.

## 10.1 INTRODUCTION

In a warm air heating system, air is used to carry heat from the point of generation to the point of use. A hydronic heating system uses a liquid medium to carry the heat. Water and steam are the two most common mediums. Other mediums are used in special applications. Water is used for most small jobs, including residences and small commercial buildings. Steam is often used for larger buildings. The systems and equipment described in this text are for water systems only.

This chapter gives a description of the total hydronic heating system, as well as the boilers, piping or distribution systems, circulating pumps, terminal units, and controls used in hydronic systems.

## 10.2 HYDRONIC SYSTEM

A **hydronic heating system** is one in which the medium for carrying the heat from the point of generation to the point of use is a liquid. In most cases, that medium is water. Water is commonly used because it carries a large amount of heat per unit. The specific heat of water is 1.00; 1 Btu of heat is needed to raise the temperature of 1 lb of water 1°F. In other words, 1 lb of water can hold 1 Btu of heat for every 1° temperature that the water is raised. The specific heat of water is higher than that of most other materials, which means that other materials have less heat-holding capacity. Table 1–1 in Chapter 1 shows the specific heat for some common materials.

Hydronic systems are used where having the heat-generating plant in one place is convenient, but the point of use is remote from that place. Transporting heat through a small pipe with water is more economical than transporting it through a large duct with air when the point of use is distant from the point of generation. High-rise and cluster-style buildings, such as campus-style school complexes, are good examples of the types of buildings in which hydronic systems are used.

The major parts of a hydronic system are the boiler

*Term defined in text.

Hydronic heating system*

**Boiler**    A closed vessel in which a liquid is heated or vaporized

**Terminal device**    The device that extracts heat from the water in a hydronic system (The device usually also distributes the heat into the spaces to be heated as warm air.)

**Pounds per square inch (psi)**    Force or pressure exerted on a surface in terms of pounds force per square inch area of surface

in which the water is heated, the piping system to carry the water to its point of use, pumps to move the water through the system, terminal devices to exchange the heat from the water to air, and controls for the boiler, pumps, and the terminal devices. Figure 10–1 is a diagram showing these major parts.

## Heating Plant

The heating plant for a hydronic system is the boiler. The **boiler** produces hot water for the entire system. A boiler may be fueled by any one of the common fuels. Its actual configuration depends on the fuel used, as well as the capacity required and the type of supply system used. Figure 10–2 is an illustration of a typical small boiler and some of the other parts of the system. This type of boiler is used primarily for small systems or small zones in a building.

## Distribution System

The hot water generated in the boiler is distributed to the point of use by a piping system. Any one of several layouts of piping could be used, such as single or double pipe or series or primary-secondary. These layouts are described in Chapter 19. The actual layout selected depends upon the type of terminal devices used and the temperature control requirements in the building. Distribution systems are described in detail in Chapters 19 and 20.

## Terminal Devices

The final step in heating a building with a hydronic system is the use of the hot water from the distribution system to heat air at the point of use. Units called **terminal devices** are used for this purpose. Several types of terminal devices are available. The ones used on a job will be selected to fit the physical and air temperature requirements for that particular job. Terminal devices are covered in detail in Chapters 19 and 20.

## 10.3   EQUIPMENT

The major pieces of equipment used in a total system are the boiler, the circulating pumps, and the terminal devices.

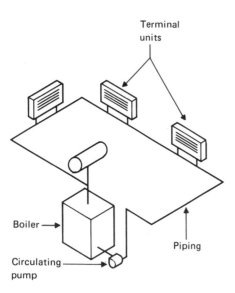

**FIGURE 10-1** _____

Schematic diagram of a typical hydronic system showing the major parts of the system

# Boiler

A boiler is a device designed to transfer heat produced by combustion, or electric elements, to water. The term *boiler* is used interchangeably for both the heat exchanger of a boiler package and for the entire unit. Boilers are available in different types for various applications. Only the main types will be covered in this text.

Four general parts make up a boiler package:

1. Cabinet or frame,
2. Heat exchanger (boiler),
3. Burner or heating elements,
4. Unit controls.

**Cabinet or Frame.**    Most smaller boilers, used for residences or small commercial buildings, are enclosed in a sheet metal cabinet. Larger boilers, used for commercial or industrial applications, are built in a frame. In both cases, the cabinet or frame holds the parts of the unit together. In boilers that are built mainly for use in residential heating systems, the cabinet is designed to be attractive as well as functional.

**Heat Exchanger.**    Boilers are grouped into classes based on working pressure, fuel or energy used for heat, materials of construction, and/or general arrangement of the tubes containing the water.

Hot water boilers are designed to operate at low or medium pressures, ranging from a low of 15 **pounds per square inch (psi)** to a high of 160 psi. Most water boilers are low-pressure type, designed to operate at no more than 30 psi working pressure.

Boilers are designed to burn wood, coal, fuel gas, or fuel oil. Electric boilers are also commonly used.

Some units are designed to use either one of two fuels. These devices are called dual-fuel units. The units can be switched back and forth from one fuel to the other depending upon fuel availability.

Boilers are constructed with either cast-iron or steel heat exchangers. The material used in a particular boiler depends on the size of the unit and its application.

The arrangement of the tubes in the boiler affects the heating unit's general efficiency. Although many different variations of the arrangements are possible,

FIGURE10-2

Small boiler used for heat in a hydronic system for a residence or small commercial application

FIGURE 10-3 _____

Water-tube boiler. The water tubes run through the areas above the fire pot where the hot flue gases can surround them.

FIGURE 10-4 _____

Fire-tube boiler. The tubes running through the water tank above the firebox increase the area of the surface exposed to the hot flue gases.

the two most common are the **water tube boiler** and the **fire tube boiler.**

In its simplest form, a boiler is a water tank with a burner under it. The heat from the burner heats the water in the tank. Most boilers used in hydronic systems are more complex, but they perform the same function. The efficiency of the boiler can be increased by increasing the surface area of the metal separating the burner from the water. The water either goes through tubes above the burner, or the hot flue gases go through tubes in the water tank. When the water goes through tubes above the burner, the boiler is called a water-tube boiler. When the flue gases go through tubes in the water tank, the boiler is called a fire-tube boiler. Each of these types, as well as the electric boiler, is described in the following sections.

*Water-Tube Boiler.*    A water-tube boiler is one in which the water is contained in tubes above the burner, or the firebox. The use of tubes in the place of a simple tank enables the efficiency of the boiler to be increased considerably. The more surface area of the water-containing tubes that is exposed to the hot flue gases, the more efficient the heat transfer will be. The tubes are connected to reservoirs, or tanks, on each end of the boiler. The reservoirs and the tubes always contain water. The hot products of combustion pass around the tubes and then leave the boiler at one end through a fitting called the **breeching.** The heat from the products of combustion is conducted through the walls of the tubes into the water. Figure 10–3 is a cutaway drawing showing the construction of a water-tube boiler.

*Fire-Tube Boiler.*    In the fire-tube boiler, the upper part of the boiler is a water tank. Tubes are connected to end breechings and pass through the water tank. The firebox is located below the water container. The water is heated by radiation and by hot flue gases. The hot flue gases also pass into a breeching at one end of the boiler and then go through tubes to the breeching at the other end. Figure 10–4 is a cutaway drawing of a fire-tube boiler.

*Electric Boiler.*    Water-tube or fire-tube boilers can be oil or gas fired. Electric boilers are a third type of boiler.

An electric boiler is basically a water tank with electric heating elements inserted in the water. When an electric current flows through the elements, the water is heated. The elements are insulated from the water electrically, so a short circuit cannot occur. Because no combustion takes place, the electric boiler is a simpler device than a gas or an oil unit. Electric boilers are more compact and often smaller than a comparably sized combustion unit. An electric boiler is shown in Figure 10–5.

**Burners.**    The burners used in fuel-fired boilers are similar to those used in warm air heating equipment. Both atmospheric and power-type burners are used. And just as in warm air equipment, the atmospheric burners are usually used in smaller units and the power burners almost always in larger ones. Figure 10–6 is a photograph of a typical oil burner used in a boiler. The general construction of the burners is the same as those described in Chapters 5 and 6. Readers should refer to those chapters for details of the burners used in boilers.

## Piping System

The piping, or distribution, system in a hydronic heating system is the part that carries hot water from the boiler to the points of use and returns cooler water from that point to the boiler. A **supply pipe** carries the hot water to the various terminal devices, and a **return pipe** brings the water back to the boiler after the water has gone through the devices. In most cases, the pipe is either iron or copper.

Unlike the boiler and other parts of the total system, the piping system is custom designed for each job; each system is unique. For this reason, describing a system specifically is impossible. Heating technicians should be able to recognize and describe some general types of systems, however. These systems are described by the number of pipes used. In a one-pipe system, the supply and return functions are included in a single loop of pipe that goes around the building or the zone to be heated. In a two-pipe system, the supply water is carried in one pipe, and the return water comes back in another. Three-pipe and four-pipe systems are used when both heating and cooling are being done by the

**Boiler, water tube**    A boiler in which the tubes contain water and steam, and the flue gases and flames heat the outside of the tubes

**Boiler, fire tube**    A boiler with straight tubes that go through the water container and through which the flue gases pass

**Breeching**    Open spaces at the end, or ends, of the boiler where the smoke and products of combustion pass from the boiler tubes to the flue outlet

**Supply pipe***

**Return pipe***

FIGURE 10-5 _____
Electric boiler

**Circulating pump**    A pump used to circulate water through a piping system

**Expansion tank**    A tank filled with air, used on a hot water system to provide a cushion for the expansion and contraction of the water in the system as the water is heated and cooled

**Pressure relief valve**    A valve that automatically opens if the pressure inside the tank it is installed on goes above the set point of the valve

one system. Each of the systems is described in detail in Chapter 19.

# Circulating Pumps

**Circulating pumps** are used in hydronic systems to move the hot water through the piping system from the boiler to the terminal devices. These pumps are usually centrifugal pumps driven by electrical motors. Most smaller pumps are built with the motor frames being part of the pump housing. Larger pumps are often indirect drive. That is, the motor is separate from the pump housing, and the pumps are connected either in line with couplings or are belt driven.

The pumps have an impeller wheel that turns at a high speed. An impeller wheel has vanes attached to a hub, and as the wheel rotates, it moves water through the pump. Water is drawn in through a suction port in the center of the wheel housing and is then thrown to the outer perimeter of the wheel by centrifugal force. A discharge port is located on the outer rim of the wheel housing. The suction side of the pump is connected to the return piping of the system, and the discharge port is connected so it provides water to the supply side.

Circulating pumps are built in various configurations. The two that are used most commonly in hydronic systems are floor-mounted and line-mounted pumps. As the names imply, the floor-mounted pump sits on the floor or a stand, and the line-mounted pumps are supported by the pipes they are connected to. A floor-mounted, in-line drive with the motor external to the pump is shown in Figure 10–7.

In most systems, the circulating pump is located in the return side of the piping system. This placement facilitates the plumbing for make-up water to the system. All systems must have a source of make-up water available, usually from the water supply used for the rest of the building. As the system operates, air separates from the water and is removed. There also may be leaks in various parts of the system. Make-up water is used to replace any volume lost so the boiler will not go dry. Putting the circulating pump on the return side of the system, where the make-up water comes in, helps to ensure a supply of water when it is needed.

FIGURE 10-6 _____

Typical oil burner used in larger boilers

FIGURE 10-7

Floor-mounted circulating pump with the motor mounted external to the pump and connected with an in-line coupling

## System Accessories

Hydronic systems also require some accessory devices in the water piping next to the boiler. Among these are the **expansion tank** and the **pressure relief valve.** Proper design of the piping arrangement for supply, return, and make-up water piping is extremely important. Improper design or installation of this piping can cause air traps to be formed in the system. An air trap will prevent the flow of water, a condition that would cause problems. A systematic arrangement of this part of the system is shown in Figure 10–8.

**Expansion Tank.**    Most hydronic heating systems are closed systems; that is, they are not open to the atmosphere. Because water expands when it is heated, provisions must be made for that expansion in the system. A small tank that is normally full of air during operation of the system is plumbed into the water system above the boiler. The water in the system expands when it is heated and is forced into the tank. The air in the tank can be compressed because it is a gas. The water in the system is not compressible in the liquid state. On an off cycle, when the water in the system cools, the water in the expansion tank flows back into the system, and the air in the tank expands.

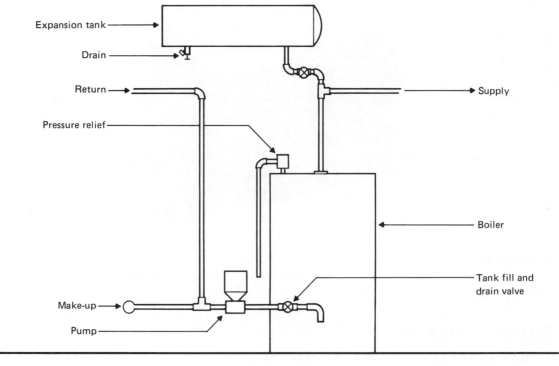

FIGURE 10-8

Diagram of the plumbing connections at the boiler for a typical heating system showing the accessories necessary for the proper operation of the system

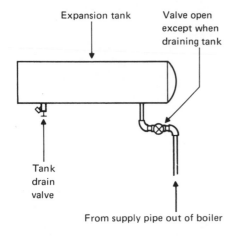

FIGURE 10-9

Expansion tank used with a hydronic system to allow for the expansion of the heated water

Use of the expansion tank allows the system to operate safely and quietly and prevents problems related to high pressure caused by temperature rise of the water. A typical expansion tank and plumbing for it is shown in Figure 10–9.

**Pressure Relief Valve.**    Every hydronic system must have certain safety features incorporated in its design to ensure safe operation of the boiler and the total system. One of the most important is a pressure-relief valve, which will open if the pressure in the boiler or the piping system goes above a predetermined safe operating pressure. Usually the valve is held closed by either a spring or a weight. If the pressure in the system exceeds the set pressure, the valve opens, and water is allowed to flow out of the system until the pressure is reduced to a safe level. The discharge port on the pressure relief valve is piped to a floor drain or other water location. A cutaway view of a pressure relief valve is shown in Figure 10–10. Some pressure relief valves will also open if the temperature of the water exceeds a safe limit.

If the burner fails to shut down for any reasons after the water has reached a **set-point temperature,** the temperature and pressure could rise to a dangerous level. If the water temperature rises, so does the pressure. In such a case, the pressure relief valve will open and vent the system. This process not only prevents the pressure from getting too high, but the operating personnel are alerted to the situation by the discharged water.

**Boiler Plumbing.**    The plumbing, or piping, connections between the boiler and the other parts of the system, such as the supply and return piping, the make-up water supply, and the expansion tank, can be complex. A typical plumbing set-up is shown in Figure 10–8. This piping must be designed and installed so that the proper water level is maintained in the boiler, the distribution system always has a water supply, and there are no air pockets in the system.

To ensure that the piping connections for any given system are correct, the boiler manufacturer's recommendations should be closely followed. Proper valves and controls should be installed according to the manufacturer's instructions.

## Controls

The control system for a hydronic heating system has two separate parts. The first part includes the controls that regulate the water temperature in the boiler. The second part includes the controls that operate the terminal devices where the heat from the water is converted to warm air for heating the spaces.

**Boiler Controls.**    The main operating control for boiler water temperature is called an **aquastat.** This control is a thermostat that senses the boiler water temperature and triggers the **cycle** of the burner to maintain the temperature as determined by the set point. The aquastat is usually a remote-bulb thermostat, meaning that the sensing bulb is located where it

**Set-point temperature**    The desired temperature at which a thermostat is set

**Aquastat***

**Cycle**    Alternating periods of time in which the burner is turned off and on

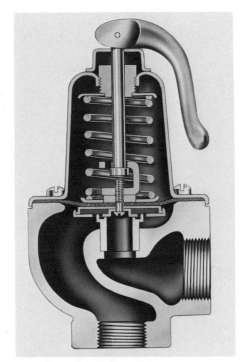

FIGURE 10-10
Cutaway view of a pressure relief valve showing the spring that holds the valve closed against water pressure up to the set point of the valve

FIGURE 10-11 _____

Motor-driven control valve for a hydronic system

will sense the water temperature, while the operating part of the control is located outside the water container. Most aquastats are temperature-pressure controls. A material in the remote bulb senses the temperature. This material exerts pressure on a diaphragm that operates a switch. If the temperature goes up on the bulb, then the pressure goes up, and the diaphragm is depressed. If the temperature on the bulb goes down, then the pressure in the bulb goes down, and the diaphragm is released.

Since the aquastat controls the temperature of the heating medium, water, this device functions as both an operating control and a limit control. The upper limit of the water temperature, as well as the lower limit, is controlled by the aquastat.

**Terminal Device Controls.**    The second part of the control system is on the terminal devices, the units that transfer the heat from the water to air. In the case of radiators or convection units, the components of the control system are usually very simple. A room thermostat operates a water valve; this valve controls the flow of water through the radiator or coil in the **convector.** The valve is motor operated. Figure 10–11 illustrates a motor-operated valve. The motor is mounted on the top of the valve body and is connected directly to the valve stem. When the thermostat calls for heat, the motor turns, and the valve is opened. When the thermostat is satisfied, the valve motor turns the opposite way, closing the valve. Sophisticated controls are not needed on a simple system because the response time of the system is very slow. The large amount of water in the system causes a time lag between the call for heat and the increase of temperature in the water to satisfy that call.

In the case of blower coil units, the thermostat may control the water flow by the use of a motorized valve, or it may control the blower in the terminal device to control the room temperature. In more sophisticated systems, both the water and the air may be controlled. Circulated air systems have a shorter response time than do radiation systems on a call for heat.

Whenever outside air is introduced into any air system ahead of a water coil, as in a **unit ventilator,** low-temperature limit controls must be used in the airstream to make sure that the air temperature will not go below the freezing temperature of the water in the system. This control is in the form of a low-temperature thermostat. This thermostat closes the outside air dampers if the temperature of the air entering the coil falls below 32°F, thus protecting the water in the hydronic system from freezing. As an added protection, some systems have a flow control device in the water piping. If the device senses that the water flow has stopped, then the damper is closed.

## 10.4 SUMMARY

A hydronic heating system is one that uses a liquid—water or steam—to carry heat from the point of generation to the point of use. Water is the liquid used in most smaller buildings and is the only one considered in this text.

The major parts of a hydronic system are the boiler, piping, pumps, terminal devices, boiler accessories, and controls. Boilers are self-contained units; the burners and controls for operation are part of the basic package. The piping system is designed specifically for each job. The accessories to the boiler and piping system, such as circulating pumps, expansion tank, pressure relief valve, and other controls and valves, are selected for each job individually. Each part of the total system has to be matched to all other parts to ensure a properly operating system.

Boilers can be categorized in several different ways. The most important are by fuel or energy used and by the general type according to the way the water is heated. Boilers are manufactured to burn any of the common fuels, and/or to use electricity for heat. Two basic types are related to how the water is heated.

In a water-tube boiler, the water is contained in tubes located over the burner. The use of tubes makes it possible to increase the area of the transfer surface, or the surface exposed to the hot flue gases. A fire-tube boiler has tubes going through the tank containing water, and the hot flue gases go through the tubes. Electric boilers have electric resistance heating elements in the water.

The burners used in boilers are similar to those

**Convector**    A heating coil enclosed in a cabinet with arrangements made for passing air through the unit to be heated by the coil

**Unit ventilator**    A blower-powered convection heating unit with a fresh air intake and damper controls

used in warm air heating units. Both atmospheric and power burners are used. Burner selection depends upon the fuel used and the size of the boiler.

A hydronic heating control system is made up of two separate sets of controls. One set maintains the temperature of the water in the boiler, and the other regulates the amount of heat given up by the terminal devices used to heat the building. The boiler water temperature is controlled by an aquastat, or thermostat, that senses the temperature of the water and cycles the burner as required. The room temperature is controlled by thermostats in the room. The thermostats operate water valves in the piping system or cycle the blower on the terminal device.

The main safety controls are the pressure relief valve, the freeze protection controls, and the high limit of the aquastat. The pressure relief valve is placed in the piping system adjacent to the boiler and is used to prevent the pressure in the system from exceeding a predetermined limit. Freeze controls, located on any blower unit that introduces outside air, limit the low temperature of the air going across the water coil in the system.

## 10.5  QUESTIONS

1. A hydronic heating system is one in which the _____ for carrying the heat to the point of use is _____.

2. Other materials are compared to water when they are listed in a table of specific heat. True or false?

3. Water has a higher specific heat than most other substances. True or false?

4. Hydronic systems are especially good when heat is _____ at a place somewhat remote from its point of _____.

5. A hydronic heating system requires a boiler to produce hot water, _____ to move the water, _____ to carry the water to its point of use, _____ to exchange the heat from the water to air, and a _____ system.

   a. terminal device    b. control    c. pumps    d. piping system

6. The two types of boilers are called _____ _____ and _____ _____ boilers.

7. Atmospheric burners are often used in gas-fired burners. True or false?

8. An expansion tank is used on a hydronic system to allow for expansion of the water when it is heated. True or false?

9. A pressure relief valve should open only in case of an emergency situation. True or false?

10. A hydronic control system has two parts. One part controls system _____ temperature, and the other part controls _____ air temperature.

11. An aquastat controls the level of the water in the boiler. True or false?

12. A motorized _____ is used to control water flow to a terminal device in some hydronic systems, and is controlled by a room _____.

13. A freeze protection control will close the outside _____ if the air temperature drops below the freezing temperature of the water in the coils.

# CHAPTER
# 11
# Unit
# Controls

## 11.1 INTRODUCTION

The heat loss of a building is directly related to the outside temperature. The function of a heating unit is to provide enough heat to maintain a comfortable inside temperature regardless of the outside temperature. Since the outside temperature constantly changes, the heat loss changes also. Therefore, the heat output of the unit must change to maintain a constant temperature in the building. Because the actual output cannot be changed, the unit is cycled on and off to provide the required amount of heat over a given period of time, usually each hour. Unit controls are used to provide this cycling of the unit.

**Unit controls** are those that are located in the heating unit and control the operation of the unit itself. Controls that operate the burner or its elements and those that operate the blowers or pumps are normally considered unit controls. Some safety controls, such as limit switches and overloads, are also unit controls.

In this chapter, the unit controls are divided into three categories: (1) operating controls, (2) safety controls, and (3) combustion safety controls. Each category is explained, and the devices used in each category are discussed.

## 11.2 UNIT CONTROLS

The proper and safe operation of a heating unit depends upon the controls used to operate it just as much as it does upon the mechanical parts. Those parts of the unit related to fuel control, combustion, and air or water distribution all have to be controlled for proper operation of the unit. The very nature of the processes used to produce heat requires that certain safety controls be used.

The controls used on heating units can be separated into three categories: (1) operating, (2) safety, and (3) combustion safety. Figure 11–1 is a wiring diagram for a typical gas furnace. The controls used are labeled and marked to show which category they belong in.

The controls in these three categories are directly

**Unit controls**\*

**Limit control**    A temperature-actuated switch that opens on temperature rise (The switch is used to open the electric circuit to the burner control when a dangerously high temperature occurs.)

\*Term defined in text.

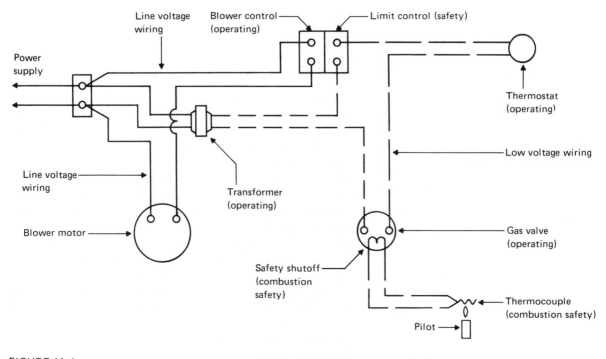

FIGURE 11-1 _____

Wiring diagram for a gas-fired furnace with the controls marked to show their category

related to the operation of the units. Other control devices may be part of the total control system; they are covered in the next chapter.

## Operating Controls

Operating controls are the controls that make the unit function. The controls include the thermostat or operator, the valve operator, and other devices that regulate the flow of fuel, steam, hot water, or electricity to produce heat. Most heating systems also have controls that activate the blower or pumps to circulate the heating medium when the burner is on. These devices are also categorized as operating controls.

## Safety Controls

**Limit controls,** air-flow switches, thermal overloads, and other devices fall into the category of safety controls. These controls are designed to protect the equip-

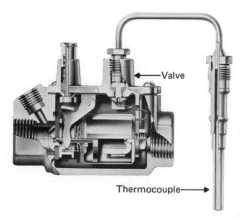

FIGURE 11-2 _____

Pilot-safety valve and thermocouple used to monitor the operation of a pilot ignition system

ment and the occupants of a building in case a unit malfunctions.

## Combustion Safety Controls

Combustion safety controls are used to prevent the operation of a unit in case of a dangerous or hazardous situation. Suppose, for example, that the pilot light was not lit on a gas furnace on a call for heat. If the gas valve opened, a gas-air mix would flood the combustion chamber, and a spark or flame could cause an explosion. On a typical gas-fired furnace with a standing pilot, an automatic pilot safety is used to monitor the pilot flame. If the pilot is not lit, the pilot safety control prevents the gas valve from opening. Figure 11–2 is a cutaway view of a pilot safety gas valve and the thermocouple used to monitor the flame.

An oil-fired furnace has no standing pilot, so the combustion safety control monitors the start-up of the unit to ensure firing. The control also continuously monitors the combustion during an on cycle to make sure the flame does not go out. The monitoring is done by a stack switch or a cad cell control.

Electric furnaces do not have combustion safety controls, but they do have limit controls on the electric power circuits to the elements. These limit controls serve the same purpose as combustion controls: they monitor the amperage in the circuit, or the temperature of the air around the element, and shut the heat off if a dangerous situation is sensed. Each circuit has two limits. One is sensitive to current flow and the other to temperature. A wiring diagram of a typical electric furnace is provided in Figure 11–3. The limits are shown wired into the circuits. If either limit senses a dangerous condition, the limit opens, and the element will not be energized.

## 11.3 CONTROLS FOR DIFFERENT FUELS AND ENERGY

All of the different heating units use the same basic control scheme. The burner, or elements, are cycled on and off to provide a specific amount of heat over a given length of time. A controller, such as a thermostat, calls

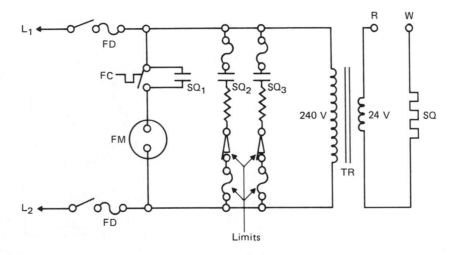

FIGURE 11-3

Primary and secondary limits in the wiring on an electric furnace

for heat when heat is needed and shuts off the burner or de-energizes the heating elements when the temperature is satisfied. Other controls provide safety functions in case of malfunctions.

These basic control functions are the same for furnaces using different fuels or electricity for the production of heat, but the application of the controls to each type of unit may vary somewhat. The particular application is usually determined by the burner or element characteristics.

## Gas Furnace Controls

A typical gas furnace control system is a 24 V electric system. A transformer reduces line voltage to 24 V for the control power. The operating control circuit runs from the transformer, through the thermostat, to the gas valve, and back to the transformer. See Figure 11–4 for a wiring diagram of this circuit.

In some larger units, the control circuit may be a line voltage circuit. The circuit in this case is the same except that a transformer is not used, and the thermostat and gas valve are rated for the higher voltage.

The blower control is also part of the operating control system. A blower control is an electrical switch similar to a thermostat. The device has a set of electric

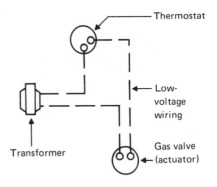

FIGURE 11-4

Low-voltage wiring from the transformer to the gas valve through the thermostat

**Transformer**    An electrical device that changes the voltage received from an electric power source

**Primary control**    A control used on oil-fired furnaces that contains the transformer, operating relays, and combustion safety relays and switches for the operation and protection of the unit

FIGURE 11-5 _____

Combination fan and limit switch used to monitor temperature in the heat exchanger and also to operate the blower

contacts that open or close on temperature rise or fall. Figure 11–5 shows a combination fan and limit control. The fan side of this control operates the blower. The blower control is located in the front part of the furnace. Its sensing element extends back into the heat exchanger section of the furnace. When the burner comes on and heat is produced, the element is heated up, closing the switch. This switch is wired in series with the blower motor, which means that all electric power to the motor goes through the switch. Figure 11–6 is a wiring diagram of the blower circuit on a typical furnace.

A unit also has safety controls that will shut the heat source off in case of a malfunction. For example, if a gas valve stuck open or a blower did not operate, the temperature in the furnace could exceed that desired for safe operation of the unit. A limit control wired in series with the gas valve in the operating control circuit will prevent this occurrence. Figure 11–7 shows a limit control and a diagram of how it is wired into a system.

The limit control is a temperature control switch that opens if the temperature reaches the set point of

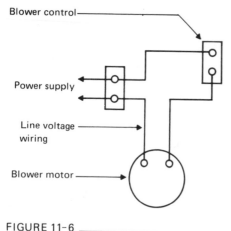

FIGURE 11-6 _____

Diagram showing the blower control wired into the circuit to the blower motor

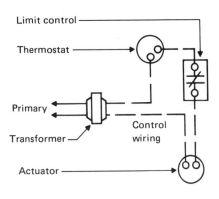

FIGURE 11-7 _____

Limit switch and wiring diagram for a typical furnace control system

the switch. The set point is a temperature between about 160°F and 200°F, the set point being the highest temperature at which the unit could operate safely. On downflow or horizontal furnaces, two limits are used. One is located in the heat exchanger, and the other is located in the unit where the air leaves the blower. The extra limit is needed in these two types of furnaces because on initial firing, the air may move by gravity in the opposite direction to the way it will when the blower comes on. In this case, the second limit will deactivate the burner or heating elements and shut off the unit. The second limit also gives protection in case of blower failure. When two limits are used, they are both wired in series in the operating control circuit, so either one can shut the burner down in case of problems. A complete wiring diagram of a typical gas-fired furnace is shown in Figure 11–8.

## Oil Furnace Controls

Most oil furnaces, like gas furnaces, have 24 V control systems. The **transformer,** used to reduce the line voltage to 24 V is part of the control package called the **primary control.** A primary control used on an oil furnace is shown in Figure 11–9. The primary control also contains the relays and other devices necessary to make

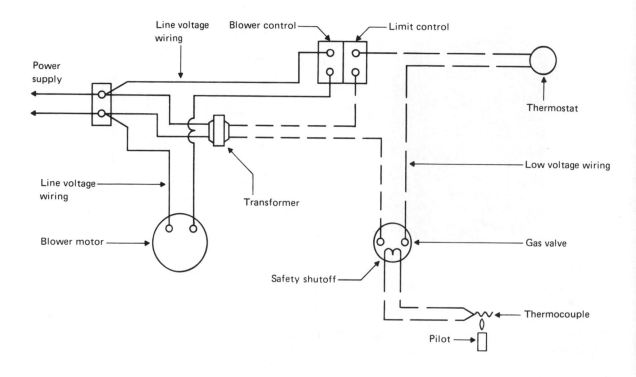

Pictorial

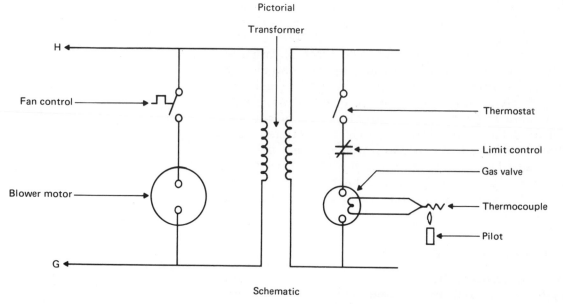

Schematic

FIGURE 11-8
Pictorial and schematic wiring diagrams for a typical gas-fired furnace identifying the control components

the combustion safety circuit. A diagram of the wiring, relays, and switches for a primary control are shown in Figure 11–10.

The primary control regulates the operation of all of the electrical components of an oil furnace. Power is brought directly to the black and white **leads** on the **control box** from the **disconnect.** An orange lead goes from the primary control to the oil burner assembly. A wire from the **neutral** terminal on the disconnect is also connected to the white lead on the burner. Two low voltage terminals are connected to the thermostat, and two more lead to the combustion safety device. Figure 11–11 shows the circuit connections.

When the thermostat calls for heat, a coil in the primary control is energized. The coil closes a set of contacts that brings on the burner and also activates the combustion safety circuits within the primary control itself.

FIGURE 11-9 _____

Primary control used to monitor the operation of an oil furnace

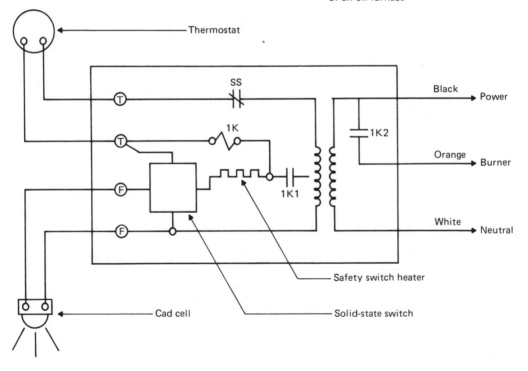

FIGURE 11-10 _____

Diagram of the internal wiring for a typical primary control

**Leads**    Electric connections at a device

**Control box**    A relay or primary control that is used to control the operation of a unit

**Disconnect**    The electric disconnect used to control the electric power to a unit

**Neutral**    The common or neutral power line on an electric circuit

**Sure-start switch**    A blower relay that is controlled by an electrical signal as well as by temperature (It will always close within a certain length of time after the electrical signal is received.)

In addition to the circuits just described, the typical oil furnace has a blower motor circuit that operates through a blower control. The hot leg, or power leg, from the power source is circuited through a blower control switch that closes when the temperature of the air in the furnace increases to a predetermined temperature. The other side of the blower control is connected to one terminal on the motor. The other terminal on the motor is connected to the neutral side of the power source. This wiring completes the blower circuit, which is the same as that used on a gas furnace.

In downflow or horizontal furnaces, a **sure-start switch** is sometimes used for blower control. In this type of control, an internal heater is wired in parallel with the thermostat terminals on the primary control so the switch will be activated by the thermostat on a call for heat. Time delay is built into this heater so the air in the system is warmed before the switch starts the blower.

The safety controls on an oil furnace are similar to those described for a gas furnace. One or two limit controls are wired in series in the power line to the primary control. See Figure 11–12 for a diagram of this wiring. If either of the limit controls opens because of high temperature, the unit will shut down.

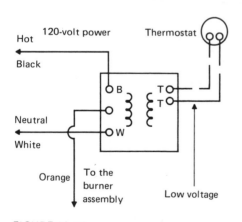

FIGURE 11-11 _____

Circuit connections to a primary control

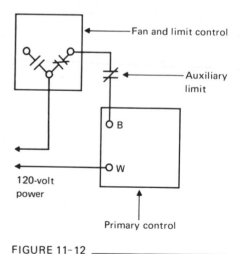

FIGURE 11-12 _____

Wiring diagram for the power to the primary control with the fan and limit control and also an auxiliary limit in series in the circuit

Combustion safety controls are described in Chapter 8, on oil furnaces. Figure 11–13 shows a complete wiring diagram for a typical oil-fired furnace, with the controls identified.

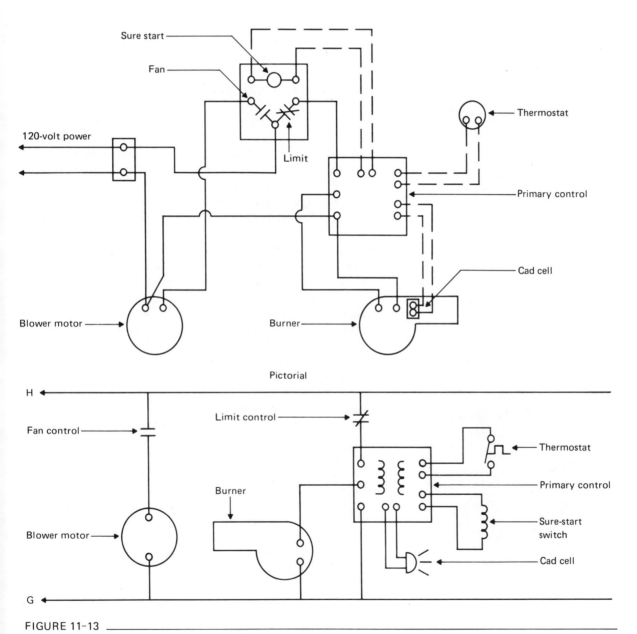

FIGURE 11-13

FIGURE 11-13
Pictorial and schematic wiring diagram of a typical oil-fired furnace

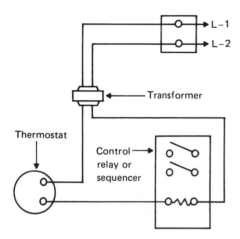

**FIGURE 11-14** _____

Electric furnace control circuit with the thermostat in series with the sequencer

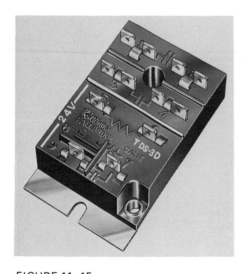

**FIGURE 11-15** _____

Sequencer for electric furnaces

# Electric Furnace Controls

The control system used on electric furnaces typically is a 24 V electrical system. The 24 V power source is a transformer. In the operating control circuit, one leg of the transformer is connected to the thermostat. The circuit goes through the thermostat and then to one terminal on a control relay coil, the magnetic coil that operates the relay. The wire from the other side of the relay coil goes back to the opposite terminal on the transformer. A wiring diagram of this circuit is shown in Figure 11–14.

When the thermostat calls for heat, the relay coil is energized, a process that closes the relay contacts and starts energizing the elements. The elements are energized through a series of relays or through a mechanical **sequencer.** A sequencer is a control device that brings the electric heating elements on in steps, one at a time. The sequencer is a series of **time-delay relays,** electrically interlocked, so that each one controls line voltage to an element and also initiates the timing sequence for the next set of contacts. A sequencer used for this purpose is shown in Figure 11–15.

The elements are sequenced on one at a time rather than all at once so the electrical load on the circuits will come on gradually. One of the most important features of the electric furnace is that the heat can be brought on in gradual steps. This sequencing closely approximates modulation and enables heat demand to match output more closely. The heat output of the unit can be varied by varying the number of elements that are energized. Solid-state controls make this operation relatively simple.

In some cases, special blower controls are used on electric furnaces to control the speed of the blower and thus regulate the volume of air across the electric elements. If such controls are used, they enable air of a constant temperature to be supplied to the building. In other words, the discharge air temperature remains constant, even though the electric elements are staged on and off. Providing these special controls is fairly expensive; moreover, the system is more sophisticated than is required for residential comfort heating applications.

The blower on an electric furnace is controlled by

a blower relay. The coil on this relay is wired in parallel with the control relay coil. When the thermostat calls for heat, the blower relay starts the blower. The relay is a time-delay relay, so the blower will not start until the air is warm enough to heat the building.

Safety controls are discussed in Chapter 9 on electric furnaces. Figure 11–16 shows a wiring diagram for a typical electric furnace.

**Sequencer\***

**Time-delay relay**    A relay in which the contacts close within a specific length of time after an electric signal is received

**Modulate**    To change capacity to meet demand

## 11.4  SUMMARY

The heat loss of a building is directly related to the outside temperature and changes as that temperature changes. The output of a heating unit, however, is constant while the unit is operating. To match the heat loss of a building, the heating unit is cycled off and on to provide an average amount of heat per hour required to offset the heat loss per hour. The cycling off and on of the unit is done by the unit controls.

The controls used to regulate the heat output of a

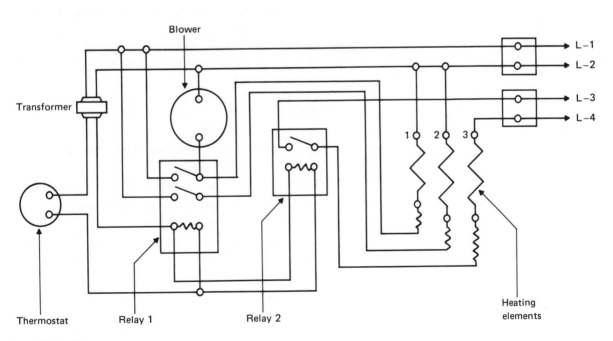

FIGURE 11-16 _____

Wiring diagram for a typical electric furnace with the controls identified

unit are called operating controls. Other controls, called safety and combustion safety controls, are also used on heating units to ensure safe operation.

The controls used on units that burn different fuels or use electric energy to produce heat are similar to each other, but their applications differ slightly depending upon the burners or the elements used.

## 11.5  QUESTIONS

1. The heat loss of a building is directly related to the temperature difference between the inside and the outside temperatures. True or false?

2. To provide a variable amount of heat, a furnace is _____ on and off. The furnace operation produces a specific amount of heat during a given _____ period.

3. The cycling of a heating unit to _____ heat output is accomplished by using _____.

4. Controls are used to provide three main functions in a heating unit. These are _____, _____, and _____.

5. Controls that make a unit functional are called _____ controls.

6. Controls that protect a heating unit and the occupants of a building against unit malfunction are called _____ controls.

7. Controls that prevent the operation of a unit in case of delayed or interrupted combustion are called _____ _____ controls.

8. Limit controls are considered to be part of which category of controls?

9. Thermostats are part of which category of controls?

10. Fan controls are part of the combustion safety control category of controls. True or false?

11. Primary controls are found on most gas-fired heating units. True or false?

12. A sure-start fan switch is used on _____ and _____ furnaces to ensure blower operation on a call for heat.

13. The heating elements in an electric furnace are brought on in steps to prevent a large electrical load from coming on the line all at once. True or false?

## 12.1 INTRODUCTION

A **control system** is the complete control package that operates a heating system. The system tells the heating equipment when to start up and when to shut down. The control system usually includes some devices that are external to the unit, as well as those that are internal. The thermostat or sensor, actuators or valves, connecting wires and/or tubes, and all the accessories that control the temperature, humidity, or any other pertinent condition are all part of the control system.

The four general types of control systems are: electrical, pneumatic, electronic, and mechanical. The electrical, pneumatic, and electronic systems are described and explained in this chapter. The mechanical control system is one in which the off-on functions are performed by mechanical devices. This control system is used so seldom in modern systems that it is not discussed in this book.

## 12.2 COMMON FACTORS IN CONTROL SYSTEMS

All control systems include components that will provide the following functions:

1. Sense conditions such as temperature, humidity, or pressure;
2. Transmit data;
3. Actuate some device to satisfy a condition;
4. Feed back information when the condition is satisfied.

Figure 12–1 is a block diagram showing the relationship of these functions in a system.

Each of the four functions can be observed in a thermostat. The thermostat measures room temperature and compares it with a set-point temperature. The information is **transmitted** by an electrical signal. If the room temperature drops below the set point, an electric circuit is completed by the closing of the contacts in the thermostat. The electric signal through the ther-

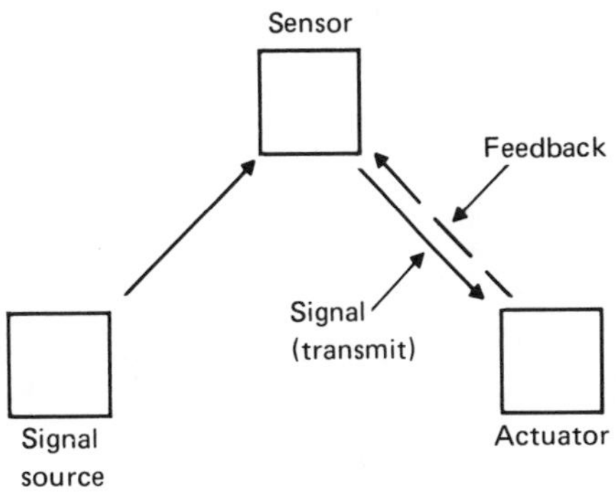

*Term defined in text.

FIGURE12-1 _____
Block diagram of the four functions of a control system

**Control system***

**Transmit**   In a control system, the signal that indicates the comparison between set point and the actual condition

FIGURE 12-2 _____

Thermostat, or sensor, used in an electric
control system

mostat actuates a gas valve in the furnace to generate heat. When the temperature in the room comes up to the set point, then the electric circuit is broken. The broken circuit is the **feedback signal** that the condition is corrected.

# Electric Control System

In an electric control system, the **sensors** are electric switches that close or open to allow an electric current to flow or not to flow. A thermostat is an example of a sensor in an electric control system. Figure 12–2 shows a typical thermostat used in an electric control system.

A thermostat has a set-point selector that can be set to the temperature desired in the space in which the thermostat is located. If the temperature in the space goes above the set point of the thermostat, an internal switch in the thermostat opens, interrupting the circuit. If the temperature in the space goes below the set point of the thermostat, the internal switch closes, completing the circuit. Figure 12–3 is a simplified diagram of wiring circuitry in a thermostat.

The thermostat, then, is a switch that controls an electric signal. The signal controls the operation of equipment.

The transmission system for an electric control system, which transmits the control signals, is the electric circuit. This transmission system includes the power source and the wires to carry the signal. The transmission system may be one circuit or more, but each electric circuit in the control system performs the transmission function.

The **actuator** in an electric control system is a solenoid, electric valve motor, or other electrically driven device. The device opens a valve when a current is flowing in a circuit and closes the valve when the current does not flow. The gas valve shown in Figure 12–4 contains a solenoid that controls gas flow. The actuator is wired into the control circuit in series with the sensor. When the sensor switch closes, current flows, and the actuator functions.

The feedback function in an electric control system is normally provided by the sensor. The thermostat opens when it is satisfied because the temperature has

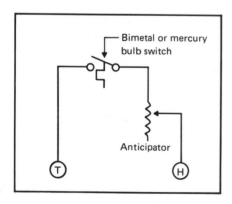

FIGURE 12-3 _____

Schematic diagram of the internal wiring for
an electric thermostat

reached the set point. The open circuit is a signal that the need for heat has been satisfied.

Other controls, such as limit controls, may be wired into an electrical control circuit to provide safety functions, but these controls are auxiliary to the operating controls and not actually part of the basic function of that part of the system.

Electric controls are basically on-off controls. The sensor is an electrical switch. When the sensor recognizes a need for heat, humidity, or whatever condition the system is controlling, an electrical switch is closed, and the unit is turned on. When the sensor indicates the controlled condition is satisfied, the switch is opened, and the unit is turned off. The actuator used in an on-off control system is also an on-off actuator.

**Multistage control** systems are also available. In a multistage control system sensor, there is more than one set of contacts. Each set of contacts is opened and closed at different control conditions, such as different temperatures. Such a sensor can control multistage burners or separate elements in an electric furnace. Each stage or set of elements is controlled by one of the sets of contacts. Multistage control makes it possible to approximate more nearly heat output to the heat loss of a building.

One other type of electrical control system is used. This is a **modulating control** system. In a modulating control system, a sensor is used that varies the electrical signal according to the conditions it is sensing. The varied signal is sent to an actuator that opens or closes a valve or other device in proportion to the strength of the signal received. This control is used with a modulating burner or other device that varies the output of the controlled unit to match the load at any condition.

## Pneumatic Control System

A **pneumatic control** system provides the same results as an electric control system does. The temperature or other conditions are controlled, but the medium of control is air rather than electricity. The main advantage of a pneumatic control system over an electrical system is that the former is inherently modulating. The sensors modulate air pressure in the transmission lines,

**Feedback signal**   A signal indicating that the set point and actual conditions are nearly the same

**Sensor**   A device in a control system that compares actual conditions with a set-point condition and sends a signal indicating that difference

**Actuator**   The device in a control system that performs a function to correct the condition causing a difference between the set point and the actual condition

**Multistage control**   A control or control system that controls the system so the output of the equipment is varied in steps or stages

**Modulating control**   A control or system that controls the output of the equipment over its operating range, in infinite increments

**Pneumatic control**   A control or control system that uses air for the control medium

Connection terminals

FIGURE 12-4 _____

Gas valve, or actuator, used in an electric control system

**Spring return**    An activator in which the "on" motion moves against a compression spring (In the "off" position,
the spring returns the actuator to its original position.)

**Field controls***

**Interface**    The point in a system where one type of control system is connected into another type of control system

and the air pressure is used to position actuators. The position of the actuators is determined by the pressure. The system is called pneumatic because air pressure is used to transmit data.

In a pneumatic system, a compressor is used to compress air and store it in a tank. The air from the tank flows to the system through a filter and a pressure reducing valve. The air leaving the pressure reducing valve is at approximately 18 psi. This arrangement provides a source of clean, dry air to the system at a constant pressure of 18 psi. Figure 12–5 is a diagram of a typical pneumatic system.

The sensors in a pneumatic control system provide the same function as thermostats in an electric system, but instead of being on-off electric switches, the sensors, or thermostats, are pressure control devices. The sensors are piped into the system between the air supply tank and the actuator, or operators, as shown in Figure 12–6.

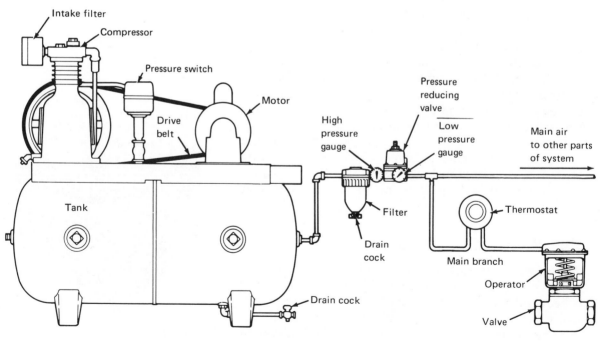

FIGURE 12-5

Diagram of a typical pneumatic control system

Small-diameter tubes (1/4 in. diameter) bring the air under pressure from the tank to the sensors. The sensor has a small orifice or nozzle from which the air can escape. A movable vane that covers the nozzle regulates the airflow. The needle valve is positioned by a bimetal element in the case of a thermostat. If the temperature surrounding the sensor goes above the set point of the sensor, then the nozzle is closed, and the air pressure in the line from the sensor to the actuator increases. If the temperature drops, the nozzle is opened, and the air pressure to the actuator is reduced. The diagram in Figure 12–7 shows how the sensor regulates the pressure.

The actuator is a pneumatic piston that is positioned by air pressure. The air leaves the controller and then goes to the actuator. Figure 12–8 shows a variety of actuators. Each has a different action.

The more pressure the sensor allows, the more the actuator is moved open; the less pressure the sensor allows, the more the actuator is allowed to move closed. The actuator is a **spring return** device. It is connected to the valve or other device being controlled. A pneumatically controlled valve is shown in Figure 12–9.

One of the advantages of a pneumatic control system as compared to an electric system is that the pneumatic system is a modulating system. The motion of the actuator is proportional to the air pressure supplied by the sensor, and the air pressure is proportional to the signal received by sensor. Pneumatic systems can be used to control heat or humidity, to position damper motors, and to regulate valves, just as an electric system can, but pneumatic systems are usually used on larger, more sophisticated jobs than are electric systems.

Pneumatic controls are usually used as **field controls** only—that is, the controls outside the unit. In most cases, the controls inside the units are electric, even though the controls external to the units may be pneumatic. When both types of controls are used, a pneumatic-electric (PE) switch is part of the system. This switch changes a pneumatic signal to an electric signal and allows a pneumatic control system to be **interfaced** with an electric one. Figure 12–10 is a diagram showing how a PE switch works.

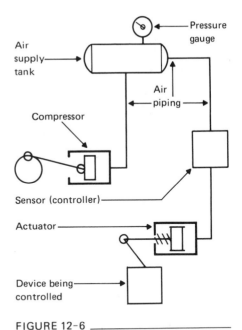

FIGURE 12-6 _____

Schematic diagram of the parts of a pneumatic control system showing the relationship between the sensor and the actuator

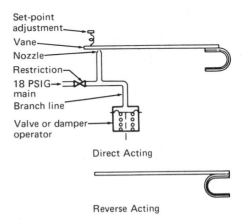

FIGURE 12-7 _____

Schematic diagram of the operating parts of a pneumatic thermostat

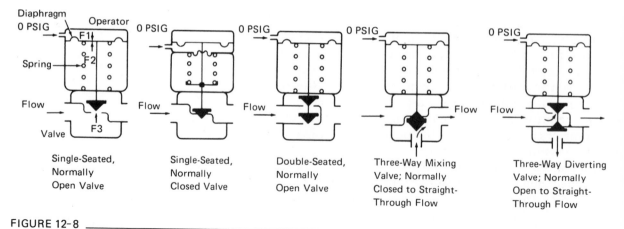

Single-Seated,
Normally
Open Valve

Single-Seated,
Normally
Closed Valve

Double-Seated,
Normally
Open Valve

Three-Way Mixing
Valve; Normally
Closed to Straight-
Through Flow

Three-Way Diverting
Valve; Normally
Open to Straight-
Through Flow

**FIGURE 12-8**

Schematic diagrams of a series of pneumatic operators illustrating the variety of actions possible

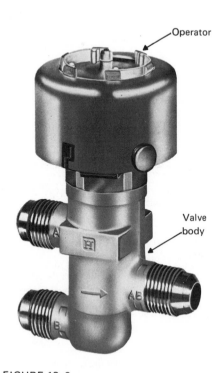

**FIGURE 12-9**

Pneumatic operator mounted on a valve

# Electronic Control System

An electronic control system, as the name implies, uses electronic devices to provide the control functions. Normally the electronic devices control electric circuits, and these circuits power the actuators in the system.

One of the major differences between an electronic control system and either the electric or the pneumatic system is the use of a control center. Electronic systems operate on low-voltage **direct current (DC)** power. But because high voltage **alternating current (AC)** is the current that is normally available on a typical job, this current has to be transformed down and **rectified** for use in the system. Rectification is the changing of the electrical power from AC to DC in a circuit. Electronic devices perform the rectification. The transformer and rectifier circuit is part of the control center, along with the other control devices and circuits. An electronic control center is shown in Figure 12–11.

A typical electronic control center has a rectifier that provides a low-voltage DC signal. The signal, used for the transmission function of the system, is circuited to the sensor of the system. The sensor is an electronic device that modifies the signal in some way. The modified signal then comes back to the control center. In the center, this signal then controls a 24 V, or line voltage, circuit that controls the actuator. The sensors in an electronic system are devices that sense temperature, humidity, light, pressure, and other conditions. A block diagram showing the relationship of the parts of an electronic control system is shown in Figure 12–12.

The control center can be designed to receive signals from more than one sensor point. The center can compare the various signals and then send out an actuator circuit signal related to the multiple signals received. For example, if both a room thermostat and an outdoor thermostat are used on a control system, the outdoor thermostat resets the indoor temperature. That is, the indoor thermostat temperature set point is automatically raised as the outside temperature gets colder. Electric and pneumatic systems can also perform this function, but not as easily as can an electronic system. In the electronic system, signals from both thermostat circuits are sensed in the one control center, and an actuator signal can be sent out from the center.

The solid-state device used to sense temperature is the thermistor. The electronically functioning part of a thermistor is a crystal, in which the resistance to an electric current changes as the temperature of the crystal changes. In a typical thermistor used in a heating control system, the resistance increases as the temperature decreases, and the resistance decreases if the temperature increases. So when a thermistor is wired in series in a circuit, temperature variations cause variations in the current flow in the circuit. The electrical signal received through the thermistor is sent to the electronic control center, where it is converted to an AC signal of high enough voltage to drive the actuators in the system.

The actuators used in an electronic system are similar in design and function to those used in an electric control system. Figure 12–13 shows a damper motor for use in an electronically controlled system.

The electronic control system combines the low-cost feature of the electric wiring found in the electric system with the advantages of inherent modulation found in a pneumatic system. When those advantages are coupled with the simplistic yet rugged features of the electronic components, it is easy to see why the electronic systems are popular and are being used in many places where electric or pneumatic systems were used previously.

**Direct current (DC)** An electrical current that always flows in the same direction in a circuit

**Alternating current (AC)** Electrical current in which the polarity changes direction cyclically, usually 60 times per second

**Rectify***

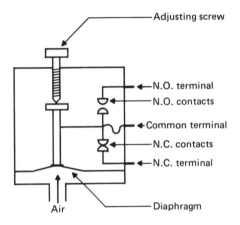

FIGURE 12-10 _____

Diagram of the working parts of a pneumatic-electric (PE) switch

FIGURE 12-11 _____

Electronic control center showing the many electronic devices that operate the unit. The terminals on the right side are used to connect into the operating system.

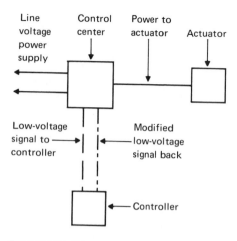

FIGURE 12-12 _____

Block diagram showing the relationship of the main parts of an electronic control system

FIGURE 12-13 _____

Modulating motor used as an operator in an electronic control system

## 12.3 SUMMARY

Just as each heating unit must have controls for proper operation, a control system is needed to integrate the different parts of a unit, or multiple units, in a building. The control system components are (1) sensors, (2) signal devices, (3) actuators, and (4) feedback devices. These parts tell the heating equipment when to start up and when to shut down. The most common systems are electric, electronic, and pneumatic.

Electric systems use electricity that is carried in normal circuits as a signal. All of the components are electrical; that is, electricity is supplied, controlled, and used as a source of power in each part of the system. Most commonly, electric controls are on-off controls; the controls turn devices on, or they turn them off. The controls usually provide no intermediate positions. Special electric controls, however, provide multi-position control or even modulation.

The pneumatic system uses air pressure as a signal. Air is compressed and stored in a tank. Small-diameter tubing carries the air to the sensor where the pressure is regulated according to the set point of the sensor. Then the modified pressure is used to position the actuator. In many cases, pneumatic controls are used to control electric signals. Pneumatic controls are usually used on larger buildings and are excellent in cases when modulation is required.

Electronic control systems are relatively new but are rapidly becoming more common. Electronic systems use solid-state devices as sensors; the signal is usually a DC electric signal that is modified by the sensor. The DC signal is changed to an AC signal in a control center. The control center then sends an AC signal to the actuators in the system to provide control. Electronic systems are becoming more popular because of the simple but rugged characteristics of the components and because modulation is so easily achieved with the system.

## 12.4 QUESTIONS

1. There are four general types of control systems. One that is seldom used is the mechanical system. The other three are _____, _____, and _____.

2. Any control system should provide four control functions. One is to sense actual conditions and compare them to a set point. The other three are _____, _____, and _____.

3. Match the words in the first column with the action in the second column that best describes each.

   1. Sensor
   2. Signal
   3. Actuator
   4. Feedback

   a. A gas valve solenoid opens the valve.
   b. The thermostat opens on temperature rise.
   c. A thermostat calls for the furnace to come on.
   d. Air pressure in a pneumatic system is changed as the temperature changes.

4. A thermostat in an electric control system can be called a temperature-actuated electric _____.

5. The transmitter in an electric control system is the electricity in the circuit. True or false?

6. Feedback in an electric control system is provided by the opening of the electrical circuit when the thermostat is satisfied. True or false?

7. An electric control system uses electricity as a transmission signal; a pneumatic system uses _____.

8. Sensors in a pneumatic system could be called air _____ _____ because of their function in the system.

9. A pneumatic actuator is a simple piston device. True or false?

10. A pneumatic actuator is a modulating control because the movement of the piston is _____ to the air pressure.

11. Electronic control systems use electric curcuits and electricity to carry signals. True or false?

12. An electronic control system can take signals from more than one point and compare them. True or false?

13. Choose the correct term for the blank space. Electronic control system actuators are similar to those used in _____ systems.

   a. electric    b. pneumatic    c. mechanical

# CHAPTER
# 13
# Heat Loss Variables and Factors

## 13.1 INTRODUCTION

When the weather is cold, heat is required to keep buildings warm. So far this text has discussed heat and the equipment that supplies it. This chapter examines what causes the need for heat in a building; subsequent chapters explain how to calculate the amount of heat needed, an operation called figuring the heat load. The **heat load** is the total Btu/h that must be supplied to keep a building at the design temperature. The **heat loss** is the Btu/h lost from the building because of temperature differences.

To figure a heat load for a building, the designer must determine what causes the load. Causes can be divided into two categories of heat loss: variables and factors. Heat loss variables include temperature difference between the inside of a building and the outside and areas of the surfaces exposed to the outside air temperature. Heat loss factors to consider are the rate of heat conduction through the surfaces of the building and cold air that infiltrates the building.

An accurate calculation of heat loss is the first step taken when a heating system is designed for any building. All of the decisions concerning the type of system to use, equipment type and size, distribution of air, and controls are dependent upon heat loss. This chapter explains what heat loss is, what causes it, and Chapter 14 explains how to use heat loss to calculate the heat load.

## 13.2 DEFINING HEAT LOSS

One of the first laws of thermodynamics states that heat always travels from a warm object to a cooler one. An ice cube in a glass of water cools the water because heat from the water is absorbed by the ice. The ice melts because it has absorbed heat, and the water is cooled because it has given up heat. See Figure 13–1.

To relate the melting of ice to the heat loss from a building, consider how the building and the spaces in it are affected by the temperature of the outside air. If the temperature of the air mass surrounding a building

**Heat load\***

**Heat loss\***

**Heat gain**  The heat in Btu/h that a building accumulates because of the design temperature difference, solar heat, and any other source

**Variable**  The elements in a heat loss calculation that are generally different for each job

**Factors**  Numerical equivalents of the amount of heat that will be conducted through a given material in Btu/h per sq ft per degree Fahrenheit

\*Term defined in text.

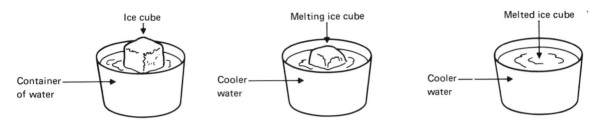

FIGURE 13-1 _____

One law of thermodynamics. Heat always flows from the warmer object to the cooler object, and, in the process, the temperature of each object changes.

is colder than the air in the building, then heat will flow from the building to the air surrounding it. See Figure 13–2. If the air outside the building is warmer than the air inside, then the heat passes into the building, and **heat gain** occurs.

To maintain a comfortable temperature inside any building when the air outside is colder than the air inside, heat must be provided at the same rate as it is lost. To determine how much heat is needed, a heat loss calculation is made.

The calculation is made by figuring the loss for those parts of the building that heat will pass through, such as walls, windows, ceilings, and floors. These losses are called conduction losses. Losses from other sources, such as air that leaks into the building through cracks or the air brought in for ventilation, also have to be figured. In all cases, the losses are proportional to the difference in temperature of the air inside and outside the building.

## Calculating the Heat Load

In calculating a heat load for any building, it is necessary to identify and consider the **variables** and **factors** that affect the loss in the building. Heat loss variables include temperature difference between the inside of the building and the outside of the building and areas of the building surface that are exposed to outside temperature. Heat loss factors include the rate of heat conduction through the surfaces of the building and the rate of infiltration of cold air into the building. These variables and factors are combined in an orderly format, and the heat loss is figured by following a series of steps.

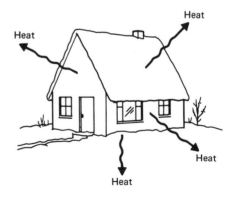

FIGURE 13-2 _____

Heat loss of a building

**Components**    The constituent parts of a structure

**Temperature***

**Design temperature***

**Outside design temperature***

**Inside design temperature***

**Design temperature difference***

**R factor***

**U factor***

—Determine the temperature difference between the inside and outside of the building.

—Consider the different types of construction materials used in the building (such as the glass in the windows, the wooden doors, or materials in the walls), and then calculate the areas of each of these.

—Figure the amount of outside air that is leaking into the building through the cracks around the opening parts of the windows and doors, as well as the air that gets into the building when the doors are opened. Figure the fresh air brought in for ventilation.

—Find the appropriate heat transfer factor for each of the building **components** and for the infiltration by using a factor table.

—Multiply the heat transfer factor for each component by the area of the component, and multiply the heat transfer factor for infiltration by the amount of crackage to determine the heat losses due to conduction and infiltration.

—Add the losses together to find the total heat load of the building.

## Terms Used in Determining Heat Loss

Some of the terms used in determining heat loss have been defined earlier in this text; others are new. To help in understanding the procedures explained later in this chapter and throughout the book, a list of the most common terms with definitions and descriptions follows.

**Temperature.**    **Temperature** is the measurement of the intensity of heat. Temperature is measured with a thermometer, which commonly is read in degrees Fahrenheit (standard) or in degrees Celsius (metric).

**Btu.**    Btu is the measurement of a quantity of heat. A Btu is defined as the amount of heat that will raise the temperature of 1 lb of water 1°F. If a common wooden kitchen match is burned, approximately 1 Btu of heat will be generated.

**Btu/h.**   Btu/h relates quantity of heat to a time period. In heat loss calculations, time is a factor, and so the term Btu/h is used to indicate the amount of heat involved in a process during 1 hr.

**Design Temperature.**   **Design temperature** is a selected or arbitrary temperature used in heat loss calculations.

**Outside Design Temperature.**   **Outside design temperature** is the temperature of the outside air for the time at which the heat loss is calculated. Design temperature tables give the lowest outside temperature that might be expected to occur in the locality for which the designer wants to figure a loss. Table A–3 in the Appendix is a table of outside design temperatures for the United States.

**Inside Design Temperature.**   The **inside design temperature** is the temperature that the designer wishes to maintain in a building. The temperature is an arbitrary one decided upon by the designer. A normal heating design temperature for buildings is between 70°F and 75°F.

**Design Temperature Difference.**   The **design temperature difference** is the difference between the inside design temperature and the outside design temperature. This variable is found by subtracting the outside design temperature for a given locality from the selected inside design temperature. Design temperature difference is the actual temperature difference used in the heat loss calculation.

**R Factor.**   An **R factor** defines the amount of resistance to heat flow through a 1 sq ft area of a given building material for each 1°F temperature difference on each side of the material. The R factor, given in Btu/h, is based on materials as tested in a laboratory. R factors are listed on insulating materials so a purchaser will know the insulating value of the material when it is purchased or used. The higher the number, the better the insulation.

**U Factor.**   U factors are the most commonly used factors for figuring heat loss. A U factor defines the

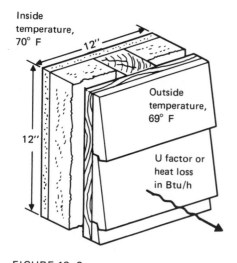

Inside
temperature,
70° F

12"

12"

Outside
temperature,
69° F

U factor or
heat loss
in Btu/h

FIGURE 13-3 _____

Pictorial representation of U factor

amount of heat, in Btu/h, that will pass through 1 sq ft of a material for 1°F of temperature difference on each side of the material. U factors are normally given for combinations of construction material such as the assorted building components that make up an outside wall.

U factors are found by dividing the sum of the R factors into the number 1. This operation is called finding the reciprocal. The formula is:

$$U = \frac{1}{R_1 + R_2 + R_3 \ldots}$$

The better the insulation value of the components, the smaller the factor.

U factors are used as multipliers in a heat loss calculation. Figure 13–3 is a pictorial representation of a U factor. Tables listing the U factors for various combinations of building components are available from the American Society of Heating, Refrigeration, and Air Conditioning Engineers, Inc.

**Heat Transfer Factor.**    A **heat transfer** factor (htf) is a factor in which the U factor for given materials has been multiplied by the design temperature difference. To determine heat loss due to conduction, the heat transfer factor is multiplied by the area of exposed surface. The same method is used to determine heat loss due to infiltration. The heat transfer factor may be selected from a factor table, or the multiplier to be used in the heat loss calculation can be obtained by multiplying the U factor by the design temperature difference. When heat transfer factors are selected from a factor table, care must be taken to make sure that they are selected for the design temperature difference of the locality in which the building will be erected. Table A–4 in the Appendix is an example of a heat transfer factor table used for calculating a room-by-room load. Table A–5 in the Appendix is an abbreviated form of Table A–4. Table A–5 is used for calculating a shell load.

**Conduction.**    Conduction is the process of heat transfer through a material.

**Infiltration.**    **Infiltration** is the term used to describe any air that leaks into a building from the outside.

Because a large part of the heat loss for any building is that caused by the leakage of cold outside air into the building, the rate of the infiltration must be determined and the heat loss caused by the infiltration must be calculated. Several methods can be used to determine the rate of infiltration. One is the **crackage** method. This method requires the estimation of the **lineal footage** of the cracks around the windows and doors and the application of an appropriate factor. Another method is to figure the amount of air in the building in cu ft and assume that that much air will infiltrate the building once each hour.

**Volume.**    The **volume,** or cubic content, of a building is figured in cu ft (ft³) or cu m (m³) and is often used in heat loss calculations. Volume is normally used when figuring the amount of air per room or zone. This variable is calculated by using the formula:

$$V = w \times l \times h$$

where:

   $V$ = volume
   $w$ = width of the space
   $l$ = length of the space
   $h$ = height of the space

## EXAMPLE 13–1

If a building is 36 ft long and 24 ft wide, with a ceiling height of 8 ft, what is the volume of the building?

**Solution**

In this case, the dimensions are given. Therefore:

$$V = w \times l \times h$$
$$= 24 \times 36 \times 8$$
$$= 6912$$

The volume is 6912 cu ft.

**Fresh Air.**    In some cases, a specified amount of outside air is brought into a building to maintain a fresh air condition for the occupants of the building. This air is called **fresh air** or **ventilation air.** The heat loss caused by this colder outside air must be figured and added to the conduction loss of the building.

**Heat transfer factor***

**Infiltration***

**Crackage**    The cracks around the opening parts of doors or windows

**Lineal footage**    The distance between two points, measured in ft

**Volume***

**Fresh air***

**Ventilation air***

**Air change***

**cfm***

**gpm***

**Dry bulb (db) temperature**    The temperature measured with a typical thermometer registering sensible heat only

**Air Change.**    The term **air change** relates to the quantity of air required to replace all of the air in a given space in a given length of time. One air change per hour for a building would be the amount of air required to replace all of the air in the building in one hour.

If a fresh air system was designed for a building that has a volume of 6912 ft$^2$, and a 1 hr air change is desired, the ventilation rate would be 6912 ft$^3$ per hour. This rate of air change would give a complete change of air in the building each hour.

**Shell Load.**    If a heat loss for a building is figured as if the entire building is one large space, this method is called a shell load calculation.

**Room-by-Room Load.**    If the heat loss is calculated for a building by figuring the losses for the individual rooms or separate zones and then adding these individual losses together to find the total load, this method is a room-by-room load calculation.

**cfm.**    Air being used as the heating medium in a given system is measured in cubic feet per minute, or **cfm.** After the heat loss in Btu is calculated for a building in which air is the circulating medium, the usual practice is to convert the Btu/h figures to cfm to determine how much air to distribute to each space in the building to offset the heat loss in the space. The term cfm may be used for the air being circulated in an entire system, or it may apply only to fresh air, infiltrated air, or any other air, when the quantity must be accounted for.

**gpm.**    When water is used as the circulating medium in a heating system, the water is measured in gallons per minute, or **gpm.** As in warm air systems, the heat loss in Btu/h is converted to gpm to facilitate the design of the piping system and the selection of heat exchangers.

## 13.3  HEAT LOSS VARIABLES

Designers calculating a heat load for a building must consider what variables will affect that heat loss. The two most important variables are the design temperature difference and the areas of the outside surfaces of the building.

## Design Temperature Difference

The first variable, design temperature difference, occurs because buildings are constructed in different geographical locations. The outside design temperature for the heat loss varies according to the location. A building constructed in the southern part of the United States has a smaller heat loss than does a building constructed in the same way in the North because of the difference in the outside temperature in the winter.

A design temperature chart or map is used to select the proper design temperature for the area in which a building is to be constructed. This map shows the winter outside design temperature according to location. An inside design temperature is chosen. The difference between the two temperatures gives the design temperature difference to use in the calculation.

**Outside Design Temperature.**    Table A–3 in the Appendix is a complete set of outside design temperature tables for the United States. These tables show the winter **dry bulb (db) temperature** for many locations in the United States and in Canada. The figures are compiled for all cities in which there are weather stations. The designer chooses the proper temperature for the area that the building will be located in.

### EXAMPLE 13–2

A designer needs to find the outside design temperature for Seattle, Washington.

#### Solution

The designer looks in the column under "Winter db" and across from the heading for Seattle. The temperature is found to be 17°F.

The design temperatures listed in the chart are somewhat higher than the lowest winter temperatures that occur in an area. Rather than design for the lowest temperature that may occur in an area, an average of the lowest temperatures that have occurred over a number of years is used. The average is a more reasonable temperature to use than the lowest that has occurred. Normally the equipment chosen to heat a building has capacity enough to carry over during colder temperatures because of heat storage in the building itself and because the coldest temperatures occur for relatively short periods of time.

**Inside Design Temperature.**    The inside design temperature is a matter of choice of the designer. Until recently, the trend was to design for 75°F or warmer inside buildings where comfort is most important. Because of the need to conserve energy and especially because of high fuel costs, the inside design temperatures now may be figured at 65°F to 70°F, depending upon the building use.

The design temperature for manufacturing plants and buildings used for storage, such as warehouses, may be much lower than that required in a residence. The choice of temperature for these buildings should be made with the use of the building in mind.

## Calculating Design Temperature Difference

The design temperature difference (DTD) to be used in the heat loss calculation is found by subtracting the outside design temperature from the inside design temperature.

### EXAMPLE 13–3

The winter db for Seattle, Washington, is found in the design table to be 17°F. If the inside design temperature of 70°F is chosen, what is the design temperature difference?

**Solution**

Subtract 17°F from 70°F:

70 − 17 = 53

The design temperature difference to be used for figuring a load for the Seattle area is 53°F.

## Exposed Surfaces

The second variable, the areas of the exposed surfaces of the building, occurs because no two buildings are ever exactly alike. The outside surfaces of windows, walls, ceilings, and other components will be different for each building. The areas of exposed surfaces for the different parts of the building through which heat can pass have to be considered. The dimensions of the surfaces for each different material combination that is exposed to outside temperatures must be determined, and then the area of each calculated.

## Calculating the Area of Exposed Surfaces

The area of a surface is figured in sq ft ($ft^2$) or sq m ($m^2$). The area of a surface is calculated by using the formula:

$$A = w \times l$$

where:

$A$ = area
$w$ = width of surface
$l$ = length of surface

### EXAMPLE 13–4

If a building is 36 ft long and has a ceiling height of 8 ft, what is the gross wall area of one side?

**Solution**

In this case, the ceiling height of the building is used as the width in the formula. Therefore:

$$A = w \times l$$
$$= 8 \times 36$$
$$= 288$$

The gross wall area is 288 sq ft.

The most common building components for which the areas have to be figured are:

1. exposed walls,
2. windows,

**Gross wall area**    The area of an outside wall, including any windows or doors located in the area

**Net wall area**    The area of a wall with the doors, windows, and any other openings subtracted from it

**Cold partition** *

**Cold ceiling** *

**Cold floor** *

3. doors,
4. ceilings,
5. floors,
6. cold partitions.

In the case of an existing building, the dimensions may have to be taken from the building itself by measurements. The best way to keep a record of the dimensions is to draw sketches of the building and record the dimensions. A floor plan should be sketched showing the exterior walls and the interior partitions and doors. Elevation sketches should be made showing the exterior doors, windows, and roof details. The dimensions of the various areas to be considered in the heat loss are shown on the sketches. Construction types and details can also be recorded on the sketches to help in selecting the proper factors for the loads. The sketches become a permanent part of the job file and can be used in the system design process, as well as for figuring the loads. In the case of a new building, plans are usually available giving dimensions.

## EXAMPLE 13–5

A typical double-hung window size is 2'4" by 5'4". Find the area.

### Solution

The dimensions of doors and windows can be figured to the nearest half-foot for purposes of finding the area. Therefore, the dimension 2'4" by 5'4" rounded off to the nearest half-foot is 2'6" by 5'6". Changing the figures to decimal equivalents makes the math simpler. In this case, 2'6" is written as 2.5 and 5'6" as 5.5. To calculate the area of the window, use the formula:

$$A = w \times l$$
$$= 2.5 \times 5.5$$
$$= 13.75$$

The area of the window is 13.75 sq ft. Working with the decimal equivalents makes the calculations of the window areas simpler, and the results are still accurate enough for purposes of determining the heat loss.

## EXAMPLE 13–6

Find the area of a wall that is 24'9" long. The ceiling height is 8'3".

**Solution**

The dimensions of walls, ceilings, and floors can be figured to the nearest whole foot for purposes of finding the area. Therefore, the dimensions 24'9" and 8'3" rounded off to the nearest whole foot become 25' and 8' respectively. To calculate the area of the wall use the formula:

$$A = w \times l$$
$$= 25 \times 8$$
$$= 200$$

The area of the wall is 200 sq ft.

Heat loss is figured on the net exposed wall only. However, in figuring the areas, it is easier to calculate the area of the entire wall, or gross wall, and then subtract the window and door areas from **gross wall area** to find the **net wall area.**

If the exposed walls of the building are made up of two or more kinds of construction, then the areas of each kind of construction must be figured separately since each kind of construction is likely to have different heat loss factors.

Closets, halls, and stairways should be treated as if they are part of the rooms into which they open.

A **cold partition** is defined as a partition that separates a heated space from a space that is not heated but is enclosed and is not exposed to the outside temperature. Such a space generally is not as cold as the outside but not as warm as the inside. An example of a cold partition is the wall between a house and an enclosed porch or vestibule.

A **cold ceiling** is the ceiling of any room that is located beneath an unheated space. This space can be an attic, the outdoors, or any other space exposed to outside temperatures.

A **cold floor** is any floor between the heated part of the house and an unheated space, such as a crawl space, a cold garage, a space under an overhang, or any other such space exposed to the outside temperature.

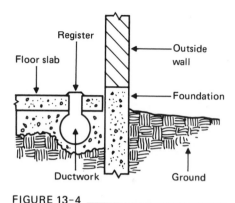

**FIGURE 13-4** _____

Duct system around the perimeter of a building under the cement slab floor

Concrete floors on the ground can pose a special problem for figuring heat losses. If the floor is in a building that is heated by an overhead system, then the floor should be treated as though it were a separation between the room and a cold area (in this case, the ground) so the area of the floor should be figured. It is necessary only to figure the outside 10 ft of the floor, however, since the heat loss through the floor farther than 10 ft from the wall is negligible.

If the concrete floor has heat ducts around the **perimeter** in the slab itself, some heat will be lost from the duct. This lost heat will warm the outside edge of the floor. See Figure 13–4 for an illustration of a duct system under a concrete slab around the perimeter of a building. In this case, the only measurement required for the floor loss is the perimeter length of the slab, which is adjacent to the outside wall.

## 13.4   HEAT LOSS FACTORS

The factors to consider when determining heat loss are the rate of heat conduction through the exposed surfaces of the building and the rate of cold air infiltration. Selecting the proper factors for each of the materials used in building construction is one of the most important steps in the calculation of heat loss. The factors are the numerical equivalents of the Btu/h that will be conducted through the building components, per sq ft of the area, per 1°F difference in temperature on the two sides of the material. There are also factors for the heat loss due to the air that infiltrates into a building through the cracks around the opening parts of the windows and doors and that gets into the building when the doors are opened.

### Factor Tables

A reasonably complete set of factors, or factor tables, should include factors for not only the different components in the building, such as walls, ceilings, and floors, but also for different types of construction for each. The set also should include factors for different amounts of insulations as used in the major construction types. The factor tables found in this text are heat transfer factor tables. In heat transfer factor tables, the U factors for various types of materials have been multiplied by the design temperature difference. U factor tables are also available. A complete set of U factor tables is available in the current manual of the American Society of Heating, Refrigeration, and Air Conditioning Engineers, Inc.

Most factor tables also have a selection of factors to be applied to door and window crackage to determine the loss due to infiltration of cold air. Some tables, however, include the infiltration factor in the conduction factor for the windows, so designers must check the particular tables being used and make sure the infiltration is included in one place or another.

# Conduction

Selecting the factor for conduction for any part of the building requires finding the description of the component in the tables that most nearly matches the construction. Under the heading for exterior walls in the factor tables are found major headings for masonry walls of various types, frame walls of different types, and possibly others.

Under each major heading, subheadings for variations are found. These variations are thickness, finish material, amount of insulation, and any other factor that affects the heat flow through the building component. All of these variations must be taken into consideration when selecting the factor so that the final loss figure will be correct for that particular construction.

The selection of the proper factors for windows, doors, walls, ceilings, and floors is made in the same way.

# Infiltration

When selecting the factor for infiltration, the type of window or door for which the loss is being figured must be identified.

No matter how tightly a building is constructed, air will leak through the cracks around windows and doors, as well as through any cracks that may exist between building components. Air at outside temper-

**Perimeter**    Circumference or distance around

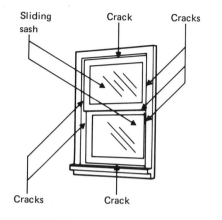

FIGURE 13-5 _____

Opening crackage for a double-hung window

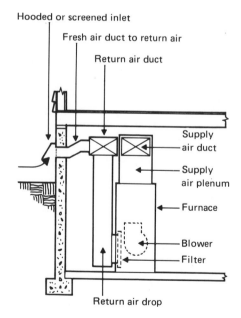

FIGURE 13-6 _____

Introduction of fresh air into the return air system

atures will leak into the building through these cracks and has to be heated if the building is to be maintained at a fixed inside temperature. The air leakage is caused by the wind pressure against the exterior surfaces of the building. The amount of leakage is determined by the size of the cracks and the velocity of the wind. In the calculations, the size of the cracks is compensated for by the factors used for infiltration, and the wind velocity is figured as 15 miles per hour. (Incidentally, a fireplace can increase the infiltration into a building appreciably because the fireplace requires combustion air to burn, and that air usually comes from crackage in the construction.)

Calculation of the heat loss for infiltration uses the actual footage of the cracks in each door or window that has an opening to the outside. The factor tables give the proper factors to use with these lineal footage figures to arrive at the heat loss occasioned by the cracks.

In measuring crackage, the lineal footage is figured for only the part of the window or door that opens. An illustration of the opening crackage for a double-hung window is shown in Figure 13–5. The cracks around the stationary, or nonopening, parts of the windows or doors do not have to be figured.

**Fresh Air.**    In some cases, fresh air will be introduced to a building as ventilation air, or **make-up air,** for an exhaust fan or a fireplace. This fresh air is introduced into the return air system, as shown in Figure 13–6.

When fresh air is introduced, the heat loss for this air has to be figured in with the heat loss caused by conduction losses. To calculate the loss, the quantity of air that is introduced is calculated in cfm. Then the loss is figured on that air by use of the sensible heat formula, as follows:

$$\text{Btu/h} = \text{cfm} \times 1.08 \times TD$$

when:

Btu/h = amount of heat required to raise or lower the temperature of the air

cfm = quantity of air in cu ft per min

1.08 = a factor including adjustments for the specific heat of air, the specific weight, and time

$TD$ = temperature difference

## EXAMPLE 13–7

If 200 cfm of fresh air is to be brought into a building as ventilation air and the temperature difference is 53°F, find the heat loss.

**Solution**

$$\text{Btu/h} = \text{cfm} \times 1.08 \times TD$$

$$= 200 \times 1.08 \times 53$$

$$= 11,448$$

The heat loss caused by the fresh air is 11,448 Btu/h.

**Make-up air**    Air that is brought into a building specifically to make up any air that is being vented

## 13.5  SUMMARY

Designers must make an accurate calculation of the heat load of a building so that they can select the proper size heating unit and design the most efficient heat distribution system. To calculate the heat load, they must consider all of the variables and factors that affect heat loss. The heat loss variables include temperature differences between the inside and outside of a building and the areas of surface exposed to outside air. Heat loss factors include the rate of heat conduction through the surface of the building and the rate of cold air infiltration.

Designers should follow certain steps when calculating a load.

—Determine design temperature difference.
—Calculate the areas of each type of construction material used in the building.
—Calculate infiltration or fresh air in cfm.
—Determine the proper factors.
—Calculate the heat loss for each component part of the building.
—Add the component losses together for the total heat load.

The design temperatures are determined by reference to an outside design temperature table and by arbitrary selection of the inside temperature. The design temperature difference is found by subtracting the outside design temperature from the inside design temperature.

The areas of each of the building components, such as windows, doors, ceilings, floors, and walls, are calculated from measurements made of the actual building or are taken from plans of the building.

Selection of the proper factors for conduction and infiltration is made from an appropriate factor table that covers the type of construction used in the building. Consideration is given to building materials, methods of application, and insulation in selecting the factors. The final result, the total heat load, is the sum of the heat losses through each of the component parts of the building and the heat required to raise the infiltrated, or fresh, air to the temperature of the indoor air.

## 13.6  QUESTIONS

1. The second law of thermodynamics states that heat always moves from a _____ object to a _____ one.

2. Heat loss occurs because of the temperature _____ between the inside and outside of the building.

3. In calculating a heat loss for a building, _____ and _____ also affect the loss.

4. Put in order the six steps to follow in calculating a heat loss:

   _____ a. Find the appropriate factor.
   _____ b. Determine the temperature difference.
   _____ c. Figure the amount of infiltration.
   _____ d. Find the total loss.
   _____ e. Find the building component losses.
   _____ f. Calculate the areas of the components.

5. Outside design temperature is the temperature of the air outside on December 31 of each year. True or false?

6. The inside design temperature is not directly related to a heat loss calculation. True or false?

7. Design temperature difference is found by subtracting _____ _____ _____ from _____ _____ _____.

8. An R factor defines the amount of resistance to heat flow. True or false?

9. A U factor defines the amount of heat that will be conducted through a specific type of construction. True or false?

10. A list of the factors that are commonly used to figure heat loss is called _____ _____.

11. A heat transfer factor is a factor in which the _____ _____ has been multiplied by the _____ _____ _____.

12. Match the words on the left with the correct descriptions on the right.

    _____ 1. Infiltration air       a. Air brought in to
    _____ 2. Ventilation                replace exhaust air
            air                   b. Leakage through
    _____ 3. Make-up air               cracks
                                  c. Air brought in to
                                     keep inside air fresh

13. The amount of infiltration and/or fresh air is figured the same way in a heat loss. True or false?

14. When the heat load for a building is figured as though the building were one large space, the load is called a _____ _____.

15. If a heat loss is figured for each individual room in a building, the load is called a _____ _____ _____ load.

16. The area of the gross wall of a building is found by multiplying the length of the building, times the width, times the ceiling height. True or false?

17. Net wall area includes windows, doors, and any other openings. True or false?

18. The total heat load of a building is the _____ of the losses of each of the component _____ of the building.

# CHAPTER 14
## Calculating Heat Loads

## 14.1 INTRODUCTION

Organizing the heat loss variables and factors in a format that makes it easy to use them for calculations and then combining them in a way to make those calculations is called figuring the heat load for a building. The heat loss variables are the design temperature difference and the areas of exposed surfaces on the building. The heat loss factors are the rate of heat conduction through the exposed surfaces of the building and the rate of cold air infiltration.

The forms used for calculating heat load are called **heat load calculation forms.** This chapter describes two types of forms that can be used for heat load calculations: the short form for calculating shell load and the long form for calculating room-by-room load.

## 14.2 HEAT LOAD CALCULATION FORMS

The first step in calculating a heating load is to identify the heat loss variables and factors that will be used for the calculations as described in Chapter 13. Then, if all of the data is properly organized, the final calculation of the load is simple. The heat load calculation form is used to record the data in an organized way. The form with the data recorded is then used to make the necessary calculations.

If the load is figured for one space in a building, or for any number of spaces treated as one, it is called a **shell load;** and a short form calculation sheet for figuring a shell load is used. This form has one column for the areas, another column for factors, and a third column for the heat loss of the separate components of the space. The component losses in the third column are added together to get the total building loss.

If the load is being figured for many rooms or many spaces, with each being treated as a separate part of the whole, it is called a **room-by-room load,** and a long form calculation sheet for figuring the room-by-room load is used. The long form has one column for the factors that remain the same for all components in the building and

*Term defined in text.

separate columns for each room and the variables related to the rooms, such as the size of windows, doors, walls, ceilings, and floors. This form allows a determination of separate loads for each room. These individual loads are added together to get the total building loss.

Any calculation form used should include space at the top to record the job name, address, date, and, most important, the design temperatures used for the load. Since the heating loads are used for sizing the equipment and designing the distribution system, the forms are valuable documents and should be kept on file for reference.

## Short Form

A typical short form calculation sheet for figuring shell load is shown in Figure 14–1 and also in Table A–6 in the Appendix. The spaces in the top section of the form are used for the statistical data for the particular job. The middle section of the form is divided into horizontal and vertical divisions for easy calculations of the loads. The bottom of the form provides space for total loads. The first column on the left identifies the various components of the building. The first several lines refer to building components. The last line is for the loss due to air that infiltrates the building or that is introduced as fresh air.

The second column, area, is for the areas of the spaces identified in the first column or for infiltration in cfm, or both.

The third column, htf, is for the heat transfer factor as taken from the factor tables. If U factors are used, the U factor must be multiplied by the design temperature difference in order to be used as a multiplier, or heat transfer factor, on this form.

The fourth column is for the subtotal Btu/h figures. Btu/h is arrived at by multiplying the areas, or cfm, times the htf. The subtotals are then added together to arrive at the total Btu/h loss, shown on the last line.

## Long Form

Whenever the heating load calculations will be used for selecting the supply and return registers and grilles

**Heat load calculation form***

**Shell load***

**Room-by-room load***

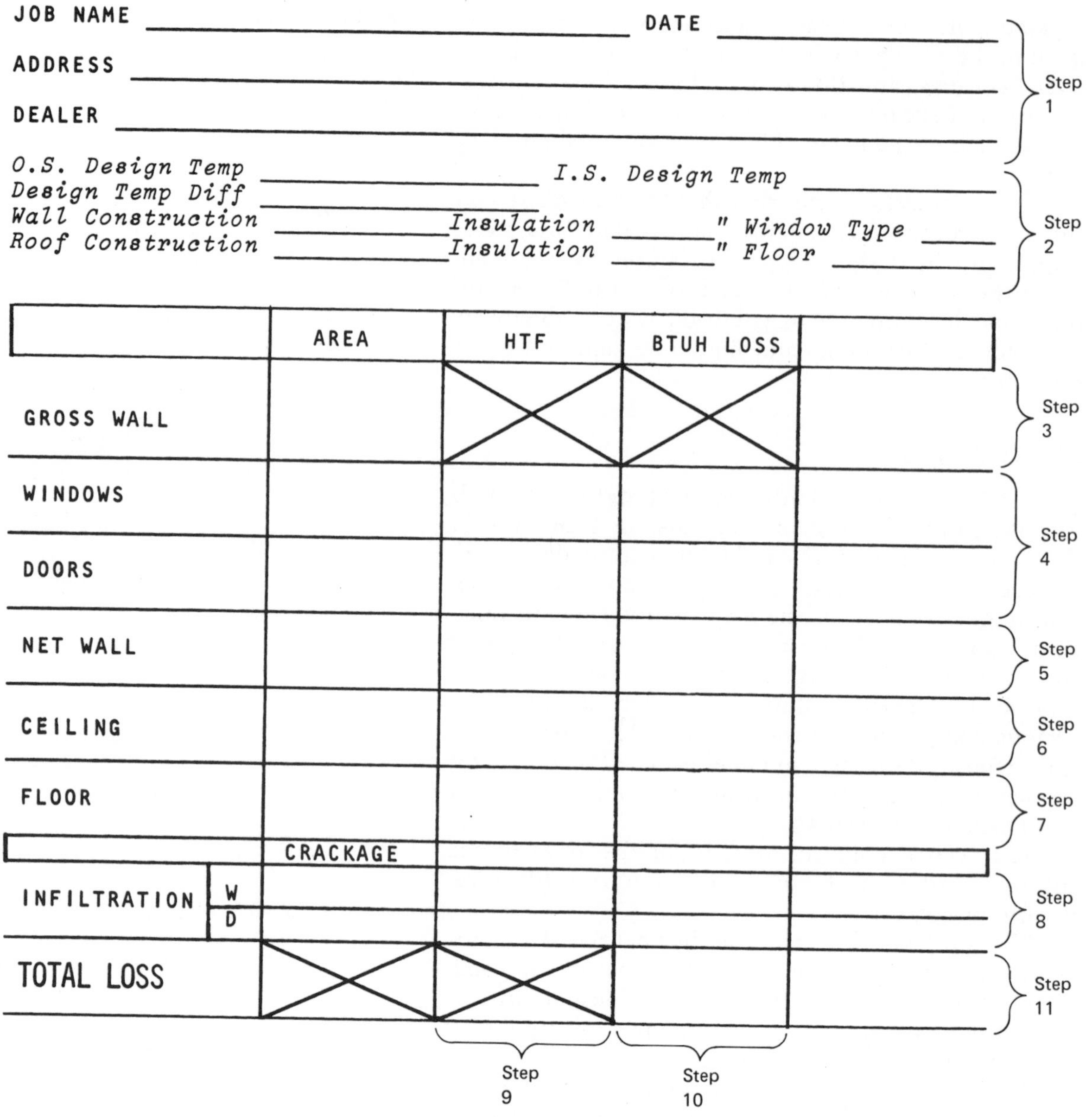

**JOB NAME** _____ **DATE** _____  ⎫
**ADDRESS** _____     ⎬ Step 1
**DEALER** _____      ⎭

*O.S. Design Temp* _____ *I.S. Design Temp* _____  ⎫
*Design Temp Diff* _____                ⎪
*Wall Construction* _____ *Insulation* _____" *Window Type* ____  ⎬ Step 2
*Roof Construction* _____ *Insulation* _____" *Floor* _____  ⎭

| | AREA | HTF | BTUH LOSS | | |
|---|---|---|---|---|---|
| **GROSS WALL** | | ✕ | ✕ | | Step 3 |
| **WINDOWS** | | | | | |
| **DOORS** | | | | | Step 4 |
| **NET WALL** | | | | | Step 5 |
| **CEILING** | | | | | Step 6 |
| **FLOOR** | | | | | Step 7 |
| CRACKAGE | | | | | |
| **INFILTRATION** W / D | | | | | Step 8 |
| **TOTAL LOSS** | ✕ | ✕ | | | Step 11 |

Step 9      Step 10

FIGURE 14-1 _____

Short form calculation sheet for figuring shell loads

and for sizing the duct system, the loads should be figured on a room-by-room basis, or at least on a zone-by-zone basis, so the supply air can be distributed to each space in the proper quantity to provide for the heating of the space.

DATE................................. BY........................

OUTSIDE DESIGN TEMP...............................

HEATING CONTRACTOR.......................................... INSIDE DESIGN TEMP............................

NAME OF JOB.......................................................... DESIGN TEMP. DIFF...........................

| | Table No. | "U" Factor | Multi-plier | Area Crack Cfm | ...uh ss | Area Crack Cfm | Btuh Loss | Area Crack Cfm | Btuh Loss | Area Crack Cfm | Btuh Loss | Area Crack Cfm | Btuh Loss |
|---|---|---|---|---|---|---|---|---|---|---|---|---|---|
| 1. Room Use and Floor Level | 1. | | | | 6. | | 7. | | 8. | | 9. | | |
| 2. Room Length and Width | | | | | | | | | | | | | |
| 3. Running Ft of Exposed Wall | | | | | | | | | | | | | |
| 4. Ceiling Height | | | | | | | | | | | | | |
| 5. Exposure Type | | | | | | | | | | | | | |
| Gross 6. Exposed Wall | | | | | | | | | | | | | |
| Windows 7. and Doors | | | | | | | | | | | | | |
| Net 8. Exposed Wall | | | | | | | | | | | | | |
| 9. Cold Partition | | | | | | | | | | | | | |
| 10. Cold Ceiling | | | | | | | | | | | | | |
| 11. Cold Floor | | | | | | | | | | | | | |
| Infiltration, 12. (Leakage) or Air Change | | | | | | | | | | | | | |
| 13. Room Sub-Total Btuh | | | | | | | | | | | | | |
| 14. Btuh Adjustment | | | | | | | | | | | | | |
| 15. Room Total Btuh | | | | | | | | | | | | | |

FIGURE 14-2
Portions of long form calculation sheet used for figuring room-by-room loads

To facilitate the calculations of loads on a room-by-room basis, a long form calculation sheet is used. Portions of the long form are shown in Figure 14–2 and the complete form is shown in Table A–7 in the Appendix.

The informational headings on the long form are similar to those used on the short form. The first column on the left-hand side of the page, in which building components are identified, is also similar. The difference in the two forms is that to the right of the column for the multiplier, the long form is divided into columns, one column for each room. A subheading across the top and separate columns for each room make it possible to figure the loss for individual rooms or spaces.

## Calculation Procedures for Finding Heat Load

The mechanics of calculating a heating load are simple. When using the short form, the designer lists the areas (A) of the building's components through which losses occur, selects the proper heat transfer factor (htf) for the design temperature difference, and then multiplies one by the other to obtain the total heat loss (H) in Btu/h. The answer gives the loss for each of the different components of the building. The component losses are then added together to get the total loss.

The formula applied for each of the components individually is as follows:

$$H = A \times htf$$

where:

$H$ = Btu/h
$A$ = area of the component
$htf$ = heat transfer factor from the factor table

Remember: For this calculation, the htf is selected for a particular design temperature difference and includes the proper U factor for the material.

## EXAMPLE 14–1

The second line on the short form is used for figuring the heat loss through the windows. Calculate the heat loss through the windows for a house with 204 sq ft of window area. Use an htf of 68.

**Solution**

$$H = A \times htf$$

$$= 204 \times 68$$

$$= 13{,}872$$

The heat loss through the windows is 13,872 Btu/h.

In the long form for calculating a heat load, the short form formula can be used or the U factors can be used. A U factor relates to the heat loss for only 1°F of temperature, and so the design temperature difference must also be included to achieve the heat transfer factor. If the U factor is used on the long form, the following formula is used for each separate component:

$$H = U \times DTD \times A$$

where:

$H$ = total heat loss in Btu/h
$U$ = U factor for the material
$DTD$ = design temperature difference
$A$ = area of the component

## EXAMPLE 14–2

On the long form calculation sheet, in Section 7, for windows and doors, the second column from the left has a place for the U factor and the third column has a place for a multiplier. The multiplier in this case is the U factor times whatever design temperature difference applies to this job.

A window in a building is 8 ft by 5 ft in dimensions, which makes it 40 sq ft in area. The window has no storm sash. Referring to Table A–5 of the Appendix, the U factor is 1.13. If, on a particular job, the design temperature difference is 60°F, what is the heat loss through the window?

**Solution**

$$H = U \times DTD \times A$$

$$= 1.13 \times 60 \times 40$$

$$= 2712$$

The heat loss through the window is 2712 Btu/h.

**Statistical data**    Identifying data of
record, such as job name and address

**Perimeter distance***

# 14.3  CALCULATING A SHELL LOAD

In a shell load calculation, the building loss is figured as though the building were one large room. Since the short form calculation sheet is used when making calculations for one space, it is ideal for figuring shell loads. The following section describes the steps for calculating the shell load for a building using the short form calculation sheet. Each step is identified with a label on the short form shown in Figure 14–1.

## Job Data

Step 1 in calculating heat load, with any form, is to fill in the proper **statistical data** for the job on the top of the form.

## Construction Data

Step 2 in calculating heat load is to fill in the proper data relative to the construction of the building and the design temperatures that will be used. It is important that a record be kept of the data used in the selection of factors in case any question arises concerning the loads.

## Gross Wall Area

Step 3 in calculating heat load on the short form is to find the gross wall area. The gross wall area is found by multiplying the distance around the building, called the **perimeter distance,** by the wall height. The resulting figure is entered in the area column, across from the gross wall designation on the form.

## Window and Door Areas

Step 4 in calculating heat load on the short form is to determine the window and door areas. To figure the areas of windows and doors, multiply the width by the height. If the dimensions are rounded off to the nearest half-foot and the totals rounded off to the nearest whole foot, the accuracy will be sufficient for the calculations. The resulting figures are entered for each in the area column, across from the proper designation.

# Net Wall Area

Step 5 in calculating heat load on the short form is to determine the net wall area. The combined area of the windows and doors is subtracted from the gross wall area to arrive at the net wall area. The net wall area is entered in the area column, across from the net wall designation.

# Ceiling Area

Step 6 in calculating heat load on the short form is to determine the ceiling area. The area of the ceiling is figured by taking the length of the building times the width of the building. This figure is entered on the form, across from the ceiling designation.

# Floor Area

Step 7 in calculating heat load on the short form is to determine the floor area. The floor area is usually the same square footage as the ceiling square footage. The exception occurs when a floor overhangs beyond the normal outside wall line or when a ceiling is of the cathedral type. In these special cases, the floor and the ceiling areas must be figured separately. The floor area can be entered in the area column, across from the floor designation.

# Infiltration or Fresh Air Loss

Step 8 in calculating heat load on the short form is to determine the infiltration losses. The most accurate way to figure the loss for infiltration air is to estimate the length of the crackage, in lineal feet, that occurs around the opening sections of the windows and doors. The length of the crack for each window and door is entered in the column under "crackage," across from either the window (W) or door (D) designation.

The length of the crack will vary according to the size and type of window. Double-hung windows have two openers to each window. The top sash can be opened downward, and the bottom sash can be opened upward. The crackage is figured for each sash separately. The

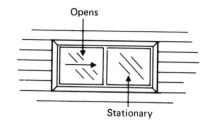

**FIGURE 14-3** _____

Horizontal slider window showing the
stationary and opening parts

two figures are added together to get the total crackage
for each window.

If the windows are horizontal sliders, usually only
one side of the window is an opener. The crackage
occurs around the opening part of the window only.
See Figure 14-3. Whatever the style of window used,
the designer should figure the crackage on the opening
cracks only.

Crackage around doors is figured in the same way
as it is for windows. There are not as many kinds of
doors as there are windows. If the crackage is being
figured for double doors, it is necessary to include both
doors only if both open.

In some cases, ventilation air, or make-up air, will
be introduced into a building. When such is the case,
the heat loss for the air is figured instead of the loss
for infiltration. The quantity of air is calculated in cfm,
and the loss is figured by the sensible heat formula.
Then this loss rather than infiltration loss is used on
the form.

## Heat Transfer Factors

Step 9 in calculating heat load on the short form is to
find the proper heat transfer factor from a set of htf
tables for each of the building components. A typical
factor table is shown in Table A-5 in the Appendix.
The factor selected for each component is entered on
the heat loss form, in the htf column, across from the
appropriate building component.

The heat transfer factor table is an abbreviated table
of factors for the most common types of building components used on residences in the western United States.
Although the table is by no means complete, it can be
used for any building component described in the left-
hand column. This column describes some of the common combinations of materials normally used as components in construction. Windows, doors, and walls
are listed here. The list also includes some choices for
different construction methods that may be employed.

The other vertical columns on the table show the
proper heat transfer factors to be used for the different
design temperature differences. The column at the extreme right shows the U factor for the type of construction shown on the left.

## EXAMPLE 14–3

A window with no storm sash is being used. The design temperature difference is 53°F, as found for Seattle, Washington, at a 70°F inside design temperature. Find the htf.

**Solution**

Using Table A–5 in the Appendix, the htf is found in the 50°F column—because this is the closest figure to 53°—and across from "glass, no storm sash." The heat transfer factor is 57. The figure 57 should be entered in the htf column for the window on the short form.

The same selection process is used for each of the other components of the construction, such as doors and walls, and the proper heat transfer factors are then entered on the form in the htf column.

The two sections at the bottom of Table A–5 in the Appendix contain factors for heat loss due to infiltration of outside air through the cracks around the openings of windows and doors. These factors give the Btu/h loss per foot of crack for the design temperature differences found at the head of the columns.

The factors for the windows are for average-fit, weatherstripped windows and are appropriate for most of the good-quality windows in use today. The door factors are separated into two categories, for two different types of door installations. The first is for an average-fit, nonweatherstripped door. The second is for a weather-stripped door, or a door and storm-door combination. The proper factors should be selected according to the type of construction used and then listed on the loss form in the htf column.

## Subtotals and Total Btu/h

Step 10 in calculating the heat load on the short form is to determine the Btu/h subtotal for each component. After the areas of all of the components of the construction are figured and entered in the second column of the short form calculation sheet and the multiplying factors are selected and entered in the third column, it is necessary only to multiply the areas times the factors to arrive at the Btu/h subtotals for each component of the construction. The subtotals are then recorded in the third column on the form.

Step 11 is to add the subtotals to find the total Btu/h loss of the building. The total Btu/h loss is also the total heat load of the building.

## EXAMPLE 14–4

Use the construction details on the floor plan shown in Figure 14–4 and the short form calculation sheet

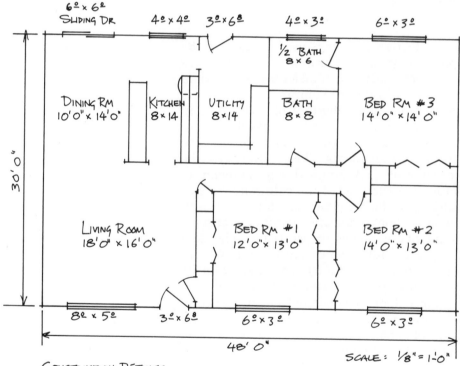

CONSTRUCTION DETAILS:

1. WINDOWS — SINGLE GLASS, NO STORM SASH
2. DOORS — W/ STORM DOOR
3. WALLS — FRAME WITH 2" INSULATION
4. CEILING/ROOF - 4" INSUL. IN ATTIC
5. FLOOR — OVER CRAWLSPACE (UNHEATED ROOM)
6. INFILTRATION — WINDOWS WILL BE HORIZONTAL SLIDERS W/ WEATHERSTRIPPING
7. CEILING HEIGHT 8'-0"

FIGURE 14-4 _____

Sketch of floor plan of a typical house with necessary data recorded for figuring a heat loss

to calculate the heat load for the residence. An illustration of a short form filled out with the data for the example is shown in Figure 14–5.

**Steps 1 and 2:** Record the job data and construction data at the top of the form. To determine the DTD,

JOB NAME _____    DATE _____

ADDRESS _____

DEALER _____

O.S. Design Temp _____−2° F_____    I.S. Design Temp ___70° F___
Design Temp Diff _____72° F_____
Wall Construction ___FRAME___ Insulation _2"_ " Window Type _____
Roof Construction ___W/C'LG.___ Insulation _4"_ " Floor ___CRAWL___

| | AREA | HTF | BTUH LOSS | |
|---|---|---|---|---|
| GROSS WALL | 1248 | | | |
| WINDOWS | 158 | 79 | 12,482 | |
| DOORS | 42 | 22 | 924 | |
| NET WALL | 1048 | 10 | 10,480 | |
| CEILING | 1440 | 7 | 10,080 | |
| FLOOR | 1440 | 10 | 14,400 | |
| | CRACKAGE | | | |
| INFILTRATION  W | 47 lin ft | 42 | 1974 | |
| D | 21 lin ft | 70 | 1470 | |
| TOTAL LOSS | | | 51,810 | |

FIGURE 14-5 _____

Completed short form calculation sheet for the house in Figure 14-4

figure the load for Salt Lake City, Utah. The outside design temperature for Salt Lake City is $-2°F$, as found on the outside design temperature chart. Use an inside design temperature of 70°F. The design temperature difference then will be 72°F:

$$DTD = 70°F - (-2°F) = 72°F$$

**Step 3:** Figure the gross wall area by multiplying the perimeter distance around the house by the ceiling height:

perimeter distance $= (2 \times 48) + (2 \times 30) = 156$

ceiling height $= 8$

gross wall area $= 156 \times 8 = 1248$

The gross wall area is 1248 sq ft. Place this figure in the gross wall area space on the form.

**Step 4:** Figure the window area by adding up the separate areas of all the windows in the house (figure the sliding glass door as a window):

window area $= 158$

Place 158 sq ft in the window area space.
Find the total door area by adding all wooden door areas (round off to the nearest whole foot):

door area $= 2 \times 21 = 42$

Place 42 sq ft in the door area space on the form.

**Step 5:** Find the net wall area by subtracting the door and window areas from the gross wall area:

net wall area $= 1248 - (158 + 42) = 1048$

Place 1048 sq ft in the space for the net wall.

**Steps 6 and 7:** Find the ceiling and floor area. To find this figure, multiply the length of the house by the width:

ceiling and floor area $= 48 \times 30 = 1440$

Enter 1440 sq ft in the spaces for both the ceiling and the floor.

**Step 8:** Find the crackage on doors and windows to determine infiltration. The length of the crackage for the windows needs to be figured for this calculation. Since the windows are horizontal sliders, one-half of each can be expected to open. Figure the lineal feet of crack for each window and for the sliding door, and add the answers up for the total crackage. Because this example is for a shell load, only half of the total crackage has to be used. Take one-half of the total crackage and place the number under "crackage," across from "windows" on the form. Do the same calculation for the wood doors:

window infiltration crackage = 94 ÷ 2 = 47

door infiltration crackage = 42 ÷ 2 = 21

Enter 47 lineal ft for the window crackage and 21 lineal ft for the door infiltration.

**Step 9:** Find the appropriate htf for each of the surfaces and the infiltration from the factor table in Table A–5 in the Appendix. Enter the data in the htf column on the form. *Note:* Although the design temperature difference is actually 72°F, the 70°F column on the factor table will give accurate enough results.

**Steps 10 and 11:** Multiply the areas in Column 2 by the factors in Column 3. The results are the component losses. Add the component losses together to arrive at the total heat loss. The total is 51,810 Btu/h, as seen in Figure 14–5.

## 14.4 CALCULATING A ROOM-BY-ROOM LOAD

The long form calculation sheet shown in Figure 14–2 is used when the designer wants to figure a load on a room-by-room basis. The procedure is similar to that used for figuring a series of shell loads. The final results are a load for each room in the building, but the format of the long form makes figuring the loads relatively easy. The following section describes the steps for calculating the room-by-room load using the long form calculation sheet. Each step is identified with a label on the form in Figure 14–6.

## Job Data

Step 1 in calculating a room-by-room load on the long form is to fill out that part of the heading on the form that is related to job information. This section is similar

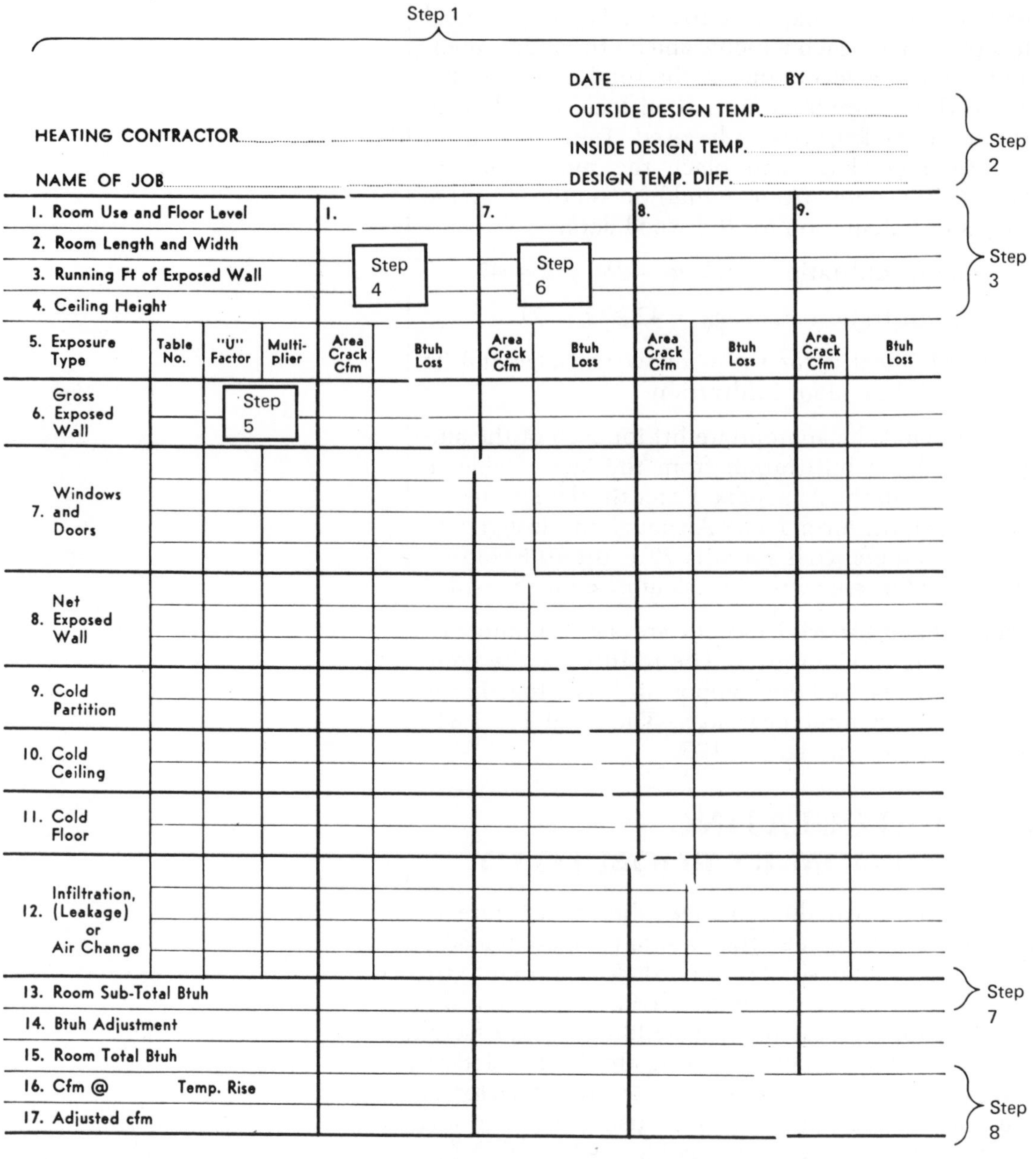

Step 1

| | | | | | DATE | | | BY | | |
|---|---|---|---|---|---|---|---|---|---|---|---|

OUTSIDE DESIGN TEMP.

HEATING CONTRACTOR                    INSIDE DESIGN TEMP.

NAME OF JOB                    DESIGN TEMP. DIFF.

Step 2

Step 3

Step 4

Step 5

Step 6

| | Table No. | "U" Factor | Multi- plier | Area Crack Cfm | Btuh Loss | Area Crack Cfm | Btuh Loss | Area Crack Cfm | Btuh Loss | Area Crack Cfm | Btuh Loss |
|---|---|---|---|---|---|---|---|---|---|---|---|
| 1. Room Use and Floor Level | 1. | | 7. | | 8. | | 9. | | | | |
| 2. Room Length and Width | | | | | | | | | | | |
| 3. Running Ft of Exposed Wall | | | | | | | | | | | |
| 4. Ceiling Height | | | | | | | | | | | |
| 5. Exposure Type | | | | | | | | | | | |
| Gross 6. Exposed Wall | | | | | | | | | | | |
| Windows 7. and Doors | | | | | | | | | | | |
| Net 8. Exposed Wall | | | | | | | | | | | |
| 9. Cold Partition | | | | | | | | | | | |
| 10. Cold Ceiling | | | | | | | | | | | |
| 11. Cold Floor | | | | | | | | | | | |
| Infiltration, 12. (Leakage) or Air Change | | | | | | | | | | | |
| 13. Room Sub-Total Btuh | | | | | | | | | | | |
| 14. Btuh Adjustment | | | | | | | | | | | |
| 15. Room Total Btuh | | | | | | | | | | | |
| 16. Cfm @    Temp. Rise | | | | | | | | | | | |
| 17. Adjusted cfm | | | | | | | | | | | |

Step 7

Step 8

FIGURE 14-6

Step-by-step procedure for figuring a long form calculation sheet

to that on the short form. The form has spaces for the name of the job, the date, and the initials of the person figuring the loads.

# Construction Data

Step 2 on the long form is to fill in the second part of the heading with the outside design temperature, inside design temperature, and the design temperature difference.

# Room Identification

Step 3 on the long form is to identify the individual rooms or spaces for which the loads are figured. The room names are recorded separately in each column, and space is provided for the dimensions for each room. Line 1 is used for recording the room identification. Line 2 is used for recording the length and width of the room. Line 3 is for recording the length of the exposed outside wall in feet. Line 4 is for recording the ceiling height of each room.

# Areas of Components

Step 4 on the long form is to find the area for each building component and list each in the proper column under the room identification headings. Below Line 4, the columns for each room are divided into two spaces. The first column provides for the recording of the area of the component part of the structure, such as walls or ceiling, as identified on the left part of the form. The second part of the column is for recording the Btu/h for each component.

The gross wall area for each room is found by multiplying the running feet of exposed wall times the ceiling height. The gross wall area is recorded on Line 6 for each room separately.

The areas for each of the components for each room are calculated and entered on the form across from the component identification, such as windows and doors.

On Line 12, the infiltration crackage, or cfm of fresh air, is listed for each room.

# Heat Transfer Factors

Step 5 on the long form is to select the U factor or the htf from an appropriate table for each of the components and enter it on the form in the second or third column from the left side.

The heat transfer factor table used with the short form method (Table A–5 in the Appendix) is also appropriate for use in the long form calculation. If more accuracy in the results is desired, a complete set of factors, such as those found in the *Manual of the American Society of Heating, Refrigeration and Air Conditioning Engineers, Inc.,* should be used.

When heat transfer factors are used, they should be entered in the column on the left headed "multipliers." If U factors are used, they should be entered in the column headed "U factors," and they should then be multiplied by the design temperature difference. The product is then entered in the column headed "multipliers."

## Component Losses

Step 6 on the long form is to determine the heat loss of each component by multiplying the area of each component by the number in the column labeled "multiplier." The product is recorded in the Btu/h loss column for that room and that component.

## Individual Room Losses

Step 7 on the long form is to add the heat losses of all of the components in each room to find the room total Btu/h loss. These calculations are recorded at the bottom of each column on Line 13.

## Total Building Loss

Step 8 on the long form is to add the room losses to find the total Btu/h loss for the building. The total Btu/h loss is also the total heat load.

### EXAMPLE 14–5

Use the construction details on the floor plan shown in Figure 14–4 and the long form calculation sheet to calculate a room-by-room load for the residence. An illustration of the long form filled out with the data for the example is shown in Figure 14–7.

**Steps 1 and 2:** Record the job data and construction data at the top of the form. To determine the design

**HEATING DATA SHEET**

JOB NO.
DATE _____ BY _____
OUTSIDE DESIGN TEMP. __-2° F__
INSIDE DESIGN TEMP. __70° F.__
DESIGN TEMP. DIFF. __72° F__

HEATING CONTRACTOR _____ ADDRESS _____

NAME OF JOB _____ ADDRESS _____

| | Table No. | "U" Factor | Multiplier | 1. L.R. Area/Crack/Cfm | Btuh Loss | 2. BR #1 Area/Crack/Cfm | Btuh Loss | 3. BR #2 Area/Crack/Cfm | Btuh Loss | 4. BR #3 Area/Crack/Cfm | Btuh Loss | 5. BATH+½B Area/Crack/Cfm | Btuh Loss | 6. UTILITY Area/Crack/Cfm | Btuh Loss | 7. KITCHEN Area/Crack/Cfm | Btuh Loss | 8. D.R. Area/Crack/Cfm | Btuh Loss | 9. HALL Area/Crack/Cfm | Btuh Loss |
|---|---|---|---|---|---|---|---|---|---|---|---|---|---|---|---|---|---|---|---|---|---|
| 1. Room Use and Floor Level | | | | L.R. | | BR #1 | | BR #2 | | BR #3 | | BATH+½B | | UTILITY | | KITCHEN | | D.R. | | HALL | |
| 2. Room Length and Width | | | | 18×16 | | 12×13 | | 14×13 | | 14×14 | | 8×14 | | 8×14 | | 8×14 | | 10×14 | | 20×3 | |
| 3. Running Ft of Exposed Wall | | | | 34 | | 12 | | 27 | | 28 | | 8 | | 8 | | 8 | | 24 | | 0 | |
| 4. Ceiling Height | | | | 8 | | 8 | | 8 | | 8 | | 8 | | 8 | | 8 | | 8 | | 8 | |
| 6. Gross Exposed Wall | | | | 272 | | 96 | | 216 | | 224 | | 64 | | 64 | | 64 | | 192 | | 0 | |
| 7. Windows and Doors — W. | | | 79 | 40 | 3,160 | 18 | 720 | 18 | 720 | 18 | 720 | 12 | 480 | 0 | | 16 | 640 | 36 | 1440 | 0 | |
| 7. Windows and Doors — D. | | | 22 | 21 | 462 | | | | | | | | | 21 | 462 | | | | | | |
| 8. Net Exposed Wall | | | 10 | 211 | 2110 | 78 | 780 | 198 | 1980 | 206 | 2060 | 52 | 520 | 43 | 430 | 48 | 480 | 156 | 1560 | 0 | |
| 9. Cold Partition | | | | | | | | | | | | | | | | | | | | | |
| 10. Cold Ceiling | 7 | | | 288 | 2016 | 156 | 1092 | 182 | 1274 | 196 | 1372 | 112 | 784 | 112 | 784 | 112 | 784 | 140 | 980 | 60 | 420 |
| 11. Cold Floor | 10 | | | 288 | 2880 | 156 | 1560 | 182 | 1820 | 196 | 1960 | 112 | 1120 | 112 | 1120 | 112 | 1120 | 140 | 1400 | 60 | |
| 12. Infiltration (Leakage) | | | 42 | 18 | 756 | 12 | 504 | 12 | 504 | 12 | 504 | 10 | 420 | 0 | | 12 | 504 | 18 | 756 | 0 | |
| 12. or Air Change | | | 70 | 21 | 1470 | | | | | | | | | 21 | 1470 | | | | | 0 | |
| 13. Room Sub-Total Btuh | | | | | 12854 | | 4656 | | 6298 | | 6616 | | 3324 | | 4266 | | 3528 | | 6136 | | 1020 |
| 14. Btuh Adjustment | | | | | | | | | | | | | | | | | | | | | |
| 15. Room Total Btuh | | | | | | | | | | | | | | | | | | | | | |
| 16. Cfm @ ___ Temp. Rise | | | | | | | | | | | | | | | | | | | | | |
| 17. Adjusted cfm | | | | | | | | | | | | | | | | | | | | | |

FIGURE 14-7

Completed long form calculation sheet for the house in Figure 14-4

temperature difference, figure the load for Salt Lake City, Utah, where the outside design temperature is −2°F. Use an inside design temperature of 70°F. The design temperature difference will be 72°F:

$$DTD = 70°F − (−2°F) = 72°F$$

**Step 3:** Fill in the individual room data for each room. Identify the rooms by name on Line 1. Insert room length and width on Line 2, running feet of exposed wall on Line 3, and ceiling height on Line 4.

**Step 4:** Multiply the running feet of exposed wall (Line 3) times the ceiling height (Line 4) to find the gross exposed wall area for each room. Enter the product in the room area column for each room on Line 6.

Follow the same procedure for each of the other

**Heat gain**    Heat added to a building by some source other than the heating unit

**Solar gain**    A gain of heat caused by the radiant energy from the sun

components of the building. In each case, the areas are related to *one room only.* Figure the areas for windows and doors, net walls, ceiling and floor, and the linear footage of cracks for windows and doors for each room. Enter these figures in the appropriate spaces on the form.

**Step 5:** Find the htf that is appropriate for each building component from the factor table. Place the factor in the column under "multiplier," across from the appropriate component. Be sure to use the factors from the 70°F column on the table.

**Step 6:** With the proper multipliers and the areas for the various components listed by room, multiply the factor times the areas for each component by room. Enter the Btu/h loss per component in the space provided.

**Step 7:** Enter the component losses for each room on the form. Add them up for each room to find the room totals.

**Step 8:** Add up the room losses to determine the building total. In Figure 14–7, the filled-out form is shown with the room totals on Line 13 and the building total in the bottom right-hand corner.

This loss is approximately 6% less than the loss figured on the same residence with the short form. The short form includes all of the inside space in the building, including the closets and walls and partitions. These spaces have been disregarded on the long form.

## 14.5  ADJUSTMENTS TO THE HEAT LOAD

After calculating the total building load, the designer should make two adjustments before using the figure to do any sizing. If there are any losses that have not been included in the original calculations, they should be added to the total. If there are any internal **heat gains** from lights, people, or equipment or external gains from solar sources, such gains should be subtracted from the total.

# Other Heat Losses

On a residential job, some losses that may not have been included on the original calculations are those caused by a fireplace and by the exhaust fans in the bathrooms and the kitchen. Most fireplace dampers are leaky even when closed. The leakage must be made up by increased infiltration to the house. The loss can be compensated for by adding 1000 Btu/h loss to the total load for each fireplace in the house. The loss caused by the exhaust fans is of the same type; that is, the air exhausted by the fans has to be made up by infiltration, which causes extra heat loss. To figure the Btu/h for this loss, take the cfm capacity of the fan and multiply the cfm by the hourly average use. Then, figure the load on that much infiltration and add it to the building total. The capacity of the fan in cfm can be found by reference to the manufacturer's literature.

The main cause of extra heat loss for commercial or industrial buildings is usually fresh air or make-up air brought in by the ventilation system. This loss should have been included in the calculations originally.

# Heat Gains

Heat gain to a building during the heating season can be substantial. Any heat gain that is constant can be deducted from the loss to find the net loss to be covered with the heating unit.

In a residence, heat gains such as **solar gain,** heat from the people in the building, or heat from the lights is considered a bonus as far as heating the building is concerned. Since there is no way to predict accurately when these gains may occur, the designer should *not* deduct them from the loss.

The internal heat gain can be calculated much more accurately in buildings used for commercial or industrial purposes because the gains are more constant. Normally a predictable number of people will be in the building, and the gain from lights and equipment is much easier to figure than in a residence. All such gains can be calculated and deducted from the loss. Variable gains, such as solar gain, cannot be predicted with enough accuracy by the methods covered in this text to deduct from the loads. On large jobs, gains are often figured by using a computer, and even the variable gains can be estimated with enough accuracy to be deducted with confidence.

## Duct Heat Loss

At this stage of the design, a preliminary layout of the duct system should be made. If any of the ducts are to run through any space that will not be heated, the heat loss through the walls of the ducts must be added to the total heat loss for the building. If the ducts will run within the heated space in the building, any loss through the duct walls will help heat the building, and no additional loss is experienced.

Depending on where the ducts run in the building, the loss through them may vary from 0% to 40% of the total loss of the building. The designer must select the proper percentage figure to use and adjust the total loss to compensate.

Three factors must be taken into consideration in selecting the percentage to use:

1. What is the design heat loss of the building in Btu/h per sq ft?
2. Is the duct in a ventilated area but exposed to outside air (such as an attic) or is it in an unheated area (such as an enclosed basement)?
3. Is the duct insulated?

Reference to a duct heat loss allowance table, such as the one shown in Table 14–1, will make the selection easy.

On the left side of the table, the two possible duct locations are listed. The designer selects the location that describes the duct placement for the particular job being considered. Then the designer selects the category under building heat loss and outside design temperature that best fits the job. The figure that corresponds to the designer's selection indicates the percentage of the total load that is lost through the ducts for that portion of the building. This percentage figure is used to find the Btu/h that should be added to determine the correct heat loss figure.

This method of correcting the total loss figure to include duct heat loss should be used only when the

TABLE 14-1
Duct Heat Loss Allowance Table

| | Building heat loss more than 30 Btu per sq ft | | Building heat loss less than 30 Btu per sq ft | |
|---|---|---|---|---|
| | Temperature below 10°F | Temperature above 10°F | Temperature below 10°F | Temperature above 10°F |
| Duct Locations | | | | |
| Ventilated area (attic or crawl space) | 40% | 30% | 20% | 10% |
| Unheated space (basement) | 25% | 20% | 10% | 5% |

duct is inside the building shell. If the duct work runs outside the building and is exposed directly to the outside air temperature, the heat loss through the walls of the duct should be figured the same as the loss through any other building component.

## 14.6  ACCURACY

A heat loss calculation for any building is only as accurate as the data used to figure it. The data for the calculations should be selected with great care and the actual calculations made carefully. If the designer thinks that adjustments should be made so that the calculations are correct, such adjustments should be made to the factors and not to the final load. The final figure should represent the actual heat loss of the building as closely as it can be figured with the data available to the designer.

## 14.7  SUMMARY

Before a system can be designed to provide heating temperature control for any building, the heat loss of the building at some predetermined set of conditions must be figured. An accurate heat loss calculation enables a designer to determine the heat load for the building, size the heating equipment for the building, and determine how much air to circulate in the building. It allows the designer to size and locate the registers and size the air distribution and return systems. Without an accurately figured heat loss, these calculations are impossible.

To figure the loss, the designer must determine the outside temperature at which the building will be heated and the inside temperature to which it will be heated. The construction of the building must be considered so that proper factors can be selected to figure the amount of heat that will travel through the building components. All of these variables must be combined on a calculation sheet in such a way that the load can be figured easily.

The heat load calculation sheets can be any format that is easy for the designer to use, but the final loads will be based on one of two basic types: shell load or room-by-room load. The shell load is figured using the assumption that the building is one large space or room. The room-by-room load is figured by determining the individual room losses, and then adding them together for the total building load.

After the basic heat load is calculated, any additional heat losses or heat gains that might not have been included in the original load must be considered. The additional heat losses can include infiltration due to fireplaces, exhaust fans, duct work, or any other source of air leakage. The heat gains can include heat increases caused by the sun, by people in the building, or by lighting in the building. The heat load should be adjusted accordingly.

The final heat loss figures should represent, as accurately as possible, the actual load of the building at the chosen design temperature. Accuracy in figuring the loads makes proper design of the complete system possible.

## 14.8   QUESTIONS

1. Figuring the heat loss of a building includes organizing the _____ and _____ and _____ them in a way to make it easy to keep track of them.

2. The building heat load is used to size the _____ _____, select and size the _____ and _____ registers, and size the _____.

3. Whenever the heating loads will be used for designing a complete system, including registers and ductwork, they should be figured on: (a) shell load basis, (b) factor tables, (c) room-by-room basis, (d) short forms.

4. Short form calculation sheets are used when the load of the entire building is wanted as one large space. True or false?

5. A heat transfer factor always includes the design temperature difference. True or false?

6. A U factor includes the design temperature difference. True or false?

7. The gross wall area includes window and door areas. True or false?

8. The net wall area is found by subtracting the _____ areas and the _____ areas from the gross wall area.

9. Infiltration is calculated by measuring the _____ around the _____ sections of windows and doors.

10. The design temperature difference is found by subtracting the _____ design temperature from the _____ design temperature.

11. _____ _____, such as fireplaces, must be added to the room totals when they have not been considered in the original calculations.

12. Solar gain, as a source of heat, should be deducted from the total heat loss of a building. True or false?

# CHAPTER
# 15
# Selection and Sizing of Heating Equipment

## 15.1 INTRODUCTION

After calculating the heat load, the heating technician must consider a number of other factors in order to select and size the specific equipment for a particular job. Proper consideration of each of these factors is necessary in order to choose the equipment best suited for the particular job.

The factors to consider in selecting and sizing the equipment are the heating capacity required, the physical arrangements possible, the fuel or energy available, the type of distribution system required, and the room available for location of the equipment in the building. Each of these factors is discussed in this chapter.

## 15.2 FACTORS TO CONSIDER

Several factors must be considered before the final selection of heating equipment for any building can be made. The major factors are:

1. Heating capacity required,
2. Type of system required,
3. Fuel or energy available,
4. Equipment location available.

The first factor to consider is the heating capacity required in the building. The heating unit selected must provide enough heat to match, or exceed, the calculated heat load so that enough heat is available to maintain the desired temperature in the building at all times. The calculated heat load also determines the design of the distribution system. The system must be designed so that the registers, grilles, ductwork, and/or piping system will distribute the heat evenly throughout the building.

The two general types of systems considered here are the hydronic system and the warm air system. The size of the building, the location of spaces to be heated in relation to the equipment room, and the degree of temperature control desired are the general factors to be considered in deciding whether to use the hydronic system or the warm air heating system. The hydronic system is usually a better choice when the spaces to be heated are located at some distance from the heating equipment. The warm air system is better for more compact spaces. If the building will have shifting loads, such as groups of people that move around during the heating period or shifting solar loads, then the warm air system provides better control than the hydronic system.

One of the major decisions to be made in the selection process is what fuel or energy to use. Cost and availability are the two main factors to consider. In some areas, certain fuels are more easily obtained than others; usually the cost for these fuels is also less..In many cases, the cost of transporting a fuel to a particular location is more than the difference in cost between two fuels. These differences have to be considered by the designer before the final selection of equipment.

In the design of a new building, space has to be provided for the mechanical equipment, such as the furnace, blower, dampers, and ductwork or piping. The initial space arrangement is made by the architect designing the building, but this space left on the plans for the mechanical systems may not be adequate. The mechanical designer usually has to fit the mechanical equipment and system into whatever space has been provided. It is necessary, then, to select a system and equipment that will fit into the space available. Some types of equipment fit into tight spaces better than others. Water piping can be worked into floor spaces, walls, and ceiling spaces more easily than air ducts can because of the smaller sizes of pipes. But water piping is very rigid and cannot be adapted to some conditions as easily as ductwork can. The entire building design has to be considered before the final selection of equipment is made.

# Heating Capacity Required

It is necessary to select a heating unit with enough capacity to heat the building it is used in. After the heat load has been calculated, two factors must be considered in determining whether a unit has the correct capacity to satisfy the load. First, the output capacity of the unit must be determined. Second, the need for extra capacity, or pickup, must be determined.

**Output Capacity.**    Fossil fuel furnaces are listed in the manufacturers' catalogs with both input and output

**Input capacity***

**Output capacity**    The amount of heat available for heating a building after the input capacity is corrected for combustion and equipment efficiency

**Setback***

**Pickup***

FIGURE 15-1 _____

Time clock used for night setback on a heating system

capacities. The **input capacity** is the amount of heat that will be given up by the fuel used in the combustion process. The **output capacity** is the actual amount of heat available for heating purposes. Because of the efficiency losses in the operation of the furnace, output is normally about 80% of the input rating. The designer must always use the output rating when sizing a furnace to match a heat load. The output of the unit selected should match or exceed the Btu/h required to heat a building at the design conditions.

**Pickup.**    When a building is to be occupied continuously, twenty-four hours a day and every day of the week, the heating system will be in operation all of the time. The equipment has to be sized to offset the heat loss of the building on a continuous basis.

If building use is periodic, perhaps only one or two days a week, such as in a typical church building, the thermostat setting should be set back to a lower temperature during the unoccupied times. This thermostat setting is called the **setback.** A time clock and appropriate controls are used for the setback. A typical time clock is shown in Figure 15–1.

Even in a setback system, the heat loss during the occupied periods of time is the same for the building as it is in an unoccupied time. The heating unit is sized to handle the total load, but it is also desirable to have extra heating capacity available to warm up the building more quickly. This extra capacity is called **pickup.** If the building has been on setback for some time, everything in it cools off. Warming the building may take several hours after the thermostat is reset to the occupied temperature. Extra capacity in the heating unit will help warm the building faster in this circumstance.

So that the heating unit has this extra capacity, it should be sized from 15% to 50% larger than that required to offset the building heat loss. The actual amount of oversizing should be determined by how much time the temperature will be set back and how much difference there is between the on-time temperature and the off-time temperature.

*Term defined in text.

# Types of Systems

The size of the building and the degree of control required will determine whether a warm air system or a hydronic system should be used to heat the building. In a building in which the equipment is centrally located but heat is needed at widely separated parts of the building, a hydronic system should be used. If the building is more compact and ductwork can be run easily to each space to be heated, or if individual equipment spaces are available to each area, then a warm air system may be more practical.

**Warm Air Systems.**    Since the final desired result for any heating system is to heat the air in a building, a warm air heating system should be used whenever possible. This system is always better when the heating equipment can be centrally located and the ductwork does not have to be too long. Figure 15–2 shows a typ-

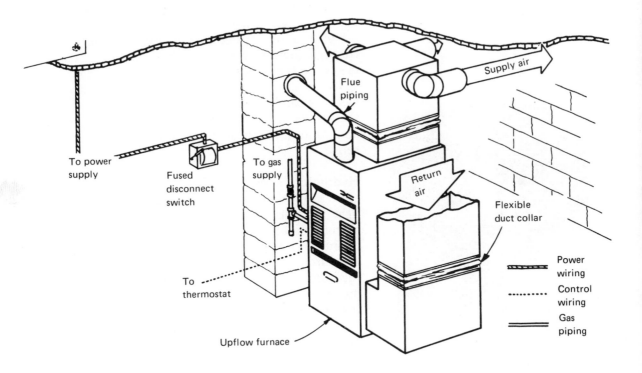

FIGURE 15-2 _____

Typical warm air furnace and ducts located in the basement of a house

**Modular system***

ical warm air furnace and ducts located in the basement of a home.

The use of a warm air system provides much more flexibility in the location of the air outlets than does a hydronic system. Commercial buildings commonly require a system that can be changed if a tenant wants to rearrange the spaces. In remodeling, it is often necessary to move the air outlets and return air registers. Because ductwork is much easier to relocate than rigid piping, the ultimate use of the building has to be considered by the designer in selecting the heating equipment.

Many commercial buildings are only one story high. Such buildings may be large in area and cover a lot of floor space, conditions not conducive to using a central warm air heating system. Instead, a **modular system** is practical. A modular system uses multiple heating units located at different strategic locations around the building, separate from each other. The units may be located in separate equipment rooms or, in many cases, located on the roof. See Figure 15–3 for a typical heating and cooling unit as used on a roof.

In roof-top systems, the ductwork drops down through the roof through special openings. This type of system is often used for retail stores where large open spaces alternate with partitioned areas. Such modular systems can be either hydronic or warm air, but since the spaces can be kept relatively small, using warm air and running ducts to distribute the air makes more sense.

**Hydronic Systems.**    Whenever it is necessary to generate heat at one point in a building and distribute the heat long distances to the points of use, a hydronic system should be used. Water carries a much greater quantity of heat per volume of medium than does air. A hydronic system employs water pipes to carry the medium. The pipes are also smaller in size than are air ducts carrying a similar amount of heat. In a large building or a multistoried one, too much space may be required for ductwork, or in some cases, the construction of the building will not allow space for the ductwork.

High-rise buildings and campus-style buildings scattered around a central heating plant are good examples of the type of buildings in which a hydronic system can be used to good advantage.

FIGURE 15-3 _____

Rooftop heating and air-conditioning unit showing ductwork that goes through the roof

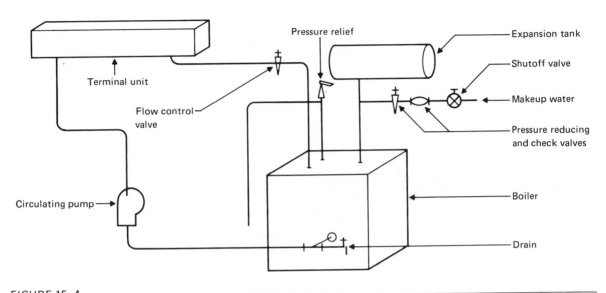

FIGURE 15-4

Schematic drawing of a hydronic heating system

In a hydronic system, a boiler in a central equipment room generates hot water. Pumps circulate that water to its point of use as shown in the diagram in Figure 15–4.

The designer must be sure that the building has appropriate spaces for the equipment and that the construction of the building will allow the distribution piping to be run.

In the spaces to be heated, terminal devices are employed to extract the heat from the water. Heated air is then used to heat the spaces. The designer should consider what type of terminal devices to use so the total system design will accommodate them. One type of terminal unit is shown in Figure 15–5.

## Fuel Types

A major decision to be made in the selection of heating equipment concerns fuel choice. In this discussion, electricity will be considered a fuel. Units that use any of the fossil fuels or electricity are available in either the warm air system or the hydronic system. The choice of which fuel to use normally depends upon the availability and cost of fuels in the area where the particular building is located. In the past, when such factors were

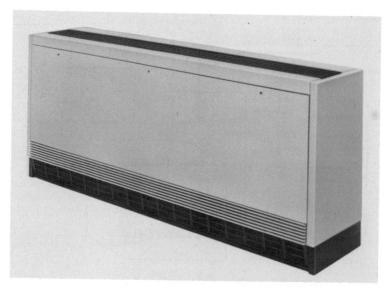

FIGURE 15-5

Convector unit used as a terminal unit in a hydronic system

reasonably stable, only the long-term effects of changes had to be considered. But now, both cost and availability are changing at so rapid a rate that it is necessary to select the fuel with great care and also to design a system so that the fuel can be changed in the future if availability or cost factors change.

**Fossil Fuels.**    In most populated areas, natural gas pipelines are common. If natural gas is available, it seems to be the most stable of all the fuels used in the United States at the present time in respect to price and availability.

Oil is an excellent heating fuel, and many of the heating plants in the United States are set up to use it. Use of oil in the United States far exceeds the domestic supply, however. Also the cost per Btu/h of heating with oil now exceeds that for either gas or electricity in most places in the United States. The future use of oil as a heating fuel probably will be limited.

Many large industrial plants are purposely located on or near sources of coal. For such buildings, the choice of which fuel to use is simple. Coal is a good source of heat. When it can be used close to the source in order to minimize the cost of transportation, and when the contamination from the products of combustion can be controlled so it is not detrimental to the environment, coal is an excellent choice for heating.

**Electric Energy.**    Since electricity is run to virtually all buildings, this form of energy is usually a good choice as a source for heating. In some areas, the cost of electricity may be greater per Btu/h than gas, oil, or coal, but electricity is always available at the building site and usually can be used for heat. In some cases, the electric utility companies are limiting the use of electric power because the demand exceeds the supply. The designer should always check with the local supplier of any fuel or energy before deciding on any particular one to use for heating a building.

## Equipment Location

Gas, oil, and coal need combustion for conversion to heat. This requirement makes it necessary to provide a way to get the fuel to the equipment, to provide combustion air, and to provide a chimney, flue, or vent to get rid of the products of combustion. The building design has to include space for fuel lines or bunkers (in the case of coal), a chimney, and combustion air ducts to and from the outside of the building. Space has to be provided for the distribution ducts or piping. The air distribution system also requires fresh air, so ducts must be provided to bring this air into the area where the heating equipment will be located.

**Fuel Availability.**    Gas or oil can be piped to the equipment room from outside meters or tanks. Because only a small amount of space is required for running the lines, the lines normally do not affect the location of the equipment room. A tank and piping for an oil-fired furnace are shown in Figure 15-6.

The electric supply lines to an electric heating system can usually be run wherever necessary in a building, and the control panels can be located where convenient. The lines are small, so the equipment room location is not significantly affected by them.

If coal is to be used as the fuel, the equipment room should be located adjacent to the coal storage area so

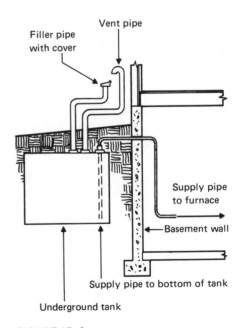

FIGURE 15-6 —————————
Buried tank and the piping installation for an oil-fired heating system

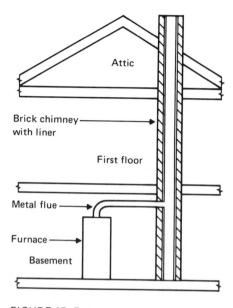

FIGURE 15-7 _____

Chimney in a multistory building

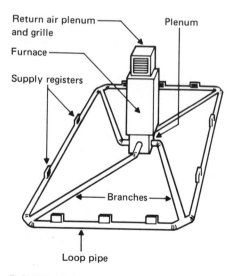

FIGURE 15-8 _____

Diagram of ductwork for perimeter loop system used with a warm air system

stoking the furnace is simplified. The heating designer and the building architect will need to coordinate their work to ensure proper location of these areas.

**Venting.**    When combustion equipment is being used, the products of combustion must be vented to the outside of the building. A vent or chimney has to be provided from inside the equipment room to the outside of the building. The vent should run vertically up through the building.

The design of the building has to be such that the chimney can be run up from the equipment room through the rooms above to the outside of the building, as shown in Figure 15–7. Since rooms and partitions on the second floor do not necessarily line up with those on the first floor, this positioning is not always easy to do. The chimney or vent is normally enclosed in an opening in the partitions or at least in a closet. If the partitions do not line up, special arrangements may have to be made to accommodate the chimney on the second, or any other, floors. It is usually necessary to locate the equipment where it is convenient to the chimney. The designer should keep this problem in mind when selecting the equipment for any job.

**Ductwork or Piping.**    In warm air systems, the supply ducts run from the equipment to the different rooms in the building. Return ducts bring the air back to the equipment. Space has to be available for these ducts. It is most desirable to have the ducts concealed in the construction of the building. Normally they are installed in ceiling spaces, furred-down sections in halls, or in special spaces called **chases** that are provided in the building. The designer should make sure that such spaces are available before selecting the heating equipment. Figure 15–8 shows the ductwork for one type of system, a perimeter loop system, which normally would be located under the floor. Other types of duct systems are shown in Chapter 16.

A hydronic system utilizes supply and return piping to transport heat. Water pipe is much smaller than air duct for a given amount of heat, so it is easier to find room for water pipes than it is for air ducts. A piping system, however, is much more rigid and inflexible in design than are air ducts. Changing a piping system is more difficult than changing a duct system. Figure 15–9 shows the piping for a simple loop system.

**Access.**    Space should be allowed around the mechanical equipment to provide for maintenance and service. Manufacturers' installation manuals show the proper amount of clearance to allow on each side of the equipment. The designer must always provide for these allowances when laying out the equipment room.

Proper access should also be planned to the equipment room. If the room is located above or below the main traffic area in a building, stairs or access ladders should be provided. Space should also be allowed for eventual replacement of parts or even of the entire unit. Doors, stairways, and lifting facilities should be large enough to accommodate these parts. The designer should ensure that such arrangements can be made for the equipment chosen for the job.

**Combustion Air.**    The designer must provide for a constant supply of combustion air to the equipment room whenever equipment is used that employs combustion for heating. The air should be brought in from the outside of the building. See Figure 15–10. If the equipment room is located on an outside wall, part of the wall may be **louvered** for combustion air and fresh air. The type of fuel used will determine the size of the opening required, but the openings and ducts should always be large enough to provide at least 50% more air than that required for complete combustion. As a general rule, openings with at least 1 sq in. per 1000 Btu/h of furnace input should be provided at the floor line and ceiling line of the furnace room.

If the equipment room is located in the central part of a building, ducts usually have to be used to bring the air into the space. If such is the case, the designer should make sure there is room for the ducts to run through the building framing or that duct chases are provided.

When electric heat is used, combustion air is not required since no combustion takes place.

**Fresh Air.**    To keep the atmosphere in the building healthy and comfortable, fresh air must be introduced into the occupied spaces. The fresh air is introduced

**Chase***

**Louver**   An opening that allows air to enter but that has overlapping outward-sloping vanes that prevent rain from entering

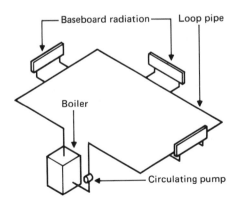

FIGURE 15-9 _____

Diagram of piping for a simple loop system used with a hydronic system

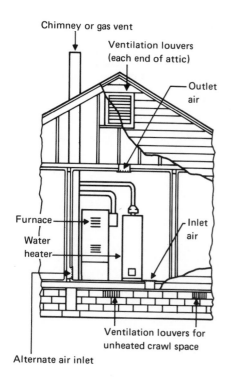

Chimney or gas vent

Ventilation louvers
(each end of attic)

Outlet
air

Furnace

Water
heater

Inlet
air

Ventilation louvers for
unheated crawl space

Alternate air inlet

FIGURE 15-10 _____

Providing ventilation and combustion air for
equipment room

into the return **plenum** of the heating system. The air is thus processed through the filters and the furnace and is introduced into the spaces with the supply air. The same amount of return air has to be exhausted or vented so that the pressure in the building remains constant.

The amount of air introduced as fresh air is determined by the number of people who will be in the building, by the amount of contamination of the air in the building, and by the cubical content of the building. This topic is covered later in the chapter.

The fresh air, like combustion air, is brought into the equipment room through ducting or louvers. Unlike combustion air, however, it is ducted directly to the return air side of the air mover. Therefore, the designer will have to make sure there is room for the ducts in the building design.

## 15.3   EQUIPMENT SIZING

After the heating unit has been selected to match the heat load and to fit into the building design, the amount of air or water that needs to be circulated in the system should be determined. To do this calculation, the Btu/h output of the heating unit is converted to cfm of air for a circulated air system or to gpm of water for a hydronic system. The heat itself is carried by the air or water; hence the need for the conversion.

## Btu/h to cfm

The formula for sensible heat can be used to convert Btu/h to cfm. Although there is always some moisture in the air being heated, the effect of that moisture is not noticeable on a heating application. The sensible heat formula is:

$$cfm = \frac{Btu/h}{TD \times 1.08}$$

where:

cfm = amount of air to circulate
Btu/h = amount of heat involved
$TD$ = temperature difference (rise or drop)

1.08 = specific heat of standard air × the specific weight × 60 min in 1 hr

The one variable that is chosen arbitrarily is the temperature rise through the furnace. The rise is normally figured at 80°F to 100°F for fossil-fired units and 50°F for electric heat units.

## EXAMPLE 15–1

Find the cfm for a gas- or oil-fired furnace of 112,000 Btu/h output (the output for a 140,000 Btu/h input furnace) and an 80°F temperature rise.

**Solution**

$$\text{cfm} = \frac{\text{Btu/h}}{TD \times 1.08}$$

$$= \frac{112,000}{80 \times 1.08}$$

$$= 1297.30$$

The amount of air that the system should be designed for is 1297.30 cfm.

**Air Circulation.**     In the choice of a temperature rise through a warm air furnace, the designer should consider how much air needs to be circulated in the building for comfort. The choice usually will be dictated by the number of people in the building or simply on an air change per hour basis.

*People Load.*     If the **people load** is high in the building, fresh air should be introduced into the circulation to keep the air in the building from becoming stale. The amount of fresh air that should be circulated can be calculated on the basis of the number of people in the building. A general rule of thumb is that the more smoking in the building, the more fresh air will be required to keep the air fresh. If the smoking is light, then approximately 5 to 10 cfm per person of fresh air should be introduced to the system. If the amount of smoking is medium, from 10 to 25 cfm should be brought in. If the smoking is heavy, from 25 to 40 cfm

**Plenum**     The box on the inlet or outlet of the furnace that connects the ductwork to the furnace

**People load**     The actual number of people in a building

TABLE 15–1 _____

Ventilation Rate Table

| Smoking | cfm per person |
|---------|----------------|
| Light   | 5–10           |
| Medium  | 10–25          |
| Heavy   | 25–40          |

Note: Figures are cfm of air per person to be brought into the return air system from outside the building.

per person will be required. Table 15–1 is a ventilation table giving guidelines to use for ventilation rates.

Remember: the fresh air is to be brought in from outside and circulated with the return air. Fresh air cfm is not the same as the total air to be circulated. A general rule is that the fresh air will be from one-tenth to one-third of the total air circulated.

*Air Change.*    A good way to determine the total amount of air that should be circulated in a building is to plan on changing the air in the building a given number of times per hour. Enough air should be circulated so that the entire amount of air in the building is replaced with fresh air a specific number of times each hour.

When the air change per hour basis is used to evaluate the amount of air needed, the following formula can be used:

$$\text{cfm} = \frac{V \times AC}{60}$$

where:

cfm = amount of air to circulate
$V$ = volume of the building
$AC$ = number of air changes per hour
60 = min in an hr

The building volume, in cu ft, is multiplied by the number of air changes desired per hour. The result is divided by 60 min. For the total air to be circulated, anywhere from 4 to 8 air changes per hour may be used. If the formula is used to determine fresh air only, then 1 air change per hour should be used.

## EXAMPLE 15–2

The floor plan in Figure 14–5 in Chapter 14 shows a residence 48 ft × 30 ft, with an 8 ft ceiling height. Find the amount of air to circulate for 6 air changes per hour in this building.

**Solution**

First, find the volume of the building:

$$V = l \times w \times h$$

$$= 48 \times 30 \times 8$$

$$= 11,520$$

The volume of the building is 11,520 cu ft. Now, use the air change formula:

$$\text{cfm} = \frac{V \times AC}{60}$$

$$= \frac{11,520 \times 6}{60}$$

$$= 1152$$

The amount of air to circulate through this building for 6 air changes per hour would be 1152 cfm.

After the amount of air to circulate has been determined by either of the above methods, the temperature rise of the air through the heating unit with that amount of air should be checked. In some cases, the amount of air will need to be adjusted to provide a proper temperature rise through the furnace.

## Btu/h to gpm

When a hydronic system is used, Btu/h must be converted to gpm. This conversion is done by dividing the Btu/h by the temperature difference multiplied by a factor of 500. The temperature difference in this case is for the temperature drop of the water as it goes through the terminal units in the system. The formula to find gpm is as follows:

$$\text{gpm} = \frac{\text{Btu/h}}{TD \times 500}$$

where:
    gpm = flow rate of the water
    Btu/h = heat loss of the building
    $TD$ = temperature difference (rise or drop)
    500 = factor derived from the weight of water per gal × the specific heat of water × 60 min per hr

## EXAMPLE 15–3

If a hydronic system was to be used to heat the building used in Example 15–1, with a heat loss of 112,000 Btu/h and a TD of 20°F (about right for a small hydronic system), find the gpm.

**Solution**

$$\text{gpm} = \frac{\text{Btu/h}}{TD \times 500}$$

$$= \frac{112{,}000}{20°\text{F} \times 500}$$

$$= 11.2$$

To carry the 112,000 Btu/h of heat, with a TD of 20°F, 11.2 gpm of water would need to be circulated in the system.

**gpm and Terminal Units.**    In a hydronic system, the heat must be extracted from the water at the point of use by some device. This device is called the terminal unit of the system.

The amount of heat extracted by a terminal unit is found by using another form of the formula shown in Example 15–3. This new formula is:

$$\text{Btu/h} = \text{gpm} \times TD \times 500$$

where:
Btu/h = heat extracted
gpm = flow rate of water
$TD$ = temperature difference
500 = factor derived from the weight of the water per gal × the specific heat of water × 60 min per hr

The heat extracted from the water by the terminal units is distributed in the building by different methods, depending upon the type of terminal units used. The different types of terminal units and methods of distribution are covered in Chapters 19 and 20.

# 15.4  SUMMARY

After an accurate heat load of a building has been calculated, the next step in designing a system is to select the heating equipment. Selecting the heating equipment that is best for a building is based on whether a warm air system or a hydronic system should be used, what fuel or energy is available, and where the equipment can be located in the building.

Since the ultimate purpose of most heating systems is to heat the air in a space, the warm air system is most commonly used. If the heating equipment has to be located some distance from the point of use, however, a hydronic system may be more practical.

Units that use any of the fossil fuels or electricity are available in either the warm air system or the hydronic system. The choice of fuel or electricity is a matter of availability, price, and convenience of use in the particular building to be heated.

Combustion equipment requires combustion air, venting of the fuel gases, and piping and/or storage of the fuel. Electric units do not need to be vented, but they require wiring and controls for the electric power. All of these factors have to be considered when selecting a system or equipment.

The quantity of air or water to be circulated has to be considered when selecting the heating equipment. The cfm of air is figured on the basis of the amount of heat to be distributed and the temperature rise of that air through the furnace. The gpm of water in a hydronic system is based on the amount of heat to be distributed and the temperature drop of that water through the system. The cfm or gpm are calculated by use of standard formulas.

When a hydronic system is used, the heat has to be extracted from the water at the point of use and transferred to the air for the actual heating of the space. The final act in designing a hydronic system, then, is to figure the cfm of air required at this point of exchange.

# 15.5  QUESTIONS

1. Name the four major factors to consider when selecting heating equipment.

2. The heat load of a building must be figured accurately so the heating unit selected will _____ or _____ that load.

3. The furnace selected to heat a building must have an input rating that matches the building loss. True or false?

4. If an oil-burning furnace has an input rating of 350,000 Btu/h, what will the nominal output rating be, as shown in the manufacturer's catalog?

5. When selecting the heating unit for some types of buildings, extra heating capacity should be provided for quick _____ after a temperature setback period.

6. In a building where the heating equipment is centrally located but heat is needed at widely separated parts of the building, a _____ system should be used.

7. If a building is compact and the spaces to be heated are close to the heating equipment location, a _____ _____ system is more practical.

8. A warm air heating system is easier to change than a hydronic system. True or false?

9. A hydronic system requires some room for piping to be run but not as much as would be required for ductwork. True or false?

10. In the selection of heating equipment, a designer must consider the _____ and _____ of fuels.

11. Electric heating units are usually easier for a designer to use because they do not require a _____.

12. Designers need not provide access to heating equipment since it never requires service. True or false?

13. All heating units need combustion air run to the equipment room. True or false?

14. How many cfm of air should be circulated in a warm air system if the furnace has a heating output capacity of 320,000 Btu/h and a TD of 85°F is desired through the furnace?

15. In planning a heating system for a restaurant, a designer determines that the capacity in the lounge area is 75 people. Lounges typically are heavy smoking areas. What would be the minimum amount of fresh air to figure for that area?

16. In a building that is 95 ft long, 60 ft wide, and has a 12 ft ceiling height, how much fresh air should be provided to give 6 air changes per hour?

17. In a hydronic system, if the heat load is 600,000 Btu/h and a TD of 20°F is desired through the system, how many gpm of water should be circulated?

# 16.1 INTRODUCTION

In a warm air heating system, the heat is generated by a furnace at one point in the building and distributed to the various parts of the building by air. A duct system carries the air where heat is wanted. The ductwork is made up of metal or fiberglass ducts laid out in a pattern that best distributes the air.

The general design of the duct system is determined by the construction used for a particular building. If the building is constructed on a concrete slab on ground, one type of system is used. If the building has a crawl space or a basement, another type of system is required. The designer must decide which system is best for each job.

After the type of system is decided upon, the final step is to size the ductwork so the proper quantity of air is delivered to each part of the building.

# 16.2 TYPES OF DUCT SYSTEMS

A warm air heating distribution system includes the supply and return duct systems. The supply duct system is usually the more complex of the two, and the name of the system describes the supply side. Three general types of distribution systems are used. They are:

1. Perimeter loop system,
2. Radial system,
3. Trunk and branch system.

These designations refer to the system's **configuration** and are normally the same whether they are installed under the floor or overhead. There are many variations of these basic configurations. However, even those configurations used in large buildings are combinations or variations of the basic systems.

## Perimeter Loop System

A **perimeter loop system** is one in which the ductwork runs around the perimeter of a building in a complete

*Term defined in text.

---

**CHAPTER**

# 16

# Warm Air Distribution and Duct Sizing

---

**Configuration**     Shape or pattern of a design

**Perimeter loop system***

**Branch ducts**    Ducts used to feed air into a loop or take air out of a trunk for distribution to registers

**Loop**    The duct that goes around the perimeter of a building, is fed by branches, and has registers located on it

**Radial system**    A duct system that has individual branches radiating out from the supply plenum, each supplying air to a register

**Trunk and branch system**    A duct system with a main trunk that supplies air to separate branches, which in turn supply individual registers

loop. The supply registers are located on the loop, and **branch ducts** carry air to the loop from the supply air plenum.

This system is most commonly used when the building has a concrete floor poured directly on the ground. The outside **loop** of the duct runs around the perimeter of the building and is under the edge of the concrete slab. Figure 16–1 illustrates a perimeter loop system.

When the ducts for a perimeter loop system are sized, the loop pipe is divided into sections with registers in each section having approximately the same cfm requirements. The loop is then sized to provide the cfm required for the section needing the greatest cfm. The loop pipe is the same size all around. The feeder pipes are sized to provide the cfm required for the sections in the area covered by the feeder.

## Radial System

A **radial system** is one in which branches run out radially from the warm air plenum to each register, as shown in Figure 16–2.

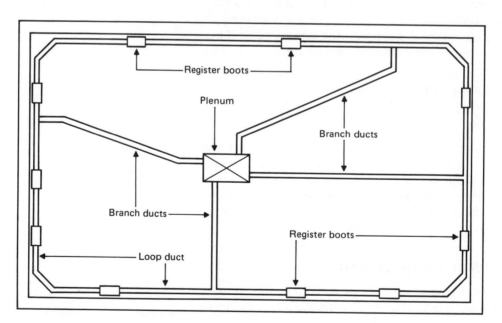

FIGURE 16-1

Perimeter loop duct system

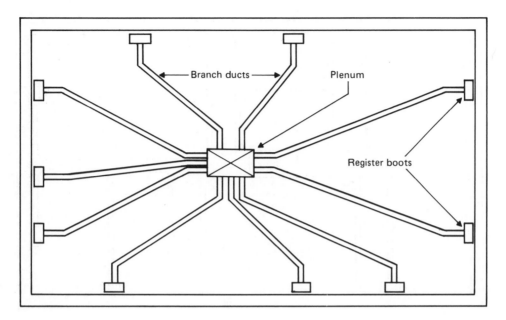

FIGURE 16-2
Radial duct system

The radial system is normally used when the duct-work is run in an attic or a crawl space of a building. This system should be used only where head room or walkways will not be affected by the ducts. It is especially adapted to buildings where the heating unit is located approximately in the center of the building, and the ductwork can run directly from the supply plenum to each supply boot as required. Each branch duct in a radial system is sized for the amount of air required by the register located on that branch.

## Trunk and Branch System

A **trunk and branch system** is one in which a main trunk, or trunks, extend out from the plenum, and branches run from the trunk to each register. Figure 16–3 is a diagram of a typical trunk and branch system.

This system is usually used when the branches can be run either up in the floor joists or in ceiling joists so they can be enclosed. The trunk that feeds the branches is run along a beam or supporting partition, where it can be enclosed as the building is finished. A

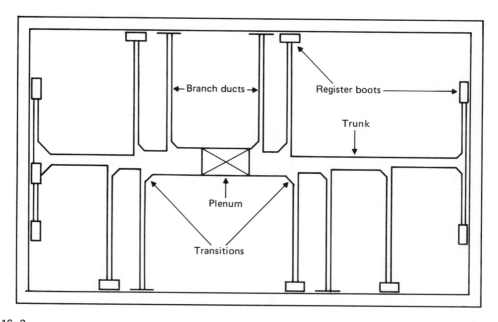

FIGURE 16-3

Trunk and branch duct system

good example of this type of system is a duct system installed in a finished basement. The ductwork is installed in the ceiling and then covered up by the ceiling finish material when the basement is finished.

The main components of a trunk and branch system are the main trunk and the branches that lead off it. The trunk is sized to carry the air needed for the part of the building that it serves, and each branch is sized to carry the air required by the registers they serve.

There are two types of trunk and branch systems: (1) an extended plenum system and (2) a graduated trunk system.

**Extended Plenum Trunk and Branch System.**    In an **extended plenum trunk and branch system,** the main trunk is kept the same size for its entire length, and each branch is a uniform size for its entire length. This system is illustrated in Figure 16–4. An extended plenum system should be used only on small jobs, where the length of the trunk is minimal. This system is difficult to **balance.** If the main trunk is not short, some branches will get more air than others.

**Graduated Trunk System.**    In the **graduated trunk system,** the size of the main duct is reduced as the quantity of air is reduced at each **branch take-off.** A branch take-off is the fitting where the branch fits onto the trunk. It is designed so air for the branch will flow smoothly from the trunk into the branch. This system is illustrated in Figure 16–3. Because of the duct reductions, the velocity of air remains the same throughout the trunk, and consequently so does the **static pressure.** The system is self-balancing. The proper amount of air will flow out of each branch take-off, regardless of how far the take-off is from the supply plenum.

**Extended plenum trunk and branch system\***

**Balance**    Adjusting an air supply system so the proper amount of air will be delivered by each register in the system

**Graduated trunk system\***

**Branch take-off**    The fitting between the main duct and a branch duct

**Static pressure**    The pressure of a moving stream of air, measured at right angles to the direction of airflow

# 16.3  LAYING OUT THE SYSTEM

After the type of system is decided upon, the next step in designing is to transfer all of the data concerning the job onto a plan of the building. The designer draws or obtains from the architect an outline plan of the foundation or basement and a plan for each floor of the building. These plans should show the walls, parti-

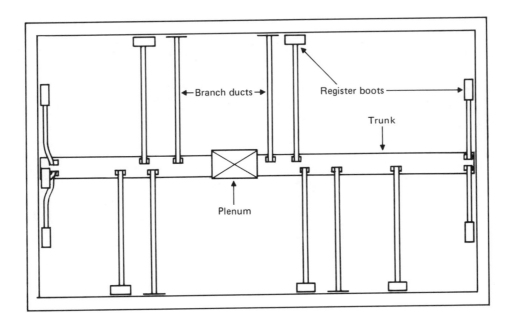

FIGURE 16-4

Extended plenum, trunk, and branch duct system

tions, windows, doors, cabinets, and any other details that will influence the location or size of registers, ducts, or other parts of the heating system. Special note should be taken of construction details in the vicinity of the equipment room. Supporting columns or beams that run where the designer needs to place ductwork should be especially noted. Such major parts of the construction cannot be moved, and the designer should avoid them in showing the location for the ducts. Piping from the kitchen or bathrooms is another example of construction details that are necessary to the building. The designer should avoid the areas where the piping goes when designing ductwork. If the plans are on tracing paper, details can be transferred from one plan to another easily.

The plans should be drawn to as large a scale as is practical, depending upon the size of the building, but no smaller than 1/8 in. to a foot for larger buildings and 1/4 in. to the foot for smaller buildings. Figure 16–5 is an example of such a floor plan for a typical residence.

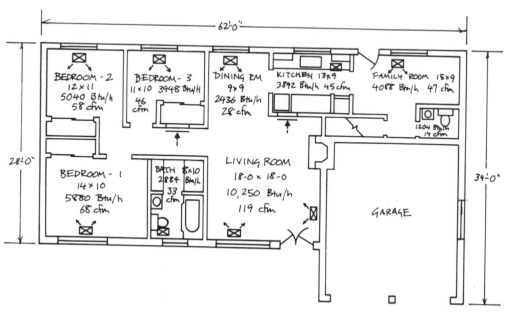

**FIGURE 16-5** _____

Floor plan of a typical residence used to prepare a duct system layout

# Design Data

Using the floor plans, the designer should make a note in each room of the Btu/h and cfm required in the room. The Btu/h and cfm are taken from the calculated heat load for the building. This information has been added to the plan in Figure 16–5.

After the Btu/h and cfm have been recorded on the plan, a tentative selection of supply registers is made. The registers are selected to provide the required amount of air to each room. This selection should be made so the registers will be compatible with the type of system decided upon.

# Register and Grille Location

Using the floor plan, the designer should mark the location of the registers. The plan in Figure 16–5 has the registers indicated with the proper symbols.

The designer uses the same procedure to select and show the location of the return air grilles on the plan. See Figure 16–5. Finally, the designer determines the correct sizes for the registers and grilles and applies this information to the plan. See Figure 16–5.

After the supply registers and return air grilles are drawn on the floor plan, a tracing of the foundation plan, or whatever floor the ductwork is going to be located on, is placed over the floor plan. The location of the registers and grilles is shown on the overlay. Figure 16–6 is an example of this drawing.

# Ductwork Layout

With the register locations marked on the plans, the ductwork can be tentatively drawn in. Now it is possible to visualize the layout of the system to see if there is any conflict with other components of the building construction, such as posts, beams, plumbing, or electrical details. Figure 16–7 shows the layout of the supply and return ducts drawn on the foundation overlay.

If there is no apparent conflict, the ductwork can be drawn in, either in single line or full scale. A single line drawing is one that shows the walls, partitions, and other major parts of the building by the use of a single line for each. Thickness of the materials is not

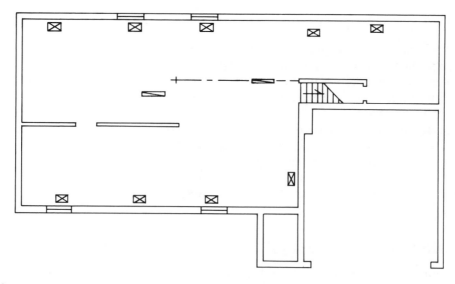

FIGURE 16-6

Overlay of the foundation or basement plan with the register and grille locations applied

indicated. The plan may be drawn approximately to scale, but individual parts are not exactly to scale. A full-scale drawing shows all parts in double lines so the thickness of each part is indicated on the plan, as well as the location in relation to other parts. The ductwork should be drawn on the plan as it will actually be placed in construction. The location of the heating unit or equipment room must also be taken into consideration when locating the duct system. After the main trunks are located on the plan for both the supply and return air systems, the branch ducts from the trunks to the register boots are drawn in, taking the same precautions pertaining to building components as used for the main trunks. Ductwork for make-up air and combustion air, if required, should also be drawn in at this time.

If the ductwork take-offs and returns to the equipment are complex, it will be difficult to show them on the floor plan. Larger-scale drawings may be needed for the equipment room. The drawings should show the supply and return plenums and the connections of the ductwork. Plan views, sections, and elevations can be used to show the details necessary.

# Equipment List and Specifications

The equipment should be listed on the plans so that the estimators and installers will know exactly what equipment is required for the job. The equipment can be shown on a schedule on the heating plan, as well as in the specifications for the job. Table 16–1 shows an equipment list for a typical job. The equipment list should include registers, grilles, dampers, controls, and any other equipment or accessories required for the heating system. Notes of any special requirements should be made on the plans to call attention to areas of special concern. Location of the thermostat should be shown on the plan.

A description of the equipment and accessories should also be made in the specifications for the job. The specifications should include the **performance standards** for the equipment and all of the components of the system, as well as physical or mechanical standards for the installation. The specifications and plans become part of the contract documents for the job and are extremely important in ensuring that the job is installed as the designer intends.

**Performance standards**    A written description of the expected results of the operation of a system, usually giving temperature, quality, velocity of air, and any other criteria necessary to describe desired performance

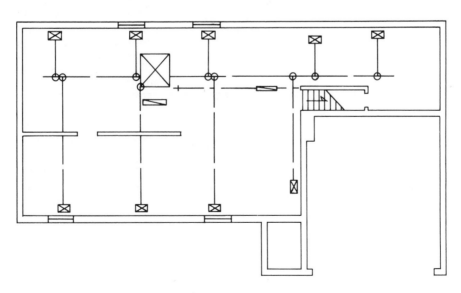

FIGURE 16-7

Tentative layout of the supply and return ducts made on the foundation overlay

**Transition**    Change in duct size

**Take-off**\*

TABLE 16–1 _____
Typical Equipment List

| Equipment | Specifications |
|---|---|
| Furnace | Shall be an upflow, gas-fired unit with 80,000 Btu/h output, and shall be capable of heating the home to 75°F when the outside temperature is 15°F. The blower shall be capable of delivering 1000 cfm of air at a total static pressure of .50 in. wg. The furnace shall be equipped with electric ignition and have a ceramic-coated heat exchanger. |
| Ductwork | Shall be of galvanized iron, installed in a proper manner and according to latest standards. Air turns shall be used in all elbows, and flexible connections shall be used at the connection to the plenums. Balancing dampers shall be installed in all branches and in the fresh air intake duct. |
| Registers and grilles | Shall be of steel construction as noted in the specifications, shall be painted with a rust-preventing undercoat, and shall match the room surfaces. |
| Air cleaner system | Shall include an electronic air cleaner on the inlet side of the furnace, which will be rated to handle the total cfm of the system with a cleaning efficiency of 80% by standard dustspot test at that cfm. |
| Humidifier | Shall be a spray-type humidifier installed in the supply duct to provide enough moisture to maintain a 50% Rh in the building at all conditions, and shall be controlled by a humidistat placed in the return duct. |

# 16.4  SIZING THE DUCTS

After the duct system is drawn in on the plan, the size of the ducts at various locations throughout the system has to be determined and the sizes marked on the plan. All **transitions** should be noted where the duct changes from one size to another. The dimensions of the ducts, with all drops, transitions, elbows, and other components of the system, are marked on the plan. It is important that all dimensions are given so the estimator can make an accurate take-off for the material and estimate the labor that will be required to install the job. A **take-off** is an accurate list of materials needed for the construction of a building or part of a building. In the case of the ductwork system, the take-off should show all equipment and materials necessary for the installation of the ductwork system. The installer will also use the plans for making up the ductwork. Details of the return air system should be shown on the plan.

The ductwork must be carefully sized to ensure that the correct amount of air gets to each room or space to provide proper heating. Any adjustment in the calculations of the cfm to be delivered to the rooms should be made before the duct is sized.

Two methods of duct sizing will be described in this text: (1) the equal velocity method and (2) the equal friction method. These methods are the basic ways to size ductwork. All other methods are an extension of them.

## Equal Velocity Method

Air noise in a duct is directly related to the construction of the duct and the velocity of the air. The equal velocity method of duct sizing is used when a designer wants to maintain a given velocity of air in each section of the system to control the sound level and still use as small a duct as possible for economic reasons. The velocity of the air in the ducts is a function of the size of the duct and the quantity of air flowing in it. This relationship is best shown by the following formula:

$$V = \frac{Q}{A}$$

where:
  $V$ = velocity of the air in ft per min (fpm)
  $Q$ = quantity of air in cfm
  $A$ = area of the duct in sq ft

## EXAMPLE 16–1

If a section of ductwork is 12 in. wide and 9 in. high, and 600 cfm of air is to pass through it, what will be the velocity of that air?

**Solution**

To start with, the dimensions of the duct are given in inches (this is the correct way to size duct). The area of the duct in square inches is:

$$A = w \times l$$

where:

$A$ = area of the duct, in the same terms as $w$ and $l$

$w$ = width of the duct

$l$ = length, or in this case height

So for this example:

$$A = 12 \times 9 = 108$$

The area 108 sq in has to be changed to sq ft to use in the velocity formula. Since there are 144 sq in. in a sq ft, then 108 sq in. divided by 144 sq in. gives the area in sq ft.

$$A = \frac{108}{144} = 0.75$$

The area of the 9 × 12 duct in sq ft is .75. The areas can be used in the velocity formula, as follows:

$$V = \frac{Q}{A} = \frac{600}{0.75} = 800$$

The velocity of the air in the duct is 800 fpm.

When the equal velocity method of duct sizing is used, a velocity table is used to find acceptable velocities for each application. Table 16–2 is an example of a velocity table. This table gives the maximum allowable velocity for the air in the ducts, for various types of installations. Reference to the table will show that the velocity of the air in the main trunks can be higher than that in the branches and that the acceptable sound level in various types of buildings is also considered.

To use the table, find the type of building use and select the acceptable, or maximum, velocity for the particular part of the system, and use that velocity in the formula. The quantity of air required through each part of the duct in the system is taken from the layout and the data sheets already prepared. The allowable

TABLE 16–2 ————————————————————————————————————————————
Recommended Duct Velocities for Various Applications

| Applications | Residences | Schools, Theaters, Public Buildings | Industrial Buildings |
|---|---|---|---|
| | *Velocities for Total Face Area* | | |
| Outdoor air intakes | 500 (800) | 500 (900) | 500 (1200) |
| Filters | 250 (300) | 300 (350) | 350 (350) |
| Heating coils | 450 (500) | 500 (600) | 600 (700) |
| | *Velocities for Net Free Area* | | |
| Air washers | 500 (500) | 500 (500) | 500 (500) |
| Suction connections | 700 (900) | 800 (1000) | 1000 (1400) |
| Fan outlets | 1000–1600 (1700) | 1300–2000 (1500–2200) | 1600–2400 (1700–2800) |
| Main ducts | 700–900 (800–1200) | 1000–1300 (1100–1600) | 1200–1800 (1300–2200) |
| Branch ducts | 600 (700–1000) | 600–900 (800–1300) | 800–1000 (1000–1800) |
| Branch risers | 500 (650–800) | 600–700 (800–1200) | 800 (1000–1600) |

Note: Maximum duct velocities appear in parentheses.

velocity is taken from the table. The duct size is figured by use of the following formula:

$$A = \frac{Q}{V}$$

when:
  $A$ = area of the duct in sq ft
  $Q$ = cfm of air
  $V$ = velocity of air in fpm

## EXAMPLE 16–2

If a branch duct is required to carry 250 cfm of air, and the velocity chart shows the allowable velocity for that part of the system to be 600 fpm, find the area of the duct.

**Solution**

$$A = \frac{Q}{V} = \frac{250}{600} = 0.416$$

The area of the duct is 0.416 sq ft. Multiplying that by 144 sq in., in a sq ft, would give 60 sq in.

The actual dimensions for square, rectangular, or round duct can be found from the area mathematically.

When rectangular duct is used, the best practice is not to make the duct height less than 8 in. and to limit the width to not more than three times the height. When a 24 in. × 8 in. duct is not large enough to accommodate the cfm, then the depth of the duct should be increased to 10 in., and so on, keeping the width at no more than three times the height. When sizing ducts, the width or horizontal dimension is always given first; the height or vertical dimension is given second.

## Equal Friction Method

The equal friction method of sizing ducts is based on the fact that as air moves through a duct, the air film moving along the surface of the duct causes a certain amount of resistance to the flow. This resistance is a function of the amount of surface exposed and the quantity of air flowing.

For air to move through a duct, there must be a pressure difference between the inlet and the outlet of the duct. The blower, or air mover, exerts a higher pressure on the air at the inlet end of the duct than exists at the outlet. This pressure difference makes the air move. The amount of difference indicates the amount of resistance in the duct. The pressure is measured with a device called a manometer. Figure 16–8 shows a manometer of the type used for measuring pressure in ducts.

The pressure, called the static pressure, is read by taking a manometer reading through a hole in the wall of the duct at a 90° angle to the airstream. The measurement is in inches of water. A more complete explanation of static pressure and the use of a manometer is given in the next chapter. Friction in the duct causes a reduction in the static pressure as the air moves through the duct. The static pressure difference between the inlet and the outlet of the duct can be used to indicate the amount of friction in the duct. Since friction, and consequently **static pressure drop,** are functions of cfm and duct size, it is possible to use static pressure drop for a given length of duct and cfm to determine duct size. Charts, called equal friction

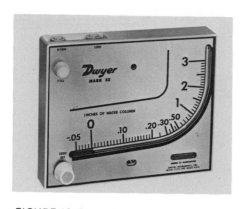

FIGURE 16-8 _____

Inclined manometer used to measure air pressure in a duct

charts, have been developed that show the relationship of static pressure drop, cfm, and duct size. Such a chart is found in Table A–8 in the Appendix. To use the equal friction chart to find the size of duct used to deliver a given cfm of air, find the static pressure (SP) drop per 100 ft of duct on a vertical line on the chart and the cfm on a horizontal line. Where the two intersect, the duct size is shown on a diagonal line coming down from the top right of the chart. The duct size is given for round duct.

To size the duct by the equal friction method, select the amount of friction desired per 100 ft of straight duct and use that amount as the base for the calculations. For residential and smaller commercial or industrial systems, a static pressure of 0.10 in. wg is about right for each 100 ft of duct. The designer, using the cfm required for a given section of duct and the static pressure drop per 100 ft of duct, uses the equal friction chart to find the duct size required.

**Static pressure drop**    The difference between the static pressure readings at two points in a duct system, which indicates the effect of resistance to airflow

## EXAMPLE 16–3

In a given application, if 250 cfm of air is required and a static pressure drop of 0.10 in. per 100 ft of duct is to be used, find the size duct required.

### Solution

The 250 cfm line is found on the horizontal scale of the equal friction chart. The static pressure drop of 0.10 in. is found on the vertical scales. Where the two cross, the duct size is found on the diagonal line coming down from the top right of the chart. In this example, the size is found to be just slightly larger than 8 in. in diameter. An 8 in. duct would be used.

If a rectangular duct is required, rather than a round one, then a round-to-rectangular conversion chart should be used to find the equivalent size of rectangular duct. A rectangular duct equivalent chart is found in Table A–9 in the Appendix. When the equivalent chart is used in Example 16–3, a 7 in. × 8 in. duct would be used in place of the 8 in. round.

# Showing the Duct Sizing on the Plan

After the duct sizes have been determined, the next step is to record them on the drawing already made of the system. The best way is to start sizing the duct from the extreme outer end of the branches or trunks. Work back toward the air mover, sizing each duct in turn. Increase the size where each branch takes off, and letter in the size on the layout. All transitions, or changes in duct sizing, should be indicated on the plan. Figure 16–9 shows the duct sizing noted on the plan.

This procedure should be used for each branch of trunk that leads off from the plenum at the unit. After each branch is sized, the quantity of air that the ducts have been sized to carry should be rechecked to be sure that the figure equals the output of the blower.

After the supply air system has been sized, the same procedure can be used to size the return air system for the job. The design of the return air system is just as important as the supply side of the system.

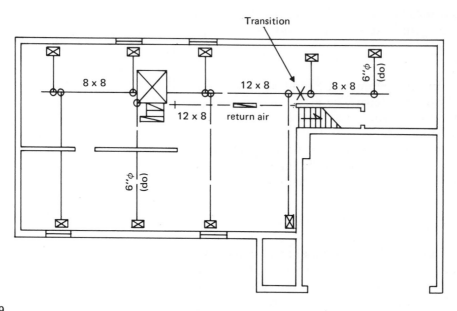

FIGURE 16-9 _____

Duct system layout showing the duct sizing and location of the transition

# 16.5  TOTAL SYSTEM FRICTION

After a duct system is sized, the designer must make sure that the heating unit used has a blower capable of delivering the cfm required against the friction in the system. This requires calculating the total friction in the system. When the equal friction method of duct sizing is used, this operation is easy. Since the duct sizing is based on a certain amount of friction per 100 ft of duct, it is necessary only to calculate the equivalent length of duct in the longest run for the supply side of the system and to make the same calculation for the return side. The two lengths are added and the sum divided by 100 (since the original figure was for friction per 100 ft of duct). This result is multiplied times the static pressure drop per 100 ft of duct used. The final result is the total pressure drop through the system.

For instance, if the actual length of the supply duct is 145 ft and the actual length of the return duct is 67 ft, first the sum of the two is found:

$$145 + 67 = 212$$

212 ft is the total length of the ductwork. This figure is then divided by 100:

$$212 \div 100 = 2.21$$

2.21 is the number of blocks of duct 100 ft long. If the ductwork was originally sized for 0.10 in. wg per 100 ft of duct, a common figure for small duct systems, then 2.21 is multiplied by 0.10 in. wg:

$$2.21 \times 0.10 = 0.212$$

0.212 in. wg is the friction drop in the duct.

To find the actual equivalent length of duct, elbows, transitions, and any other part of the ductwork that is not straight must be converted into the equivalent length of straight duct. The equivalent lengths for many different types of fittings are found in the Appendix in Table A–10.

To use the equivalent length tables, choose the fitting from the table that most nearly matches the fittings in the duct system. Then add the equivalent length given in the table for that fitting to the straight duct length. This calculation is made for each fitting in the system, and the total equivalent length is used to determine the total friction in the duct.

## 16.6  SYSTEM INSTALLATION PROCEDURES

The designer should be involved in the installation of the heating system. Unless the system has been planned in great detail and coordination with all of the other building component designers has been possible, there will be a need for some flexibility in the installation of the ductwork and equipment. Quite commonly, beams and girders, water and sanitary piping, or electrical wiring will run through the same areas that the heating system designer has chosen for the ductwork. In such cases, the designer and the contractors installing the systems must work together to decide which system can best be changed so that both can be installed. Usually the designer or contractor can rearrange the work to accommodate the installation of both.

The installers of the systems must know exactly what the designer intends as far as performance is concerned. Slight changes in duct or pipe sizing can make a great difference in performance, and yet the installer may not appreciate that fact. If inspections are included in the design of the system, then the designer should make sure that the installer follows the original plan as far as possible.

## 16.7  SUMMARY

After the loads are figured and the tentative selection of equipment is made, the next step in designing a heating system is to select the distribution system to be used, lay out that system on a plan, size the ducts for air distribution, and check the blower capacity against the system requirements.

Three main types of air distribution systems are used: (1) perimeter loop, (2) radial, and (3) trunk and branch.

The perimeter loop system has a loop of duct that runs around the perimeter of the building, usually under the floor. This system is basically an under-floor system, and it is almost always used in a concrete slab-on-grade construction. The main advantage of this system is that the loop duct warms the concrete slab around the outside perimeter of the building, making the concrete floor more comfortable.

In the radial system, branch ducts radiate out from a supply plenum, and each feeds a register. The branches are usually of small size since they each feed only one register. This system is generally installed in either a crawl space under a building, or an attic space.

In a trunk and branch duct system, the trunk is the main distribution member, taking the air off the supply plenum, and branches lead off the trunk to the registers. Two types of trunk and branch systems are used: the extended plenum system and the graduated trunk. Each has advantages and disadvantages, but the extended plenum system should be used only on small systems. The graduated trunk system can be used on any size system.

The ducts should be carefully sized to ensure the proper delivery of air required into each part of a building. This work can be done by the use of an equal friction chart. Proper sizing also makes it easy to calculate the total resistance to airflow so the blower can be properly sized for the delivery of air.

When the heating system is being installed, the designer should be involved in the job to ensure that the components go in as planned. Sometimes changes have to be made to the system to facilitate the installation of other components of the building. If changes are made, they must not alter the basic function of the system.

# 16.8 QUESTIONS

1. The simplest and best format to use to show the details concerning a heating system is a set of _____ and _____.

2. The three types of duct distribution systems used for residential and smaller commercial buildings are (a) _____ , (b) _____, and (c) _____.

3. In a perimeter loop distribution system, the loop runs around the outside _____ of the building.

4. In a radial distribution system, each branch carries approximately 50% of the total air of the system. True or false?

5. The two main parts of a trunk and branch system are the _____ and the branches that run to individual _____.

6. There are two types of trunk and branch duct systems: _____ and _____.

7. An extended plenum system is a good system to use regardless of the size of a building. True or false?

8. In a graduated trunk duct system, the main trunk is _____ in size, as the _____ of air is reduced at each take-off.

9. Performance standards define the conditions that will be maintained in a building by the system. True or false?

10. The fitting used where a duct changes from one size to another is called a _____.

11. The two principal methods of duct sizing are _____ and _____.

12. The equal velocity method of duct sizing is used when a designer wants to control the air _____ in a system.

13. What would the velocity of air be in a 20 in. × 10 in. duct if 1000 cfm of air was flowing in it?

14. What would be the dimensions of a square duct to carry 1275 cfm of air at 800 fpm?

15. Static pressure is that pressure exerted by the air in a duct, at right angles to the direction of flow of the _____.

16. Static pressure drop is the difference between the static pressure readings at two points in a duct system. True or false?

17. An equal friction chart shows the relationship among static pressure drop, _____, and _____ in a duct system.

18. The blower furnishing air to a duct system must be able to deliver the required air against the _____ in the system.

19. To calculate total friction loss of a system, the _____ length of duct is found by adding the actual length of duct to the _____ length for each fitting.

## 17.1 INTRODUCTION

In designing a warm air heating system, supply registers and return air grilles are selected, located, and sized to ensure that the correct amount of warm air is distributed to each space in the building as determined by the heat load calculations. The required cfm for each room is used to select the registers and grilles. This chapter describes the various supply registers and return air grilles available and outlines factors to consider when selecting, locating, and sizing the registers and grilles.

## 17.2 SUPPLY REGISTERS

The **supply registers** are the components of the warm air heating system that actually provide air to the rooms or spaces to be heated. A supply register is made up of a **diffuser** or face, a frame to hold it, and a damper to control the quantity of air passing through it. A typical supply register is shown in Figure 17–1.

Register type, location in the room, and size are the most important factors to consider in the selection of the supply register for a system. The style of the register and the location in the space will determine how effectively the air is distributed. The size of the register will determine the amount of air that will be delivered. Each of these important factors is discussed in the following sections.

### Selecting Supply Registers by Location

There are three basic types of supply registers categorized by their location in a room:

1. Ceiling,
2. Sidewall,
3. Perimeter.

**Ceiling registers,** as the name implies, are located on the ceiling of a room and are designed to deliver the air both horizontally and vertically. Figure 17–2 is an illustration of a typical square ceiling register.

*Term defined in text.

# CHAPTER
# 17
# Warm Air Registers and Grilles

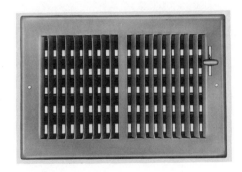

FIGURE 17-1 _____

Typical supply register used on a warm air system

---

**Supply register***

**Diffuser**   A face grille, used on a register, with vanes that cause the air to spread as it leaves the face

**Ceiling register***

FIGURE 17-2
Ceiling register

FIGURE 17-3
Typical sidewall register

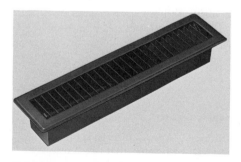

FIGURE 17-4
Floor register used in perimeter floor systems

**Sidewall registers** are located on a wall of a room, usually in one of two general locations: high on the wall near the ceiling or low near the floor. In a typical sidewall register, the air is diffused horizontally and vertically by the vanes in the grille. A sidewall register is shown in Figure 17–3.

**Perimeter registers** are located around the outside walls of the building, either on the floor or the ceiling. Perimeter registers deliver the air either vertically up or vertically down, depending on their location. A perimeter floor register is shown in Figure 17–4.

**Ceiling Registers.**    Ceiling registers are used when the space to be heated is too large to heat adequately from the walls or along the walls only, or when the amount of air required is greater than can be provided from the wall locations. When registers are placed at the proper locations around the ceiling of a room, air can be circulated evenly regardless of the size of the room.

Ceiling registers can be either square or round in shape. Figure 17–5 shows a round ceiling register.

The ductwork for ceiling registers is located above the register. The shape of the register is normally determined by the construction of the ceiling or the shape of the room. Aesthetics, or looks, are also considered in the choice.

Although the air pattern is basically vertically down, the face of the register is designed to spread the air so that it will cover a wider area. A diffuser type of register face accomplishes this purpose. Figure 17–6 shows the diffuser from a register. The curved vanes direct the airflow.

FIGURE 17-5
Round ceiling register

Care must be taken in sizing and locating ceiling diffusers so the air from them will not strike the occupants of the space at too high a velocity or before it has had time to mix with the room air. Air of too high a velocity or too low a temperature will be uncomfortable.

**Sidewall Registers.**    Sidewall registers are used when the design of the building is such that the registers can be located in the walls but the discharged air will not strike any occupant until it has mixed with the room air. The registers are located to deliver the air basically in a horizontal pattern, but the register face is designed so the airstream is also spread over a fairly large horizontal area. Figure 17–7 is an illustration of a sidewall register. The vanes are adjustable so that the desired air distribution pattern can be attained.

Location and sizing of sidewall grilles is critical to ensure comfort. Low sidewall registers should be used only when it is possible to direct the air so it does not enter the occupied space at too high a velocity for comfort.

**Perimeter Registers.**    The most commonly used register system for small commercial and residential systems is the perimeter floor or ceiling system. In this system, narrow slot-type diffusers are located in the floor or ceiling around the perimeter of the building so the air will cover the windows, doors, and any other area of high heat loss.

The air from perimeter registers is directed vertically up or down, but it is also spread by vanes in the face of the register. Since the registers are located around the walls of the room, the air usually will not strike any occupant of the space until it has had a chance to mix with the room air. The air distribution pattern, or spreading pattern, for a perimeter register is shown in Figure 17–8.

# Selecting Supply Registers by Air Pattern

The style of supply register selected for any distribution system will be determined by the type of register that will deliver the air where it will best cover the area of

Sidewall register*

Perimeter register*

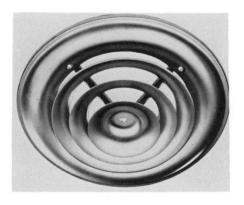

FIGURE 17-6
Diffuser used on the face of a ceiling register to deflect the air

FIGURE 17-7
Sidewall register

**Throw***

**fpm**    Foot per minute; a designation for velocity used in air distribution systems

**Spread***

**Noise level***

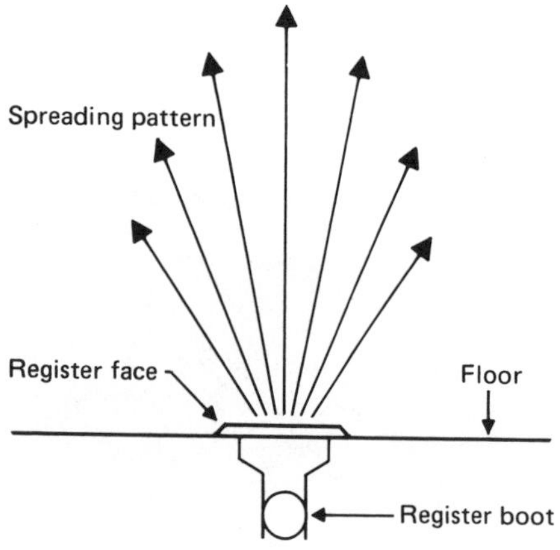

FIGURE 17-8 _____
Spreading pattern of a perimeter register

heat loss and provide the amount of air required. Four factors have to be considered:

1. cfm,
2. Throw,
3. Spread,
4. Noise.

Cfm, the air required in the room, has been discussed in previous chapters. **Throw** is the distance the air will travel directly away from the register before it reaches a terminal velocity of 50 **fpm** (ft per min). **Spread** is the width of the space that the air will cover, usually in a fan shape, before it reaches a terminal velocity of 50 fpm. **Noise level** of a register is related to the sound generated by the air passing through the vanes in the register face. Each of these factors has to be considered in the selection process.

**cfm.**    The first consideration in selecting the supply registers is based on the fact that air is the medium for carrying the heat where it is wanted in the building. The proper cfm has to be delivered by each register in the system to ensure proper distribution of heat. The amount of air that any register is designed to deliver is shown in a register manufacturer's catalog. Table 17–1 is a page from a register catalog. The table shows how cfm is related to register size.

The amount of air that a register will deliver is determined by the opening size of the register. The velocity of air out of the opening is determined by the amount of air. Manufacturers' catalog selection sheets show the different types of registers with the cfm that each size will deliver and the velocity of the air out of the face. They also show the spread and throw, which will be discussed later. The following example shows how to use the register catalog data for register selection.

## EXAMPLE 17–1

A room with the dimensions of 12 ft × 24 ft is found to need 250 cfm of air to heat it. It is decided that a ceiling diffuser will be used. What size register will deliver the required cfm?

TABLE 17–1

Page from a Register Catalog

## specifications—series 95    ONE WAY PATTERN

| SIZE IN INCHES | VEL* / PRESS† | 400 / .044 | 450 / .056 | 500 / .068 | 550 / .083 | 600 / .100 | 650 / .114 | 700 / .130 |
|---|---|---|---|---|---|---|---|---|
| **PLAN VIEW** | | | | | | | | |
| **CORE NO. 111** | | | | | | | | |
| 6 x 6 | TOTAL CFM | A 100 B | A 112 B | A 125 B | A 137 B | A 150 B | A 162 B | A 175 B |
| | CFM/side | 100 | 112 | 125 | 137 | 150 | 162 | 175 |
| | Throw, ft. | 6-11 | 6-12 | 7-14 | 7-15 | 9-17 | 9-18 | 10-20 |
| 9 x 9 | TOTAL CFM | 224 | 252 | 280 | 308 | 336 | 364 | 392 |
| | CFM/side | 224 | 252 | 280 | 308 | 336 | 364 | 392 |
| | Throw, ft. | 7-13 | 7-15 | 9-17 | 9-18 | 10-20 | 11-22 | 12-24 |
| 12 x 12 | TOTAL CFM | 400 | 450 | 500 | 550 | 600 | 650 | 700 |
| | CFM/side | 400 | 450 | 500 | 550 | 600 | 650 | 700 |
| | Throw, ft. | 7-15 | 9-17 | 10-19 | 10-21 | 11-23 | 12-25 | 13-27 |
| 15 x 15 | TOTAL CFM | 624 | 702 | 780 | 858 | 936 | 1014 | 1092 |
| | CFM/side | 624 | 702 | 780 | 858 | 936 | 1014 | 1092 |
| | Throw, ft. | 9-17 | 10-19 | 10-21 | 11-23 | 13-26 | 14-28 | 15-30 |
| **CORE NO. 112** | | | | | | | | |
| 6 x 9 | TOTAL CFM | 150 | 169 | 188 | 206 | 225 | 244 | 262 |
| | CFM/side | 150 | 169 | 188 | 206 | 225 | 244 | 262 |
| | Throw, ft. | 6-12 | 7-14 | 7-15 | 9-17 | 10-19 | 10-20 | 11-22 |
| 9 x 12 | TOTAL CFM | 300 | 338 | 375 | 413 | 450 | 488 | 525 |
| | CFM/side | 300 | 338 | 375 | 413 | 450 | 488 | 525 |
| | Throw, ft. | 7-14 | 8-16 | 9-18 | 10-20 | 10-21 | 11-23 | 12-25 |
| 9 x 15 | TOTAL CFM | 375 | 422 | 469 | 516 | 563 | 610 | 656 |
| | CFM/side | 375 | 422 | 469 | 516 | 563 | 610 | 656 |
| | Throw, ft. | 7-15 | 9-17 | 9-18 | 10-20 | 11-22 | 12-24 | 13-26 |
| 12 x 15 | TOTAL CFM | 500 | 562 | 625 | 688 | 750 | 813 | 875 |
| | CFM/side | 500 | 562 | 625 | 688 | 750 | 813 | 875 |
| | Throw, ft. | 8-16 | 9-18 | 10-20 | 11-22 | 12-24 | 13-26 | 14-28 |
| 12 x 18 | TOTAL CFM | 600 | 675 | 750 | 825 | 900 | 975 | 1050 |
| | CFM/side | 600 | 675 | 750 | 825 | 900 | 975 | 1050 |
| | Throw, ft. | 8-16 | 9-18 | 10-20 | 11-22 | 12-25 | 13-27 | 14-29 |
| **CORE NO. 113** | | | | | | | | |
| 6 x 9 | TOTAL CFM | 150 | 169 | 188 | 206 | 225 | 244 | 262 |
| | CFM/side | 150 | 169 | 188 | 206 | 225 | 244 | 262 |
| | Throw, ft. | 6-12 | 7-14 | 7-15 | 9-17 | 10-19 | 10-20 | 11-22 |
| 6 x 12 | TOTAL CFM | 200 | 225 | 250 | 275 | 300 | 325 | 350 |
| | CFM/side | 200 | 225 | 250 | 275 | 300 | 325 | 350 |
| | Throw, ft. | 7-13 | 7-15 | 8-16 | 9-18 | 10-20 | 10-21 | 11-23 |
| 6 x 15 | TOTAL CFM | 250 | 281 | 312 | 344 | 375 | 406 | 438 |
| | CFM/side | 250 | 281 | 312 | 344 | 375 | 406 | 438 |
| | Throw, ft. | 7-13 | 7-15 | 9-17 | 10-19 | 10-20 | 11-22 | 12-24 |
| 6 x 18 | TOTAL CFM | 300 | 337 | 375 | 413 | 450 | 487 | 525 |
| | CFM/side | 300 | 337 | 375 | 413 | 450 | 487 | 525 |
| | Throw, ft. | 7-14 | 8-16 | 9-18 | 10-20 | 10-21 | 11-23 | 12-25 |
| 6 x 21 | TOTAL CFM | 350 | 394 | 437 | 481 | 525 | 570 | 612 |
| | CFM/side | 350 | 394 | 437 | 481 | 525 | 570 | 612 |
| | Throw, ft. | 7-14 | 8-16 | 9-18 | 10-20 | 11-22 | 12-24 | 13-26 |
| 6 x 24 | TOTAL CFM | 400 | 450 | 500 | 550 | 600 | 650 | 700 |
| | CFM/side | 400 | 450 | 500 | 550 | 600 | 650 | 700 |
| | Throw, ft. | 7-15 | 9-17 | 10-19 | 10-21 | 11-23 | 12-25 | 13-27 |
| 9 x 21 | TOTAL CFM | 524 | 590 | 655 | 720 | 786 | 852 | 917 |
| | CFM/side | 524 | 590 | 655 | 720 | 786 | 852 | 917 |
| | Throw, ft. | 8-16 | 9-18 | 10-20 | 11-22 | 12-24 | 13-26 | 14-28 |
| 9 x 24 | TOTAL CFM | 600 | 675 | 750 | 825 | 900 | 975 | 1050 |
| | CFM/side | 600 | 675 | 750 | 825 | 900 | 975 | 1050 |
| | Throw, ft. | 8-16 | 9-18 | 10-20 | 11-22 | 12-25 | 13-27 | 14-29 |

## specifications—series 95    TWO WAY PATTERN

| SIZE IN INCHES | | 400 / .044 | 450 / .056 | 500 / .068 | 550 / .083 | 600 / .100 | 650 / .114 | 700 / .130 |
|---|---|---|---|---|---|---|---|---|
| **PLAN VIEW** | | | | | | | | |
| **CORE NO. 211** | | | | | | | | |
| 6 x 6 | TOTAL CFM | A 100 B | A 112 B | A 125 B | A 137 B | A 150 B | A 162 B | A 175 B |
| | CFM/side | 50  50 | 56  56 | 62  62 | 69  69 | 75  75 | 81  81 | 88  88 |
| | Throw, ft. | 5-10  5-10 | 6-11  6-11 | 6-12  6-12 | 7-13  7-13 | 7-15  7-15 | 8-16  8-16 | 9-17  9-17 |
| 9 x 9 | TOTAL CFM | 224 | 252 | 280 | 308 | 336 | 364 | 392 |
| | CFM/side | 112  112 | 126  126 | 140  140 | 154  154 | 168  168 | 182  182 | 196  196 |
| | Throw, ft. | 6-11  6-11 | 7-13  7-13 | 7-14  7-14 | 8-16  8-16 | 9-17  9-17 | 10-19  10-19 | 10-20  10-20 |
| 12 x 12 | TOTAL CFM | 400 | 450 | 500 | 550 | 600 | 650 | 700 |
| | CFM/side | 200  200 | 225  225 | 250  250 | 275  275 | 300  300 | 325  325 | 350  350 |
| | Throw, ft. | 7-13  7-13 | 7-15  7-15 | 8-16  8-16 | 9-18  9-18 | 10-20  10-20 | 10-21  10-21 | 11-23  11-23 |
| 15 x 15 | TOTAL CFM | 624 | 702 | 780 | 858 | 936 | 1014 | 1092 |
| | CFM/side | 312  312 | 351  351 | 390  390 | 429  429 | 468  468 | 507  507 | 546  546 |
| | Throw, ft. | 7-14  7-14 | 8-16  8-16 | 9-18  9-18 | 10-20  10-20 | 10-21  10-21 | 11-23  11-23 | 12-25  12-25 |
| 18 x 18 | TOTAL CFM | 900 | 1012 | 1125 | 1240 | 1350 | 1462 | 1575 |
| | CFM/side | 450  450 | 506  506 | 562  562 | 620  620 | 675  675 | 731  731 | 788  788 |
| | Throw, ft. | 7-15  7-15 | 9-17  9-17 | 10-19  10-19 | 10-21  10-21 | 11-23  11-23 | 12-25  12-25 | 13-27  13-27 |

*Neck velocity fpm. Jet velocity will approximate neck velocity x 2.
†Total pressure (sum of velocity and static pressures) inches of water.

Reprinted with permission from Lima Register Co.

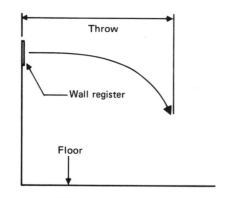

FIGURE 17-9 _____

Air throw from a wall register

**Solution**

Refer to the catalog page for ceiling diffusers (Table 17–1). The catalog page includes a register with a two-way pattern. The boldface numbers in the bottom section give the cfm for each register. The nearest cfm for this example is 252, provided by a 9 in. × 9 in. register. The register will deliver 126 cfm out of each side. The throw will be from 7 ft to 13 ft on each side. This register will deliver the wanted air in a way that will cover the room completely.

The designer may choose one or more registers for each area in a building, depending on the amount of air required. The choice of the number of registers is based on the amount of air and the air distribution pattern desired.

**Throw.**    An important consideration in selecting the registers is how far and in what direction the airflow will travel after it leaves the register. The distance the air travels before its velocity becomes too slow to be effective is called throw. Figure 17–9 is an illustration of throw.

The throw for a given register depends upon the velocity of the air at the register face. The velocity is a function of the quantity of air and the size of the register. The face area and the throw of the register are shown on the selection charts.

Some register catalogs show two face areas. One is the total area; it is based on the size of the opening in the register through which air can pass. The other area is the net face area; it is the total area less the area taken up by the vanes or any other obstructions in the face of the grille.

Registers must direct the air to where it is needed in the space. A terminal velocity of 50 fpm is considered the point at which throw is ended.

**Spread.**    Most registers have a grille, or curved vanes, on the face to spread the air as it leaves the register. The amount of this spread is important in determining the air pattern that will be achieved. An illustration showing air throw and spread from a floor register is provided in Figure 17–10. Spread is the width of the

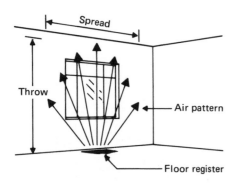

FIGURE 17-10 _____

Air throw and spread from a floor register

fan of air that comes out of a register. The perimeter of the pattern is considered to be where the air has dropped to a velocity of 50 fpm or less.

Good design requires that the conditioned air cover the areas of greatest heat loss, such as windows, doors, and cold outside walls. Therefore, the designer should select the registers with the proper spread to cover these areas with a blanket of air.

**Noise.**    The air passing through the face of any register goes past the vanes in the face that are used to direct it. These vanes are relatively small. If the quantity of air is too great, they vibrate, like the strings on a harp. If this vibration becomes too great, noise is produced. To control the amount of noise, the designer must select registers large enough so the velocity of air will not create vibrations.

Most register manufacturer's selection charts are marked in such a way that they indicate how much air can be delivered by each size of register without excess noise. Table 17–2 is a copy of a register manufacturer's catalog sheet for floor registers. The chart is separated into two sections. All register selections for any given cfm on the left side of the double vertical lines will be quiet in operation; those on the right side may be noisy.

# 17.3  RETURN AIR GRILLES

A **return air grille** has two parts: a decorative grille or face and the frame that holds it and connects it to the ductwork. The face may have curved or louvered vanes to make it more attractive or, more likely, to deflect the view through it. A typical return air grille is shown in Figure 17–11.

If the supply air system is properly designed and the supply registers are sized and located so they will do a good job of supplying the air to take care of the heat loss, then the function of the return air system is to provide a way for that air to get back to the air mover. The air mover, or blower, can deliver only as much air on the supply side as it takes in on the return side. Therefore the return air grilles have to be located so that the air, as supplied by the system, can find its way to them.

**Return air grille***

FIGURE 17-11
Typical return air grille

## TABLE 17–2

### Catalog Sheets for Registers Indicating Limits at Which They Will Operate Quietly

| NOMINAL SIZE | FREE AREA SQ. IN. | | 3045 | 4565 | 6090 | 7610 | 9515 | 11415 | 13320 | 15220 | 17125 | 19025 | 20930 | 22830 | 24735 | 26635 | 30440 | 34245 | 38050 | 45660 |
|---|---|---|---|---|---|---|---|---|---|---|---|---|---|---|---|---|---|---|---|---|
| | | Heating BTU/h | 3045 | 4565 | 6090 | 7610 | 9515 | 11415 | 13320 | 15220 | 17125 | 19025 | 20930 | 22830 | 24735 | 26635 | 30440 | 34245 | 38050 | 45660 |
| | | Cooling BTU/h | 855 | 1280 | 1710 | 2135 | 2670 | 3200 | 3735 | 4270 | 4805 | 5340 | 5870 | 6405 | 6940 | 7470 | 8540 | 9605 | 10675 | 12810 |
| | | C.F.M. | 40 | 60 | 80 | 100 | 125 | 150 | 175 | 200 | 225 | 250 | 275 | 300 | 325 | 350 | 400 | 450 | 500 | 600 |
| 2¼" x 10" | 19 | T.P. Loss | .013 | .022 | .035 | .045 | .060 | .092 | .120 | .150 | | | | | | | | | | |
| | | Vert. Throw (ft.) | 3.0 | 4.5 | 5.5 | 6.5 | 8.5 | 10.5 | 13.0 | 15.5 | | | | | | | | | | |
| | | Vert. Spread (ft.) | 6 | 8 | 10 | 12 | 15 | 18 | 23 | 26 | | | | | | | | | | |
| | | Face Velocity | 309 | 463 | 617 | 771 | 964 | 1159 | 1352 | 1545 | | | | | | | | | | |
| 2¼" x 12" | 21 | T.P. Loss | .009 | .015 | .027 | .037 | .050 | .080 | .105 | .134 | | | | | | | | | | |
| | | Vert. Throw (ft.) | 3 | 4 | 5 | 6 | 8 | 10 | 12 | 14 | | | | | | | | | | |
| | | Vert. Spread (ft.) | 6 | 8 | 10 | 11 | 14 | 17 | 22 | 25 | | | | | | | | | | |
| | | Face Velocity | 280 | 420 | 565 | 705 | 880 | 1050 | 1230 | 1400 | | | | | | | | | | |
| 2¼" x 14" | 24 | T.P. Loss | .006 | .010 | .021 | .031 | .042 | .070 | .093 | .121 | .150 | | | | | | | | | |
| | | Vert. Throw (ft.) | 3 | 4 | 4.5 | 5.5 | 8 | 9.5 | 11 | 12.5 | 14 | | | | | | | | | |
| | | Vert. Spread (ft.) | 6 | 8 | 9 | 11 | 14 | 16 | 19 | 22 | 25 | | | | | | | | | |
| | | Face Velocity | 245 | 365 | 490 | 610 | 760 | 915 | 1065 | 1220 | 1370 | | | | | | | | | |
| 4" x 10" | 32 | T.P. Loss | | .008 | .021 | .026 | .032 | .045 | .062 | .084 | .110 | .134 | .163 | | | | | | | |
| | | Vert. Throw (ft.) | | 3 | 4 | 5 | 7 | 8.5 | 10 | 11 | 12 | 13 | 14 | | | | | | | |
| | | Vert. Spread (ft.) | | 6 | 8 | 9 | 12 | 14 | 17 | 19 | 22 | 24 | 26 | | | | | | | |
| | | Face Velocity | | 265 | 355 | 445 | 555 | 665 | 775 | 890 | 1000 | 1120 | 1220 | | | | | | | |
| 4" x 12" | 39 | T.P. Loss | | .004 | .010 | .016 | .023 | .033 | .042 | .058 | .075 | .089 | .107 | .128 | .159 | | | | | |
| | | Vert. Throw (ft.) | | 2 | 3 | 4 | 6.5 | 8 | 9 | 10 | 11 | 12 | 13 | 14 | 15 | | | | | |
| | | Vert. Spread (ft.) | | '4 | 6 | 8 | 12 | 14 | 16 | 18 | 20 | 23 | 24 | 26 | 28 | | | | | |
| | | Face Velocity | | 220 | 295 | 370 | 460 | 555 | 645 | 735 | 830 | 925 | 1020 | 1110 | 1200 | | | | | |
| 4" x 14" | 46 | T.P. Loss | | | .006 | .010 | .016 | .021 | .028 | .039 | .051 | .060 | .080 | .101 | .124 | .137 | .167 | | | |
| | | Vert. Throw (ft.) | | | 3 | 4 | 6 | 7 | 8 | 9 | 10 | 11 | 12 | 13 | 14 | 15 | 16 | | | |
| | | Vert. Spread (ft.) | | | 6 | 8 | 11 | 12 | 14 | 16 | 18 | 20 | 22 | 24 | 26 | 28 | 30 | | | |
| | | Face Velocity | | | 255 | 320 | 395 | 475 | 555 | 635 | 715 | 790 | 870 | 950 | 1025 | 1110 | 1270 | | | |
| 6" x 10" | 52 | T.P. Loss | Series 40 Only | | | .009 | .014 | .019 | .027 | .035 | .044 | .054 | .064 | .078 | .090 | .104 | .135 | .171 | .210 | .314 |
| | | Vert. Throw (ft.) | | | | 4 | 5.5 | 6.5 | 7.5 | 8.5 | 10 | 11 | 12 | 13 | 14 | 15 | 17.5 | 19 | 21.5 | 26 |
| | | Vert. Spread (ft.) | | | | 8 | 9.5 | 11.5 | 13.5 | 15.5 | 17 | 19 | 21 | 23 | 25 | 27 | 31 | 34.5 | 38.5 | 46 |
| | | Face Velocity | | | | 278 | 348 | 417 | 487 | 556 | 626 | 695 | 765 | 834 | 904 | 973 | 1112 | 1251 | 1390 | 1668 |
| 6" x 12" | 59 | T.P. Loss | Series 40 Only | | | .007 | .011 | .015 | .021 | .027 | .035 | .042 | .050 | .060 | .071 | .083 | .107 | .134 | .165 | .242 |
| | | Vert. Throw (ft.) | | | | 4 | 5 | 6 | 7 | 8 | 9 | 10.5 | 11.5 | 12.5 | 13.5 | 14 | 16.5 | 18.5 | 20.5 | 29 |
| | | Vert. Spread (ft.) | | | | 7 | 9 | 11 | 12.5 | 14.5 | 16 | 18 | 20 | 21.5 | 23.5 | 25 | 29 | 32.5 | 36 | 43 |
| | | Face Velocity | | | | 245 | 307 | 368 | 429 | 491 | 552 | 613 | 675 | 736 | 797 | 859 | 981 | 1103 | 1227 | 1472 |
| 6" x 14" | 66 | T.P. Loss | Series 40 Only | | | .005 | .009 | .012 | .017 | .022 | .028 | .034 | .041 | .050 | .058 | .067 | .088 | .110 | .132 | .194 |
| | | Vert. Throw (ft.) | | | | 4 | 5 | 6 | 7 | 8 | 8.5 | 9.5 | 10.5 | 11.5 | 12 | 13 | 15 | 17 | 19 | 23 |
| | | Vert. Spread (ft.) | | | | 7 | 8.5 | 10 | 12 | 13.5 | 15 | 17 | 18.5 | 20.5 | 22 | 24 | 27 | 30.5 | 34 | 41 |
| | | Face Velocity | | | | 219 | 274 | 329 | 384 | 439 | 494 | 549 | 604 | 659 | 713 | 768 | 878 | 988 | 1098 | 1317 |

Reprinted with permission from Lima Register Co.

The return air grilles should be located as near to the center of the building as possible. This placement keeps the friction in the return air duct as small as possible. If a return grille is located in the center of a zone, care should be taken to make sure that all of the air from the supply registers can get back to it. If any of the supplies are located in rooms that may be cut off from the return grille location, then door grilles or wall grilles should be provided to allow that air to reach the location of the return grille.

Return air grilles are sized to handle the quantity

of air in the zone they will serve at a low enough velocity to be quiet in operation. The acceptable velocity for a variety of applications is shown in Table 17–3.

When more than one return air grille is used for a zone, each grille should be sized to handle the amount of air required for its part of the zone at a face velocity below that shown on the chart. Door grilles, **transfer grilles** (wall grilles), or any other return air grilles should be sized in the same way. The return air system must be sized to carry at least the amount of air that the supply side will furnish. If the air handler cannot get air back from the building, it will not be able to supply it.

## 17.4  OUTSIDE AIR AND EXHAUST GRILLES

Outside air and exhaust grilles, or louvers, should be sized in the same way as the other registers or grilles in a building. Each should be sized for the cfm required to flow through it and for a velocity of air that will be quiet and acceptable for each location.

Choosing the location of the fresh air and exhaust grilles should be carefully done. Fresh air grilles must

**Transfer grille**  A grille installed in an opening between two rooms that allows air to pass between the rooms

TABLE 17–3

Return Air Grille Selection Chart

| Product Number and Size | Free Area Sq. In. | Maximum Recommended C.F.M. | Product Number and Size | Free Area Sq. In. | Maximum Recommended C.F.M. | Product Number and Size | Free Area Sq. In. | Maximum Recommended C.F.M. | Product Number and Size | Free Area Sq. In. | Maximum Recommended C.F.M. |
|---|---|---|---|---|---|---|---|---|---|---|---|
| 18H- 6 x 4 | 16 | 65 | 18H-14 x 10 | 74 | 390 | 18H-18 x 18 | 216 | 900 | 18H-24 x 16 | 256 | 1060 |
| 18H- 8 x 4 | 21 | 89 | 18H-14 x 12 | 113 | 468 | 18H-20 x 6 | 77 | 318 | 18H-24 x 18 | 288 | 1200 |
| 18H- 8 x 6 | 32 | 135 | 18H-14 x 14 | 132 | 545 | 18H-20 x 8 | 104 | 432 | 18H-24 x 20 | 320 | 1330 |
| 18H- 8 x 8 | 43 | 180 | 18H-16 x 4 | 43 | 180 | 18H-20 x 10 | 130 | 540 | 18H-24 x 24 | 384 | 1600 |
| 18H-10 x 4 | 26 | 107 | 18H-16 x 6 | 65 | 270 | 18H-20 x 12 | 156 | 648 | 18H-30 x 4 | 78 | 324 |
| 18H-10 x 6 | 39 | 162 | 18H-16 x 8 | 87 | 360 | 18H-20 x 14 | 182 | 755 | 18H-30 x 6 | 118 | 486 |
| 18H-10 x 8 | 52 | 216 | 18H-16 x 10 | 109 | 450 | 18H-20 x 16 | 209 | 870 | 18H-30 x 8 | 158 | 654 |
| 18H-10 x 10 | 65 | 270 | 18H-16 x 12 | 130 | 540 | 18H-20 x 18 | 235 | 977 | 18H-30 x 10 | 199 | 828 |
| 18H-12 x 4 | 31 | 130 | 18H-16 x 14 | 152 | 630 | 18H-20 x 20 | 262 | 1090 | 18H-30 x 12 | 239 | 990 |
| 18H-12 x 6 | 47 | 192 | 18H-16 x 16 | 174 | 725 | 18H-24 x 4 | 63 | 258 | 18H-30 x 14 | 279 | 1159 |
| 18H-12 x 8 | 63 | 258 | 18H-18 x 6 | 71 | 294 | 18H-24 x 6 | 95 | 390 | 18H-30 x 16 | 319 | 1325 |
| 18H-12 x 10 | 79 | 324 | 18H-18 x 8 | 95 | 396 | 18H-24 x 8 | 127 | 528 | 18H-30 x 18 | 360 | 1500 |
| 18H-12 x 12 | 96 | 396 | 18H-18 x 10 | 119 | 491 | 18H-24 x 10 | 159 | 660 | 18H-30 x 20 | 399 | 1660 |
| 18H-14 x 4 | 37 | 150 | 18H-18 x 12 | 144 | 600 | 18H-24 x 12 | 191 | 792 | 18H-30 x 24 | 479 | 1990 |
| 18H-14 x 6 | 56 | 228 | 18H-18 x 14 | 167 | 696 | 18H-24 x 14 | 224 | 930 | 18H-30 x 30 | 600 | 2500 |
| 18H-14 x 8 | 75 | 312 | 18H-18 x 16 | 192 | 797 | | | | | | |

Reprinted with permission from Lima Register Co.

be located so contaminated air from chimneys, flues, or vents will not be drawn into the building. Exhaust grilles must be located so exhaust air will not recycle into the fresh air grilles or be blown into any occupied space.

## 17.5   REGISTER PLANS

Layouts of the distribution and return air system should show the register sizes. The type of register to be used at each location, the cfm to be delivered by the register, and the size of the register should be lettered in by the register symbol or indicated on a separate schedule. These data will be used in figuring the cost of the job, in laying it out, and also in balancing the system for proper airflow.

## 17.6   SUMMARY

The supply registers in a warm air heating system introduce the heated air into the space to be heated, control the quantity of heated air, and provide the proper distribution of heated air into the space to be heated. The selection, location, and size of the registers is of paramount importance in the design of a good heating system. The return air grilles are used to collect the air from the building and return it to the blower so a continuous supply of air can be provided.

Supply registers come in three main types: ceiling, wall, and perimeter. Ceiling registers are used when having the air distributed through the center area of the room is desirable. Wall registers are used when the air can be distributed from one or more sides of a room and still cover the entire space. Perimeter registers are used to blanket the outer surfaces of a building with warm air to offset the greater heat losses that occur there.

Registers are also designed to spread the air over a wide area. The throw and spread of that area is considered when a selection is made of a register for a particular application.

Supply registers are sized to supply the amount of air required at a given location at a high enough velocity to direct it where it is needed. The selection should also be made so that the velocity will not be so great that air noise at the register face becomes a problem.

Return air grilles are located in a building so they return all of the air that the supply side distributes. A supply blower will put out only as much air as it can get back from the return side of the system. The return air grilles must be located so that air from each room or space with a supply register can get to them. They are sized for quantity of air and velocity at the face. The face velocity must be low enough so the grilles are quiet.

Outside air or exhaust air grilles, or louvers, are located and sized in the same way as supply registers and return air grilles.

# 17.7  QUESTIONS

1. Registers are selected, located, and sized to provide the correct amount of air for each room as determined by the _____ _____ of the room.

2. A supply register includes a _____, a frame, and sometimes a _____ to control the quantity of air.

3. The three basic types or styles of registers, as determined by physical location, are: (a) _____, (b) _____, and (c) _____.

4. Ceiling registers are a good choice for a very large room. True or false?

5. Ceiling registers usually use a _____ for the face so the air will be spread horizontally to some extent.

6. Low sidewall registers work well when they blow the air into the occupied space at high velocities. True or false?

7. The most commonly used register for residential and small commercial buildings is the _____ register.

8. Perimeter distribution systems are designed to cover the _____ surfaces of a building because that is where the greatest _____ _____ occurs.

9. The four major factors to consider when selecting supply registers are: (a) _____, (b) _____, (c) _____, and (d) _____.

10. The distance air travels at right angles to the face of a register before it slows to 50 fpm is the _____.

11. The effective width of the envelope of air out of a register is the _____.

12. When the sound produced by air coming through a register face is loud enough to be unpleasant, it is called _____.

13. A return air grille is used to draw the air where it is wanted. True or false?

14. An air mover in a heating system always puts out a little more air than it gets back from the return air system. True or false?

## 18.1 INTRODUCTION

The installing contractor often is unable to install a warm air heating system in exactly the way that the designer draws it up on the mechanical plan. Unforeseen complications, such as changes in construction or conflicts with plumbing or electric systems, can require the ductwork to be changed. When such changes are made in the system, there will be more or less friction than was originally planned. Because the duct has been sized to provide a specific amount of air into each room or zone based on the predicted friction in the system, these changes will alter the actual amount of air flowing. Adjustments must be made to the system to compensate for these changes. Also, since the blower and the registers are not factory adjusted for a specific amount of air, they must be adjusted on the job.

An air balance for a warm air heating system includes adjusting the blower so it puts out the total air required for the system, making sure all branches in the system have the proper amount of air flowing in them, and adjusting each register so that it has the amount of air flowing out of it that was calculated when the heat load was converted to cfm. Each of these operations is explained and described in this chapter.

## 18.2 TERMS RELATED TO AIRFLOW

Some specific terms related to airflow, blowers, and duct systems are important to an understanding of airflow and measurement of air in a duct system.

### Friction

As air flows through a duct, there is always friction between the walls of the duct and the film of air in contact with them. See Figure 18–1.

There is also resistance to the airflow in any fitting in the system that requires the air to change speed or direction, such as **take-offs, reducers,** or elbows. All duct resistance to airflow can be termed **friction.** Friction can be measured by the effect on the static pressure in the system.

*Term defined in text.

# CHAPTER
# 18
# Air Balance

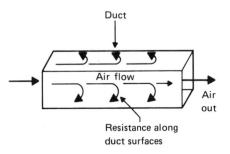

FIGURE 18-1 _____

Resistance caused by friction between the air flowing in a duct and the walls of the duct

**Take-off**　A duct fitting designed to allow air to flow from a trunk to a branch

**Reducer**　A section of air duct where the size of the duct becomes smaller

**Friction***

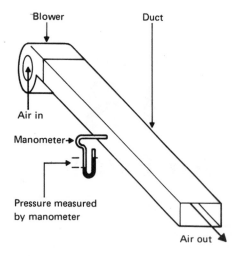

FIGURE 18-2 _____
Measuring static pressure

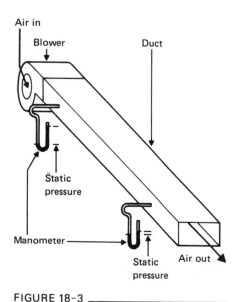

FIGURE 18-3 _____
Measuring static pressure drop

## Static Pressure

Static pressure (sp) describes the amount of air pressure in a duct as measured at right angles to the flow of the air. It can be defined as the pressure that would have a tendency to burst the duct. Figure 18–2 illustrates how static pressure is measured.

## Static Pressure Drop

The difference between the static pressure at two different points in a duct, or between the static pressure in the duct and the pressure in a room, is called the static pressure drop. The resistance in a duct causes a static pressure drop to occur as the air passes along the duct (disregarding changes in velocity), so static pressure drop can be used to measure friction. Figure 18–3 demonstrates how static pressure drop is measured.

Static pressure drop can be represented by the following formula:

$$SPD = P_1 - P_2$$

where:

$SPD$ = static pressure drop

$P_1$ = static pressure at one point in a duct

$P_2$ = the static pressure at a point downstream from $P_1$

## EXAMPLE 18–1

The static pressure read at the blower discharge on a typical duct system is found to be 0.06 in. wg. This pressure is compared to the pressure at the discharge of the duct, which will be 0.00 in. wg (room pressure). Find the static pressure drop in that section of duct.

### Solution

Use the following formula:

$$SPD = P_1 - P_2$$

where:

$SPD$ = static pressure drop in the duct

$P_1$ = pressure at blower discharge (in this example, 0.06 in. wg)

$P_2$ = pressure in the room at duct discharge (in this example, 0.00 in. wg)

Therefore:

$$SPD = 0.06 - 0.00 = 0.06$$

The static pressure drop in the duct is 0.06 in. wg. This measurement indicates the amount of friction in the duct.

## Manometer or Water Gauge

The static pressure in a duct is measured with a manometer, a device that measures air pressure. Basically, a manometer is a U-shaped tube containing water or some other liquid. If a fluid other than water is used, the scale is calibrated to give a reading corresponding to one for water. See Figure 18–4 for a diagram representing a manometer.

When a manometer is used, one end of a rubber tube is connected to either one of the upright legs of the U. The other end of the tube is connected to a hole in the side of the duct. The pressure exerted by the air against the inside wall of the duct is transferred through the tube and causes the water in the one leg of the U to be depressed. A corresponding rise will be observed in the water in the other leg.

The difference between the level of the water in the two legs of the U is an indication of the pressure exerted by the air in the duct. The pressure is measured in inches and so is called the static pressure in inches of water (in. wg). See Figure 18–5.

## Measuring Static Pressure to Determine Friction

To measure the static pressure drop in a duct, the water gauge pressure is read at each end of the duct. The difference between the two readings indicates the amount of friction in the duct. The pressure reading in the upstream part of the system must be higher than in the downstream part or the air would not move. The pressure is read by placing the tube at right angles to the flow of air and flush with the inside of the duct so that only the pressure exerted against the side of the duct is read.

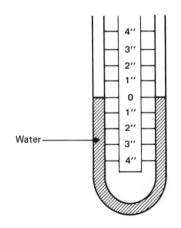

FIGURE 18-4
U-tube manometer

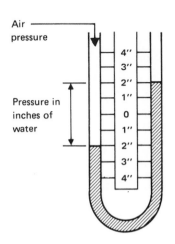

FIGURE 18-5
Using a manometer to measure air pressure

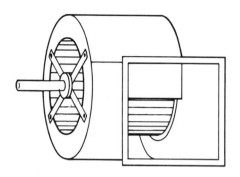

FIGURE 18-6 _____

Typical forward-curved, belt-drive blower

## 18.3   CHECKING THE QUANTITY OF AIR IN THE SYSTEM

The first step in performing an air balance is to check the cfm output of the blower and adjust the output as necessary to deliver the amount of air required for the system. This operation should always be done with all balancing dampers and register dampers in the system set wide open.

When selecting the heating unit for a job, the system designer chooses a unit with a blower with the capacity to provide the amount of air needed for the job. Blower characteristics are such that any given blower will deliver air over quite a wide range of cfm and for different static pressure drops. When starting an air balance, the installer must determine the actual cfm output for the conditions on the particular system being balanced.

The air moving capacity of the blower used in a typical low pressure system is a function of the size of the blower wheel and of the speed of rotation of the wheel. The blowers most often used for heating systems are double-inlet, forward-curved type. Double inlet means that air will enter the blower from both sides. Forward curved relates to the angle at which the blades of the wheel are inclined to the direction of the air discharge. A typical blower is shown in Figure 18-6.

Manufacturers provide data sheets with their equipment that include blower performance charts. These charts give the output of the blowers in cfm, for different static pressure drops, and blower revolutions per minute (rpm). These data are used to make the original selection of blower for the job. A copy of one such blower table is shown in Table 18-1.

There are two common methods of determining the cfm output of the blower. One is to measure the temperature rise of the air through the unit and then use the sensible heat formula to find the cfm mathematically. The other is to find the rpm of the blower, measure the static pressure drop in the duct system, and then find the cfm from the blower table. Both of these methods are explained in the following sections.

## Using the Temperature Rise Method

To determine the cfm output of a blower with the temperature rise method, find the Btu/h output of the unit

TABLE 18-1
Blower Data from a Furnace Manufacturer's Catalog

## DRIVE KITS

| Usage | Drive Kit Model No. | Motor hp | Motor Pulley (in.) & Groove | **Blower Pulley (in.) & Groove | *Rpm Range | Belt | Shipping Weight (lbs.) 1 Package |
|---|---|---|---|---|---|---|---|
| Heating | DK 1/3-5 (BM-5790) | 1/3 | 1/2 x 3-3/4 — OA | 1 x 7 — A | 590 — 835 | 4L410 | 17 |
| 3-1/2 & 4 tons | DK 1/2-6 (BM-5791) | 1/2 | 5/8 x 4-1/8 — OA | 1 x 7 — A | 690 — 935 | 4L420 | 22 |
| | DK 3/4-7 (BM-5792) | 3/4 | 5/8 x 4-1/8 — OA | 1 x 7 — A | 690 — 935 | 4L420 | 27 |
| 5 tons | DK-2007 (BM-7523) | †1 | 5/8 x 4-3/4 — OA | 1 x 7 — A | 840 — 1085 | 4L430 | 38 |

*At 1725 rpm motor speed.
†Fan relay included in kit.
**Factory installed in furnace package and not included in drive kit.

## G11E-200V BLOWER PERFORMANCE

STATIC PRESSURE EXTERNAL TO UNIT (Inches Water Gauge)

| Air Volume (Cfm) | 0 | | .10 | | .20 | | .30 | | .40 | | .50 | | .60 | | .70 | | .80 | | .90 | | 1.00 | |
|---|---|---|---|---|---|---|---|---|---|---|---|---|---|---|---|---|---|---|---|---|---|---|
| | RPM | BHP | RPM | BHP | RPM | BHP | RPM | BHP | RPM | BHP | RPM | BHP | RPM | BHP | RPM | BHP | RPM | BHP | RPM | BHP | RPM | BHP |
| 1400 | 460 | .23 | 530 | .29 | 600 | .35 | 665 | .41 | 725 | .46 | 775 | .51 | 830 | .57 | 880 | .66 | 920 | .72 | 965 | .79 | 1005 | .86 |
| 1600 | 520 | .31 | 590 | .38 | 650 | .43 | 715 | .49 | 765 | .56 | 815 | .62 | 865 | .71 | 910 | .79 | 955 | .86 | 995 | .93 | 1035 | 1.0 |
| 1800 | 580 | .40 | 650 | .46 | 705 | .53 | 760 | .61 | 810 | .69 | 860 | .78 | 905 | .86 | 950 | .94 | 990 | 1.02 | 1030 | 1.09 | --- | --- |
| 2000 | 650 | .51 | 705 | .58 | 760 | .67 | 815 | .77 | 860 | .86 | 905 | .95 | 950 | 1.03 | 990 | 1.10 | --- | --- | --- | --- | --- | --- |
| 2200 | 715 | .65 | 765 | .74 | 820 | .86 | 870 | .96 | 910 | 1.04 | 950 | 1.12 | --- | --- | --- | --- | --- | --- | --- | --- | --- | --- |
| 2400 | 780 | .85 | 825 | .95 | 875 | 1.07 | 915 | 1.14 | --- | --- | --- | --- | --- | --- | --- | --- | --- | --- | --- | --- | --- | --- |
| 2600 | 840 | 1.02 | 885 | 1.13 | --- | --- | --- | --- | --- | --- | --- | --- | --- | --- | --- | --- | --- | --- | --- | --- | --- | --- |

NOTE — All cfm is measured external to the furnace with the air filter in place.

Reprinted with permission from Lennox Industries Inc.

**Tachometer**  A meter for reading the speed of rotation of a moving wheel

by reference to the manufacturer's data (the output can be doublechecked by finding the fuel use rate and calculating the actual output). Then find the temperature rise through the unit by taking the dry bulb temperature of the air entering the furnace and the dry bulb temperature of the air leaving the furnace, find the difference between the two, and use the sensible heat formula, as follows:

$$\text{cfm} = \frac{\text{Btu/h}}{(T_O - T_I) \times 1.08}$$

where:

cfm = quantity of air in cu ft per min

Btu/h = heat content of the air in Btu per hr

$T_O$ = temperature of the air at the outlet of the furnace

$T_I$ = temperature of the air at the inlet of the furnace

1.08 = constant that includes the sensible heat of air, the specific weight, and 60 minutes in an hour

## EXAMPLE 18–2

If the Btu/h output of a furnace is found to be 200,000 Btu/h, the return air temperature is 76°F and the supply plenum air temperature is 154°F, find the cfm.

**Solution**

$$\text{cfm} = \frac{\text{Btu/h}}{(T_O - T_I) \times 1.08}$$

$$= \frac{200,000}{(154 - 76) \times 1.08}$$

$$= 2374$$

The airflow through the furnace is 2374 cfm.

## Using the Blower rpm Method

The second method of checking the cfm output of a heating unit is by determining the blower rpm, measuring the static pressure in the system, and then using the manufacturers' blower tables to find the cfm. The rpm of the blower is most easily found by using a **tach-**

**ometer.** This device gives the rpm of the motor directly. If a tachometer is not available, the rpm can be determined for direct-drive motors by reading the rpm data on the motor nameplate. If the blower is a belt-drive blower, the rpm is a function of the motor speed and the motor and blower pulley sizes. Figure 18–7 shows a furnace with a belt-drive blower.

To find the rpm of the blower from the motor rpm and pulley sizes, use the following formula:

$$\text{rpm}_b = \text{rpm}_m \times \frac{pd_m}{pd_b}$$

where:
    $\text{rpm}_b$ = rpm of the blower
    $\text{rpm}_m$ = rpm of the motor
    $pd_m$ = diameter of the motor pulley
    $pd_b$ = diameter of the blower pulley

The motor rpm can be found by reference to the motor nameplate, and the pulley sizes can be measured. The diameters used should be the effective diameter across the belt where it fits on the pulley.

## EXAMPLE 18–3

In a particular application, the motor speed on a belt-drive blower is determined to be 1725 rpm, the pulley diameter on the motor is 3-1/2 in., and the blower pulley diameter is 6 in. Find the blower rpm.

**Solution**

$$\text{rpm}_b = \text{rpm}_m \times \frac{pd_m}{pd_b}$$

$$= 1725 \times \frac{3.5}{6}$$

$$= 1006.2$$

The rpm of the blower in this application is 1006.2.

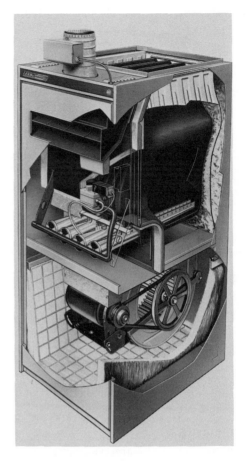

FIGURE 18-7 _____
Furnace with a belt-drive blower

After the rpm of the blower is determined, the total static pressure drop in the system is found by reading the static pressure at the blower inlet and also at the blower outlet. The pressure reading at the inlet will be a negative number since this is the static pressure drop

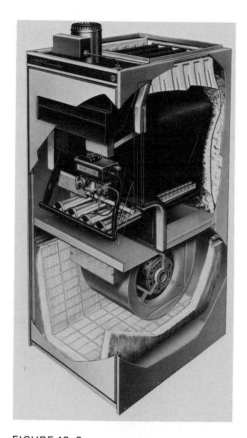

FIGURE 18-8

Furnace with a direct-drive blower

in the return air side of the system. The reading taken on the outlet side will be positive since this is on the supply side of the system. To find the total static pressure drop, these two figures should be added together mathematically *without regard to the signs.* The total is the total static pressure drop in the system. Reference can then be made to the manufacturers' blower charts, and the cfm can be found by comparing the rpm of the blower to the static pressure drop in the system.

## EXAMPLE 18–4

Find the static pressure drop for a blower with an rpm of 1006.2. The static pressure in the supply side of a system is 0.25 in. wg by measurement. The pressure in the return side is −0.15 in. wg.

### Solution

Add the two water pressures together without regard to the signs:

$$0.25 + 0.15 = 0.40$$

The total static pressure drop is 0.40 in. wg. This static pressure is used with the rpm to find the cfm from the blower performance chart.

## 18.4   ADJUSTING THE BLOWER FOR PROPER CFM

If when the blower output is checked, the total cfm is not correct for the system, adjustments must be made to get the correct amount. The blower speed will need to be changed.

The characteristics of the blowers used in most heating units are such that the quantity of air moved, in cfm, is directly related to the rpm of the blower wheel. To increase the volume of air being delivered, the speed of rotation of the wheel is increased. To decrease the volume of air, the speed is decreased. Because the volume of air is directly related to the speed of rotation, the relationship can be shown mathematically, as follows:

$$\frac{cfm_1}{cfm_2} = \frac{rpm_1}{rpm_2}$$

where:

cfm$_1$ = original blower output
cfm$_2$ = new blower output
rpm$_1$ = original blower speed
rpm$_2$ = new blower speed

The desired blower speed for the corrected cfm can be taken directly from the manufacturer's blower performance chart. The blower rpm is then set at this speed.

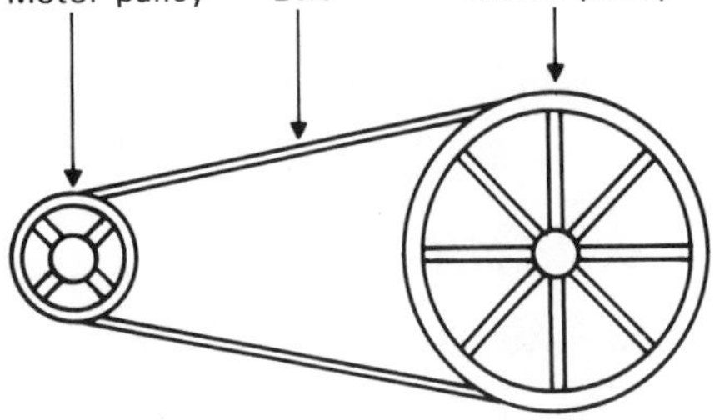

FIGURE 18-9

Typical pulley-belt arrangement

## Changing cfm on a Direct-Drive Blower

If a blower is direct drive—that is, the blower wheel is mounted directly in the motor shaft, as seen in Figure 18–8—the blower speed is the same as the motor speed. The speed of the motor can be changed by rearranging the electrical leads to the motor. The speeds related to the different electrical connections are shown in the manufacturer's data sheets and usually on the motor nameplate. Since the speeds are constant at each different step, the closest speed to give the desired cfm must be used.

## Changing cfm on a Belt-Drive Blower

To change the cfm on a unit with a belt-drive blower, the rpm of the blower is changed by changing pulley sizes. The blower pulley diameter is usually two to three times the diameter of the motor pulley. Figure 18–9 shows a typical pulley-belt arrangement. Because of the size difference, the blower turns more slowly than does the motor. When the speed of the blower must be changed, the size of the motor pulley is changed.

Sometimes adjustable motor pulleys are used. When

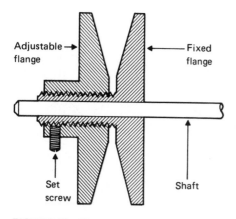

Adjustable → flange

Fixed flange

Set screw

Shaft

FIGURE 18-10 _____

Adjustable pulley used for changing the speed of rotation of a blower wheel

this is the case, the speed of the blower can be changed by changing the diameter of the adjustable pulley. To make this change, a set screw on the pulley is loosened, and the outer part of the pulley is turned in or out on a threaded hub. See Figure 18–10. Turning the outer part of the pulley moves it in closer to the inner part or moves it farther away. The effective diameter of the pulley is changed by allowing the belt to ride closer in to the hub or causing it to ride farther out. When more adjustment is needed than the adjustable pulley will give, a motor pulley of a different size must be chosen.

The speed of rotation of the blower wheel is directly related to the speed of rotation of the motor that drives it, but since the two are connected by the belt that runs between the two pulleys, the speed of the blower is indirectly proportional to the diameter of the two pulleys. Mathematically stated, this is:

$$\frac{rpm_m}{rpm_b} = \frac{pd_b}{pd_m}$$

where:

   $rpm_m$ = revolutions per minute of the motor
   $rpm_b$ = revolutions per minute of the blower
   $pd_b$ = diameter of the blower pulley
   $pd_m$ = diameter of the motor pulley

This formula is the same as that used to find the original speed of the blower. Here, the elements are transposed to show the rpm as a ratio of the inverted pulley diameters.

Since the new rpm has already been determined from the blower performance chart, the only calculation necessary is to find the new motor pulley size for the required rpm. To do this operation, the formula is transposed to solve for $pd_m$, as follows:

$$pd_m = pd_b\left(\frac{rpm_b}{rpm_m}\right)$$

where:

   $pd_m$ = diameter of the motor pulley
   $pd_b$ = diameter of the blower pulley
   $rpm_b$ = revolutions per minute of the blower
   $rpm_m$ = revolutions per minute of the motor

## EXAMPLE 18–5

In Example 18–3, the following data were determined:

$pd_m = 3.5$

$pd_b = 6$

$rpm_m = 1725$

$rpm_b = 1006.2$

Use these data as a starting point. If a new blower speed of 1260 rpm was required, as found from a blower performance chart, find the proper motor pulley to give this speed.

**Solution**

$$pd_m = pd_b\left(\frac{rpm_b}{rpm_m}\right)$$

$$= 6\left(\frac{1260}{1725}\right)$$

$$= 4.38$$

The motor pulley diameter should be changed to 4.38 in. to provide 1260 rpm, which will then increase the cfm to that required.

This change will also make a change in the static pressure drop in the system. If the change is not too great, then the blower performance chart will still be accurate enough for proper operation of the system.

After the motor pulley size is changed, the cfm should be checked again by the temperature rise method to make sure the cfm is acceptable.

# 18.5  BALANCING SYSTEM BRANCHES

Most small duct system layouts are quite simple. The system has a main trunk and individual branches to each register. The individual branches are sized to carry

**Pitot tube**   A tube used in measuring airflow that gives a total pressure reading and a static pressure reading simultaneously

**Inclined manometer\***

**Velocity pressure**   The pressure exerted by a moving column of air in a direct line of flow

**Total pressure**   The pressure exerted by a moving column of air in a direct line of flow (Total pressure includes both velocity pressure and static pressure.)

enough air to supply only one register. Balancing is done at the register damper. In larger systems and in better quality small systems, balancing dampers are installed in the branches.

If the branch serves only one register, then the balancing damper is used in the same way as a register damper. Many larger systems have multiple register takeoffs from each branch; many such branches may take off from the trunk. In such systems, the airflow in each branch must be regulated if the entire system is to remain in balance. The balancing dampers are found in the branch ducts just after the takeoff from the trunk.

## Measuring Airflow in Branches

All of the balancing dampers in the branches should have been open when the blower was adjusted for design airflow. It is now necessary to take each branch duct in turn and measure the airflow in it and then adjust the dampers for the correct flow according to the design. Various instruments are used in this operation.

**Measuring Instruments.**   The most common method of measuring airflow in a duct is by use of a manometer and a special tube, called a **Pitot tube,** that is used to measure the velocity pressure of the air in the duct. The manometer used is the same as described earlier in this chapter, only the bottom part of the U is elongated and sloped. This shape allows the vertical height reading of the water to be spread over a longer distance, and the scale for reading it is larger and more accurate. This type of a manometer is called an **inclined manometer.** See Figure 18–11.

The special tube used with the manometer for measuring velocity is the Pitot tube. The Pitot tube gives pressure reading related to the velocity of the air in the duct. The reading obtained is called the **velocity pressure** ($P_v$) reading. We have already discussed static pressure ($P_s$). Static pressure plus velocity pressure equals **total pressure** ($P_T$). Velocity pressure is that pressure exerted by the air in a direct line with the airflow. Because of the nature of air pressure, velocity pressure cannot be measured directly. But it is possible to measure static pressure and total pressure. Since velocity

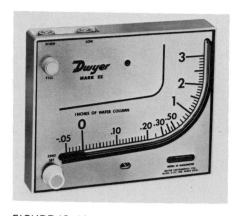

FIGURE 18–11 _____
Inclined manometer

pressure is the difference between these two pressures, it can be found from them by the use of this formula:

$$P_v = P_T - P_S$$

where:

$P_v$ = velocity pressure
$P_T$ = total pressure
$P_S$ = static pressure

A Pitot tube does this operation automatically.

A Pitot tube is constructed so that one tube is inside another. The inner tube extends to the end that is aimed into the airstream and is open on the end. See Figure 18–12 for a diagram of a Pitot tube. The diagram shows the construction, location of the pressure taps, and points where each pressure is read. This tube measures total pressure because it reads both velocity pressure and static pressure simultaneously. The outer tube is closed at the end but has several holes around its outside perimeter so that it reads only static pressure.

The two tubes from the Pitot tube are connected to the manometer. See Figure 18–13. When they are connected in this way, the pressure exerted in the tubes opposes each other. The result is the velocity pressure reading, shown directly on the manometer scale. The velocity pressure can then be used to find the actual velocity by formula.

**Calculating the cfm.**    Three steps are required to find the quantity of air flowing in a duct. First, the cross-sectional area of the duct must be found in sq ft. Second, velocity pressure is converted to velocity in fpm. Third, the area and the velocity are used to find the cfm.

In the first step, the area of the duct is found by multiplying the width of the duct in in. times the height in in., and dividing the results by 144 inches (number of in. in 1 sq ft):

$$A = \frac{w \times h}{144}$$

The second step is to use the following formula to find the velocity of air in the duct from the velocity pressure:

$$\text{fpm} = 4005\sqrt{P_v}$$

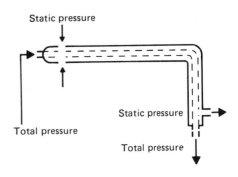

FIGURE 18-12
Diagram of a Pitot tube

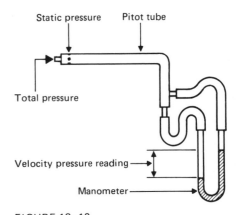

FIGURE 18-13
Tubes of the Pitot tube connected to the manometer in order to get velocity pressure reading

where:

fpm = velocity of the air in ft per min
4005 = constant for standard air
$P_v$ = velocity pressure from manometer reading

The third step is to convert the velocity to cfm using this formula:

$$Q = Av$$

where:

Q = quantity of air in cfm
A = area of the duct in sq ft
v = velocity of the air in fpm

## EXAMPLE 18–6

A 24 in. × 10 in. duct provides air to a small zone in a building. Upon measurement by manometer and Pitot tube, the velocity pressure $(P_v)$ is found to be 0.048 in. wg. Determine the cfm.

**Solution**

First, find the cross-sectional area of the duct:

$$A = \frac{w \times h}{144}$$

$$= \frac{24 \times 10}{144}$$

$$= 1.67$$

The cross-sectional area of the duct is 1.67 sq ft.
  Second, find the velocity of air in the duct:

$$fpm = 4005\sqrt{P_v}$$

$$= 4005\sqrt{0.048}$$

$$= 877.5$$

The velocity of air in the duct is 877.5 fpm.
  Third, find the cfm (The area (A) of the duct must be in sq ft.):

$$Q = Av$$

$$= 1.67 \times 877.5$$

$$= 1465$$

The quantity of air actually flowing in the duct is 1465 cfm.

# Adjusting the cfm

If the existing cfm calculated for a branch is different from the design requirements, the adjusting damper needs to be changed to provide the correct amount. Finding the proper position may be a matter of some experimentation, but by checking the velocity pressure as the damper is changed, the correct cfm can be obtained.

## EXAMPLE 18–7

Assume in Example 18–6 that the required cfm is 1330. The 1465 found in the example would be too much air. The damper should be closed somewhat to reduce the flow.

### Solution

To simplify this procedure, first find the desired $P_v$ for the 1330 cfm using the formula:

$$P_{v1} = P_{v2}\left(\frac{\text{cfm}_1}{\text{cfm}_2}\right)$$

where:
  $P_{v1}$ = velocity pressure desired for new cfm
  $P_{v2}$ = velocity pressure for existing cfm
  $\text{cfm}_1$ = desired cfm
  $\text{cfm}_2$ = existing cfm

Therefore:

$$P_{v1} = 0.048\left(\frac{1330}{1465}\right)$$

$$= 0.044$$

The desired velocity pressure is 0.044 in. wg. The damper should be regulated until that reading is found on the manometer.

# 18.6 BALANCING THE SUPPLY REGISTERS

After the blower has been adjusted so that the correct amount of air is provided to the system and the branch adjustments are made, the next adjustment is that of the individual supply registers. Each register should be

**Anemometer***

**Velometer***

set so that it distributes the proper amount of air to the room or space that it serves. This balancing is done by adjusting the balancing dampers in the branch ducts or in the registers themselves.

The airflow out of each supply register can be measured with any one of several instruments especially designed for this purpose. One is a special fitting that is used with a manometer. The fitting goes over the register face, and the manometer gives the velocity of the air in fpm. A second type is an **anemometer.** This device measures airflow by measuring the effect of the air on an electronic sensing device. Another type is a **velometer,** which reads the velocity of the air directly.

## Instruments for Measuring Air at Supply Register Faces

When a manometer is used for measuring air quantity, a special fitting is used on one of the tubes. This fitting is placed over the register face so that the air passes through it. The fitting converts velocity pressure of the air from the register to static pressure. The manometer is calibrated to read in fpm.

An anemometer is an electronic device in which an electric current flows through a thermistor or similar sensing element. This sensing element is held in the airstream. The flow of air across the sensor affects the electric current. The effect is measured by a meter, calibrated to give a reading in fpm or cfm. Figure 18–14 shows a technician reading cfm from a register with an anemometer.

A velometer has a propellor rotated by a flow of air or a vane deflected by the flow. The propellor or vane is connected to a needle through a series of connecting links. The needle shows the velocity of the air on a calibrated scale. The velometer gives the velocity in fpm. Figure 18–15 shows a typical velometer.

One other meter used for checking cfm out of a register is shown in Figure 18–16. This device is a deflecting vane type of meter, but a scale on the indicating dial makes it possible to dial in the register size and read the cfm directly. The meter is used by placing it in the front of the register, and the air from the register flows through the meter.

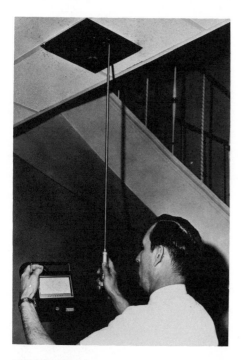

FIGURE 18-14

Using an anemometer to determine cfm from a register

# Finding the cfm

After the velocity of the air at a register face has been found, the quantity of air is calculated by using the velocity and face area of the register in the following formula:

$$Q = Av$$

where:

$Q$ = quantity of air in cfm
$A$ = area of the net face opening in sq ft
$v$ = velocity of air in fpm

The free area of the register opening is given in the manufacturer's catalog. The balancer should have access to the proper performance chart for each type of register that is balanced.

FIGURE 18-15 _____
Deflecting vane velometer used to find fpm at a register face or in any airstream

## EXAMPLE 18–8

Find the cfm through a 2 in. × 14 in. floor register with a velocity found in the manufacturer's catalog to be 760 fpm. The face area of the register is 23.6 sq in.

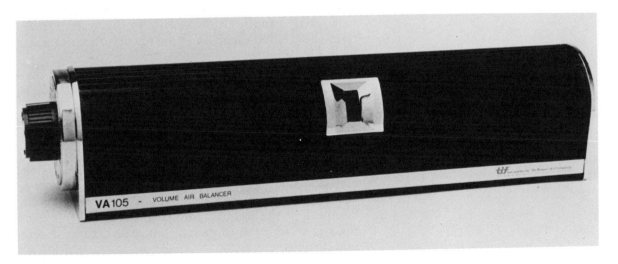

FIGURE 18-16 _____
Velometer used over the face of a register to find the cfm flowing in a direct reading

**Solution**

The area in sq in. needs to be changed to sq ft:

$$A = \frac{\text{sq in.}}{144}$$

$$= \frac{23.6}{144}$$

$$= 0.1639$$

The face area of the register in sq ft is 0.1639. This figure, with the velocity 760 fpm, is used in the velocity formula:

$$Q = Av$$

$$= 0.1639 \times 760$$

$$= 124.56$$

The volume of air out of the register is 124.56 cfm.

## 18.7  SUMMARY

After an air distributing system has been designed and installed, that system must be balanced for proper operation. An air balance is the process of adjusting the various air handling components in the system to deliver the desired quantity of air to each room or space in the building.

The three main components of the system that need checking and adjusting are the blower, the branch ducts, and the registers.

To start an air balance, all dampers in the system should be opened to ensure unrestricted flow of air. Then the blower is adjusted so it will provide the amount of air required as determined when the system was designed. Next, the airflow is adjusted in each of the branch ducts in the system using the balancing dampers in the ducts. Third, the airflow out of each register is adjusted to match that required in the original design.

# 18.8   QUESTIONS

1. Three separate operations must be performed in providing an air balance. They are: (a) _____, (b) _____, and (c) _____.

2. The total amount of air put out by the blower must equal the sum of the cfm that is circulated to each _____ or _____.

3. The flow of air in the branch ducts is regulated by balancing _____ in the duct.

4. Friction in a duct system can be said to be equivalent to _____.

5. The difference between the static pressure at two points in a duct is called _____.

6. A manometer is used to measure the air pressure in a duct system. True or false?

7. The air moving capacity of the type of blower used in most warm air heating applications is a function of the _____ of the blower wheel and the _____ of rotation.

8. If a furnace with a heating capacity of 96,000 Btu/h is operating with a temperature rise of 87°F through it, how many cfm of air is it handling?

9. If a belt-drive blower in a furnace is driven by a motor that is rated at 1725 rpm and has a 3 in. diameter motor pulley and a 7 in. diameter, blower pulley, what is the rpm of the blower wheel?

10. A static pressure reading taken in a duct at a blower inlet will be positive compared to room pressure. True or false?

11. If a blower delivers 680 cfm when the wheel is rotating at 720 rpm, at what speed would the wheel have to turn to deliver 780 cfm?

12. If a belt-drive blower wheel is found to be turning at 767 rpm, the motor pulley has a 4 in. diameter, and the blower pulley has a 9 in. diameter, what size should the motor pulley be changed to if a blower wheel rpm of 800 is desired? (Hint: find the motor rpm first.)

13. The speed of rotation of a multispeed direct-drive blower is changed by changing the _____ leads on the motor.

14. A manometer and Pitot tube are used to measure _____ _____ in a duct.

15. Air pressure measured at right angles to the direction of airflow in a duct is called _____.

16. The sum of static pressure and velocity pressure is called
_____ _____.

17. A Pitot tube, as used to measure airflow in a duct, measures
_____ with the tube that extends out of the end,
and _____ _____ with the other tube.

18. What would the velocity of air in a duct be if the velocity
pressure was found to be 0.043?

19. If a section of 36 in. × 10 in. duct had an air velocity of
863 fpm, how many cfm would be flowing in the duct?

20. If a register has a gross face area of 120 sq in. but the
net area is 70% of the gross, what would the net face
area be in sq ft?

21. If a register has a free face area of 0.583 sq ft and a face
velocity of 450 fpm, what quantity of air in cfm is coming
out?

## 19.1 INTRODUCTION

A hydronic heating system is one in which heating is accomplished by the use of liquids. In most cases, the heat is generated at a central location in a building. Then water carries the heat for use in other points in the building. A piping system carries the water. At the points of use, air is heated by the hot water and then used for heating.

The many similarities between the design of the warm air heating system and the hydronic heating system will become apparent as the hydronic system is described. In this chapter, the major components of a hydronic system are described and their function in the system explained.

## 19.2 HEATING LOADS

The first step in designing any heating system is to find out how much heat must be furnished to offset the heat loss of the building. The heat load is found by accurately calculating the losses, as already described in Chapters 13 and 14.

Each part of a hydronic system is selected and sized to provide the correct amount of heat to offset the heat loss. The calculations must be made carefully and accurately. For design purposes, the heat load should be calculated for individual rooms or zones in the building, not for the building as a whole.

## 19.3 MAJOR COMPONENTS

The major components of a hydronic heating system are the boiler, the distribution system, the circulating pumps, the terminal devices, and the control system. See Figure 19–1 for a diagram showing these components in a typical system. Each of these components is described in this chapter. The boiler and the controls are covered in detail in Chapter 10, however, and so are only mentioned briefly here.

# CHAPTER
# 19
# Hydronic System Components

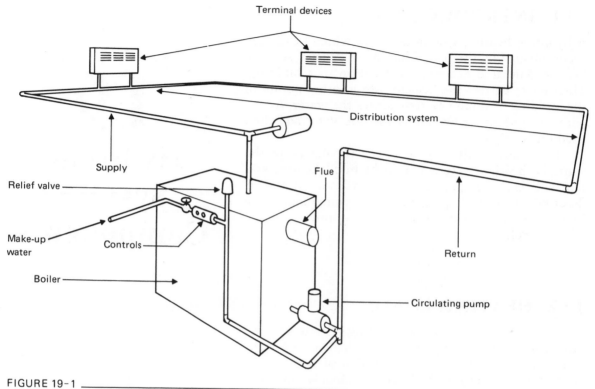

FIGURE 19-1

Major components of a hydronic heating system

## 19.4  BOILER

The boiler is the part of the system where the hot water is generated for the ultimate heating of the spaces in the building. The construction and performance of the major types of boilers are covered in Chapter 10. Readers may want to review that description.

## 19.5  DISTRIBUTION SYSTEM

The second major component in the system is the distribution, or piping, system. The **distribution system** is the part of the system in which the hot water is transported from the boiler to the various points of use.

Several basic piping layouts can be used for hydronic distribution systems. The major types are:

1. One-pipe, series loop system,
2. Primary-secondary system,
3. Two-pipe, direct return system,
4. Two-pipe, reverse return system,
5. Three- and four-pipe system.

Each of these is described in the sections that follow.

## One-Pipe, Series Loop System

A **one-pipe, series loop system** is one in which a single pipe runs entirely around the building or a zone within a building. The terminal units are piped in series so that all of the water goes through each unit. Figure 19–2 illustrates a one-pipe, series loop system.

This system is simple and inexpensive compared to other systems. There are two major objections to its use: (1) All of the water for the entire system flows through each terminal device. Therefore, each device must be sized large enough to carry the total flow. (2) The temperature drop *(TD)* of the water through each unit affects all other units from that point on in the system. This system is practical only for the smallest jobs where the *TD* through the entire system is not too great.

## Primary-Secondary System

A **primary-secondary system** is similar to the series system in that a single pipe carries the heating water around the building or zone, but the heating units are not in series on the loop. They are in parallel with it. An example of a typical primary-secondary system is shown in Figure 19–3.

In a primary-secondary system, each terminal device is connected to the main pipe with a supply and return connection. Although the *TD* of each individual unit affects the water temperature in the main line, the entire water supply does not flow through each unit. The main pipe is sized for the total gpm for the building or zone, but the secondary piping to each unit is sized to carry the gpm for that unit only.

The use of this system allows smaller piping for

*Term defined in text.

**Distribution system***

**One-pipe, series loop system***

**Primary-secondary system***

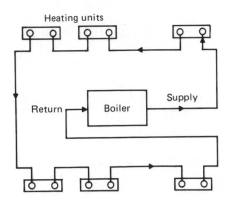

FIGURE 19-2 _____
One-pipe, series loop distribution system

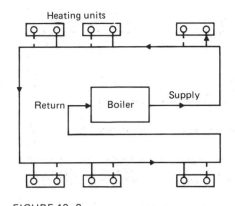

FIGURE 19-3 _____
Primary-secondary distribution system

**Two-pipe, direct return system**\*

**Two-pipe, reverse return system**\*

**Three- or four-pipe system**\*

FIGURE 19-4 _____

Balancing valve used to control the flow of
water in a branch or secondary loop of a
distribution system

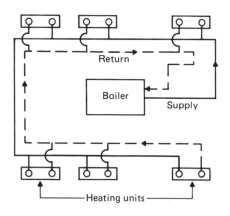

FIGURE 19-5 _____

Two-pipe, direct return supply and return
system

each unit and also makes it easier to provide proper temperature control in each separate zone served by the units.

The one-pipe, primary-secondary system is best used in small to medium-sized buildings where more flexibility of control is needed than can be attained with a series loop. If the building is too large to heat with one loop, then multiple loops can be used. In this case, each loop is treated as a separate piping system.

This system requires a method of ensuring a flow of water through each of the terminal devices and a way to balance that flow with the heat demand for the particular load. To accomplish this purpose, automatic or manual balancing valves are installed on either the supply or the return for each individual unit. Such a valve is shown in Figure 19–4.

## Two-Pipe, Direct Return System

On larger jobs, especially when the designer desires to keep the pipe sizing to a minimum, a two-pipe system should be used. A **two-pipe, direct return system** incorporates a supply pipe and a return pipe. The return pipe runs parallel with the supply line but returns the water directly back to the boiler from each terminal device. Figure 19–5 illustrates a typical two-pipe, direct return supply and return system.

The water from the first unit out of the supply pipe is the first in to the return and the first back to the boiler. In this system, the return water is running in the opposite direction to that in the supply line.

## Two-Pipe, Reverse Return System

In a **two-pipe, reverse return system,** the main supply pipe runs from one heating unit to another. The return pipe runs parallel with it from the first unit on. See Figure 19–6 for an example of this system.

The hot water is taken from the supply line, through the secondary piping to the unit, and then to the return pipe from the outlet of the unit. The water flow in the return pipe then continues in the same direction as that in the supply pipe. This system makes it possible to decrease the size of the supply pipe as it goes from unit to unit. The return pipe size increases as it goes around the loop. The heating units are sized for only the quantity of water required for their individual load since they are actually in parallel with the main piping.

This system is the more desirable of the two-pipe systems to use because all of the water in the system flows entirely around the system and the friction in the piping is always the same for each loop.

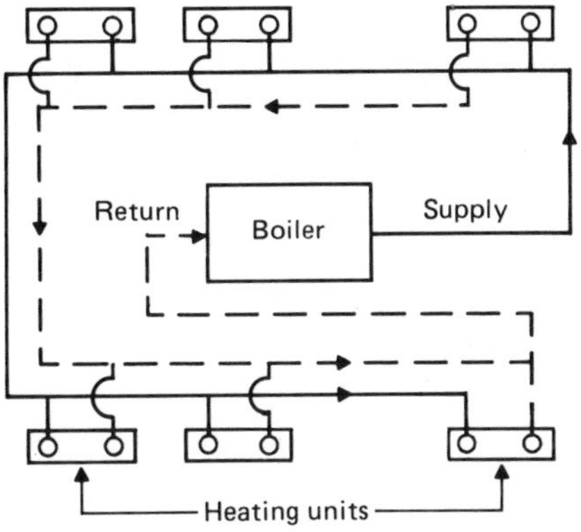

## Three- or Four-Pipe Systems

On a large job where the hot water is distributed for heating and chilled water is used for cooling in the building, a three- or four-pipe system can be used. In either a **three-** or **four-pipe system,** one supply pipe is used for carrying the heating water and a different pipe for cooling water. One or two return lines can be used. These systems are normally used only on jobs requiring heating in some zones at the same time as cooling is needed in others. See Figure 19–7 for an illustration of how a three-pipe system is used for heating and cooling at the same time.

## Piping Arrangements

When a hydronic system is designed, the routing and location of supply and return water piping is an important consideration. In most buildings, it is impor-

FIGURE 19-6 _____
Two-pipe, reverse return system

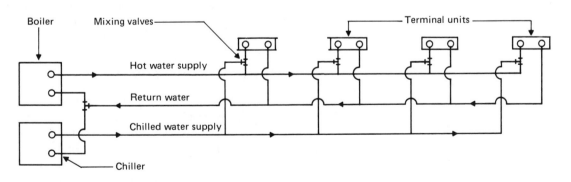

FIGURE 19-7 _____
Three-pipe supply and return system used on a heating-cooling application

**Centrifugal pump***

**Feet of head**   Term related to water pressure; pressure of a fluid expressed in terms of the height of a column of the fluid that it would support

**Water-to-air heat exchanger***

**Radiator***

**Convection coil***

FIGURE 19-8 _____

Typical line-mounted circulating pump used on a hydronic system

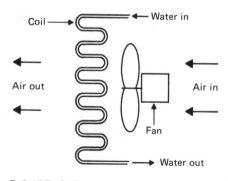

FIGURE 19-9 _____

Water-to-air heat exchanger

tant to locate the piping so that it is concealed; yet various components of the system will require service. Access is needed to these components. The piping used is generally fairly heavy and needs to be located so that it can be supported by the structure of the building. Designers must consider all of these points when locating the terminal units so the piping will fit into the general design of the building.

## 19.6  CIRCULATING PUMPS

In a typical hydronic system, pumps circulate the water through the various piping loops. If a system has more than one zone loop, each loop will have a pump. Such a system will be used when a building has separate zones, and each zone needs individual control.

The pumps used are centrifugal pumps. In **centrifugal pumps,** water is drawn in at the center of a housing in which an impeller wheel is rotated at high speed. The water is thrown by centrifugal force to the outer edge of the wheel. A discharge port is placed on the housing at one point on the outer perimeter. The water is forced through the port by the wheel.

Many types of circulating pumps are available. They are designated by physical arrangement as related to the way they are connected to the water pipes and by size related to the volume of water they will pump; some of the physical arrangements are line mounted or floor mounted. Figure 19–8 illustrates a typical line-mounted circulating pump.

Capacities of such pumps are given in gpm of water that can be circulated against a specified resistance to flow. The resistance is measured in **feet of head.**

## 19.7  TERMINAL UNITS

Heat exchangers must be employed to use the heat in the water in a hydronic system. The heat exchangers extract heat from the water and transfer the heat to the air. The air is then used for the heating medium. Figure 19–9 illustrates a **water-to-air heat exchanger.**

The heat exchanger in the hydronic system is generally in the form of a **radiator** or a **convection coil.** In both the radiator and the convection coil, the hot water flows through the inside of the elements and air is heated as it passes over the outside of the element.

In some cases, coils are built in to the walls, floors, or ceilings of a building. The building surfaces become a source of radiant heat. More often, the coil is installed in a cabinet located in the space to be heated. Whatever method is used for the final exchange of heat from the water to the air, the device used to make that exchange is called the terminal unit of the system. Figure 19–10 illustrates some typical terminal devices used for hydronic systems.

## Selection Criteria

Some basic design considerations go into the selection of the type of terminal unit to use for a particular job. These considerations are:

1. Aesthetic requirements,
2. Physical limitations,
3. Heat distribution requirements,
4. Sensitivity of control requirements,
5. Piping arrangements,
6. Fresh air requirements.

There may be other considerations for special applications, but the factors in the preceding list should be considered for every job.

**Aesthetic Requirements.**    If the terminal device is being used in a warehouse or a manufacturing area or any part of a building where appearance is not important, relatively simple equipment can be used. Either direct radiation or unit heaters could be chosen. **Radiation panels** or **unit heaters** are usually suspended from the ceiling or some overhead part of the building structure. Because appearance is not important, the piping and control valves can be exposed. The designer's main concern is to get the units high enough so they will be out of the way, yet low enough and in the right locations so the warm air will be directed where it is needed.

In areas where appearance is important, such as offices, the terminal devices are usually chosen to blend in to the surroundings. Coils can be enclosed in cabi-

**Radiation panel**    Heating panel that emits radiant energy

**Unit heater**    Self-contained heating unit, with a heat exchanger and blower in one cabinet, used with ductwork

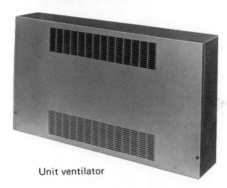

Baseboard convection unit

Unit ventilator

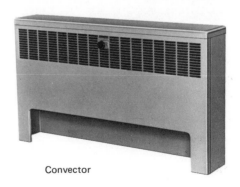

Convector

FIGURE 19-10

Selection of terminal devices used with hydronic heating systems

FIGURE 19-11 _____

Enclosed convection unit

nets that are neatly finished and painted a neutral color. These cabinets fit along the walls, either close to the floor or low enough on the wall to be inconspicuous. Figure 19–11 illustrates an enclosed convection unit of this type. The unit is both attractive and functional.

School rooms and large office spaces commonly have terminal units that fit under the windows. This arrangement makes it possible to bring in fresh air through the outer wall. Heat is brought into the room at its perimeter, where the heat loss is the greatest. This type of unit is called a unit ventilator. The units are normally enclosed in metal cabinets and painted to match the rooms in which they are used.

Exposed cast-iron radiators of the type previously used in many school rooms are seldom employed anymore. About the only use for the open-type radiator is in factories or manufacturing buildings.

**Physical Limitations.**    Most buildings are not designed with the placement of the mechanical equipment in mind. Designers must not only put the equipment in the best place for heating the building but be sure the equipment will fit. In most cases, the best location for the terminal units is along the outside walls, since that is where the greatest heat loss occurs. If the general arrangement of the rooms makes such placement impossible, then the designer should fit the units where they will do the best possible job of heating the space and still fit into the space available.

**Heat Distribution Requirements.**    The designer must bear in mind that the space will actually be heated by air although the heating medium is water. The terminal units are simply devices that use hot water to heat air, which is then circulated through the space to be heated. The terminal devices, or heating units, should be located so the heated air will circulate through the space as evenly as possible and, especially, should cover any areas of high heat loss, such as windows, doors, and/ or outside walls, with a blanket of warm air.

**Sensitivity of Control Requirements.**    A hydronic heating system provides one of the most important functions of a comfortable heating system—that is, a source of heat that remains constant. This source of heat is the mass of hot water in the system. After the water is heated, it gives up heat relatively slowly. This slowness can be a disadvantage when the temperature in the building changes rapidly, as it might in a building where the number of people varies or in a building where there are large windows on the south or west side. Because of the hot water in the system, heat can radiate from the terminal units after the thermostat in the room or zone where the units are located is satisfied. Because of the mass of water in the system, a hydronic system is relatively slow to provide heat. To compensate for this slowness of response, controls should be used on the components of the system that can sense changes in temperature rapidly and that can change the water flow in the units and the air flow through them as needed. These specific controls are covered in general later in this chapter and in detail in Chapter 20.

**Piping Arrangement.**    The particular piping arrangement to be used in a building must be considered when selecting the terminal units. The location of the main pipes in the system may make it impossible to use certain types of units. The general layout of the piping system will certainly influence the location of individual terminal units. Details concerning the terminal units and their selection are found in Chapter 20.

**Fresh Air Requirements.**    One of the prime requirements for comfort in a building is that the air be kept fresh by the addition of clean outside air. Many of the types of terminal units used with hydronic systems are basically radiant heat units. These units have no provision for the introduction of fresh air to the building through the unit. The designer should make the terminal unit selection with the above facts in mind. If the building will have a fairly large people load, then the terminal units should be a type that will introduce fresh air into the rooms. Unit ventilators are one such type. If the building will be used primarily for storing goods, and people are not a big part of the load, then units that merely heat the air without introducing fresh air are acceptable.

# Types of Units

Several general types of terminal units, and many variations of each type, are available. The main types are:

1. Direct-radiation elements,
2. Radiators,
3. Convection coil units.

All of these types are used to some extent, and each has advantages and disadvantages. Each is discussed in the sections following.

**Direct-Radiation System.**　A **direct-radiation system** is one in which the heating elements—in this case, hot water coils—are built in to the walls, floors, or ceilings of a building. The coil warms that part of the construction, and the hot surface radiates heat into the space. The heat travels in the form of rays. When these rays strike any object in the room, the radiant energy becomes heat energy.

This form of heating has some disadvantages. The surface temperature must be fairly high to emit enough rays for warmth; therefore, a lot of coil surface is required. A second disadvantage arises from the coil location; the coil is embedded in the building structure. If a leak occurs, the structure is usually damaged before the leak is found. Finally, the piping is difficult to get to for repairs.

Radiation systems are used in some locations, but such systems will not be covered further in this text.

**Radiators.**　When hot water heating systems were first used, they usually employed only radiators as a means of heating the air in a building. Cast-iron radiators are an example. See Figure 19–12.

The radiators heat the air by convection, direct radiation, and natural draft. The warm air moving across the surface of the radiator is the only air mover in the system. Simple radiator systems are seldom used in hot water systems today, except in commercial or industrial applications. Circulated air systems are more common.

**Convection Coil Units.**　The basic component used in most hydronic heating units to heat the air is the

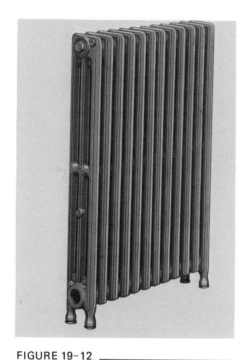

FIGURE 19–12 _____
Cast-iron radiator

convection coil. When a convection coil is used, air is forced across the coil to carry the heat away from the coil and into the space to be heated, as shown in Figure 19–13.

Two basic types of coils are used. The first is a **bare copper** or **steel coil.** The size of the tubing varies, depending upon the amount of heat transfer required. The air is simply blown over the coil in a direction at right angles to the tubing. The amount of heat extracted from the water within the coil is dependent upon the water temperature, the thickness of the coil material, the metal in the coil, and the amount of air moving across the coil.

The second type of coil used, and the more common one for comfort heating systems, is the **finned-tube coil.** Figure 19–14 illustrates a finned-tube coil. Aluminum fins are pressed onto the tubing in such a way that the fins extend the surface of the heated area and increase the ability of the coils to heat the air passing over them.

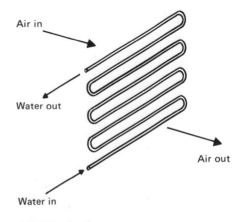

Air in

Water out

Air out

Water in

FIGURE 19-13 _____
Bare coil heating unit with air as the heating medium

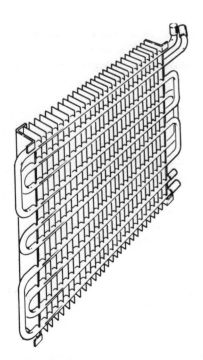

FIGURE 19-14 _____
Finned-tube coil

**Duct heater***

**Blower coil***

**Convector***

**Forced air convector***

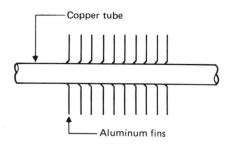

FIGURE 19-15 _____

Fins pressed on the coil to ensure good heat transmittal

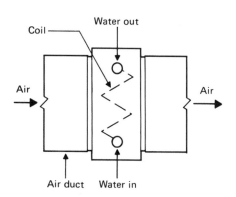

FIGURE 19-16 _____

Diagram of a typical duct heater application

The coil is still installed in the airstream so the tubes are at right angles to the air flow, but the aluminum fins are parallel with the flow. See Figure 19-15.

*Duct Heaters.*    If a central air handling system is being used in which the air is distributed to various parts of the building through ductwork, coils can be installed in the ducts so the air from the central blower passes through the coils. Such coils are called **duct heaters.** A diagram of a duct heater application is shown in Figure 19-16.

*Blower Coils.*    Sometimes it is necessary to furnish heating in areas that are not provided with air from a central source blower. In this case, blower coils are used. A **blower coil** unit is a self-contained blower and heating coil installed in a common cabinet. Figure 19-17 illustrates a typical blower coil unit. Blower coil units usually have the controls necessary for their operation as part of the package. Quite often, such units have filters in the cabinet also.

*Convectors.*    A **convector** is a unit with a heating coil encased in a metal cabinet. A return air inlet is located at the bottom of the unit, and a warm air outlet is located at the top. See Figure 19-18 for a typical convector unit.

Cooler air enters at the bottom of the unit. The air is heated by the coil and rises. The motive power, which moves the air upward and out of the convector, is provided by the natural draft effect of the rising warm air. Some convectors, called **forced air convectors,** contain blowers to move the air.

The outer metal cabinet on the convector improves the draft effect and the appearance of the unit. The cabinet also helps direct the air up.

# 19.8  CONTROL SYSTEM

The last major component in the hydronic system is the control system. A hydronic control system has two parts. The first part includes those components that control the burner and maintain the water temperature; the second part controls the operation of the terminal devices. There are many different combinations for each part of the system and also many ways to interlock the two.

The general scheme and operation of a control system has already been covered in Chapter 10. Some general observations are given here concerning the special characteristics of hydronic control system design.

Because of the large amount of water used in a total system, a hydronic system reacts slowly to temperature changes within a building. To compensate for this slowness, controls should be used that anticipate changing loads as much as possible. The usual method is to use outdoor reset controls that raise the water temperature in the boiler when the temperature of the outside air goes down.

Low voltage controls are recommended on the terminal devices. Such controls are more sensitive to temperature changes and can provide better anticipation. When rapid load changes are anticipated, as in a building with a lot of window area or shifting people loads, the controls should always be chosen to control the air across the coils, as well as the flow of water in the coils.

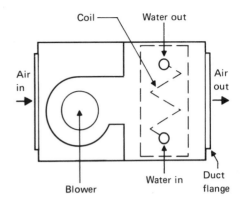

FIGURE 19-17 _____

Diagram of a blower coil unit

## 19.9  SUMMARY

A hydronic heating system is one in which the heat for a building is generated in a central location and then is carried to different areas in a water system to heat separate parts of the building.

The major components of a hydronic heating system are the boiler, distribution system, circulating pumps, terminal devices, and the controls. Two basic types of boilers are used. These are described in Chapter 10. Several types of distribution systems are used. These types are usually described by the arrangement of the supply and return pipes and by the way the terminal devices are connected in the system. Pumps are used in pipe loop for circulation of the water through the system and are described by the method used for mounting them in the system and by the volume of water that they can distribute. The control system has two parts. One part controls the burner and regulates the temperature of the water available to the system. The second part controls the terminal devices and regulates the heat output of each.

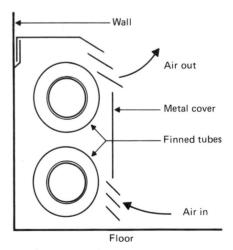

FIGURE 19-18 _____

Typical convector unit

## 19.10  QUESTIONS

1. The science of heating and/or cooling with liquids is called _____.

2. The first information required when a heating system is designed is the _____ _____ of the building.

3. The five major parts of a hydronic heating system are: _____, _____, _____, _____, and _____.

4. Even in a hydronic system, the desired final result is to heat _____.

5. The component used to move the heated water through the piping system of a hydronic system is called a _____ _____.

6. On a large hydronic heating system, more than one circulating pump may be used. True or false?

7. Since a hydronic system uses water as a medium, fresh air is not required in the building. True or false?

8. A hydronic heating system has a rapid or slow response time to changing load conditions. Circle the underlined word that makes the sentence correct.

9. An outdoor reset control helps a hydronic system compensate for changing external loads. True or false?

10. The five major types of hydronic distribution systems are: (a) _____, (b) _____, (c) _____, (d) _____, and (e) three- and four-pipe systems.

11. In a one-pipe, series loop system, all of the water goes through each _____ _____.

12. A one-pipe system in which the terminal units are plumbed-in parallel with the main pipe is called a one-pipe, _____ _____ system.

13. All of the system water does not flow through each terminal unit in a one-pipe, primary-secondary system. True or false?

14. A two-pipe system should be used when a designer wants to make pipe sizes larger. True or false?

15. The two-pipe system in which the water from the first terminal unit in the supply line is the first back to the boiler is called a _____ _____ system.

16. The two-pipe system in which the supply pipe gets smaller as it goes around the loop and the return line gets larger is called a _____ _____ system.

17. Match the terms in the first column with the definitions in the second column.

    1. One-pipe, series
    2. One-pipe, primary-secondary
    3. Two-pipe, direct return
    4. Two-pipe, reverse return

    a. Return, same direction
    b. Return, opposite direction
    c. Units in parallel
    d. Units in series

18. The three basic types of hydronic system terminal devices are: (a) _____, (b) _____, and (c) _____.

19. Terminal devices can be selected without any consideration of the other parts of the system. True or false?

20. A direct radiation system heats the air in a space. True or false?

21. Radiators heat by a combination of _____, _____, and natural draft.

22. Name the two main types of coils used in hydronic coil units.

23. Fins are used on coils to extend the surface area. True or false?

24. Match the terms in the first column with the descriptions in the second column.

    1. Duct heater
    2. Blower coil
    3. Convector
    4. Unit heater
    5. Unit ventilator

    a. Introduces fresh air
    b. Circulates air by gravity
    c. Connects to ductwork
    d. Remote blower
    e. No ductwork

# CHAPTER

# 20

# Hydronic System Design and Component Sizing

## 20.1 INTRODUCTION

For a hydronic heating system to operate properly, all parts must be sized correctly, and each part must work with the others. This chapter discusses the procedures to follow when designing a system; sizing the terminal devices and the supply and return pipes; calculating the total system friction; selecting the circulating pump, piping accessories, and controls; and determining the general layout of the equipment room.

## 20.2 SYSTEM DESIGN

The general layout of a total hydronic system and a description of the individual parts were described and explained in Chapter 19. Systems may vary, but the general type of component for each part will be the same. Each part of a system should be chosen to provide the best possible performance in that system for the application. Manufacturers' specification sheets or catalogs are the best source of information for each separate component.

## 20.3 DESIGN STEPS

Certain steps should be followed when laying out a plan for a heating system. These can be called the design steps. They are:

1. Obtain or draw a building plan;
2. Figure the heat load;
3. Locate the boiler and terminal devices on the building plan;
4. Make the piping layout on the building plan;
5. Size all components;
6. Select the control system;
7. Show the details of the boiler room piping on the building plan.

If each of these steps is completed in sequence, the design process will go smoothly, and the finished product will operate properly.

The first step is to obtain a plan of the building in

which the system is to be installed. The plan may be a simple one-line drawing or a complete set of plans. It should show the walls, partitions, windows, doors, and any cabinets or other detail that will affect the heating system. The plan should be drawn to scale. All major dimensions should be shown on it. See Figure 20–1 for a sample of the drawing. The designer will use this plan to show the heating system, equipment, and piping arrangement.

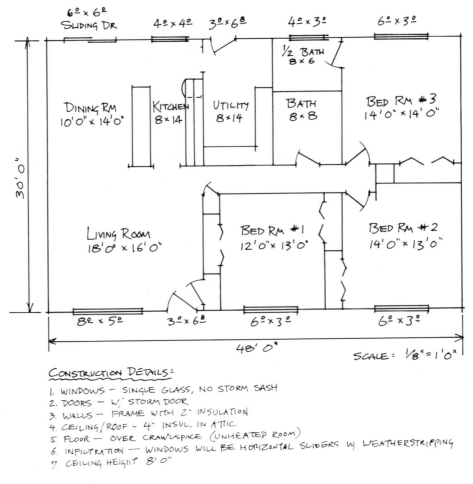

CONSTRUCTION DETAILS:

1. WINDOWS — SINGLE GLASS, NO STORM SASH
2. DOORS — W/ STORM DOOR
3. WALLS — FRAME WITH 2" INSULATION
4. CEILING/ROOF - 4" INSUL. IN ATTIC
5. FLOOR — OVER CRAWLSPACE (UNHEATED ROOM)
6. INFILTRATION — WINDOWS WILL BE HORIZONTAL SLIDERS W/ WEATHERSTRIPPING
7. CEILING HEIGHT 8' 0"

FIGURE 20-1 _____

Floor plan of a typical residence with all data recorded for the layout of a hydronic system

The next step is to figure the heat load for the building. Before the components can be selected for the system, the designer must know how much heat needs to be furnished to offset the heat loss of the building. Chapter 14 explains how to figure a heat load. Each individual part of a hydronic system is sized to provide the amount of heat required to each part of a building. The calculations must be accurate to ensure that the final design will be adequate. For design purposes, the heat load should be calculated for individual rooms or zones.

After the heat load is figured, the designer should write on the plan the load for each room. The terminal devices can also be selected and shown on the plan at this time, as shown in Figure 20–2.

Next, the boiler should be selected and its location shown on the drawing. If the boiler is placed on a floor other than the one that the terminal devices are on, then the location of the terminal devices should be

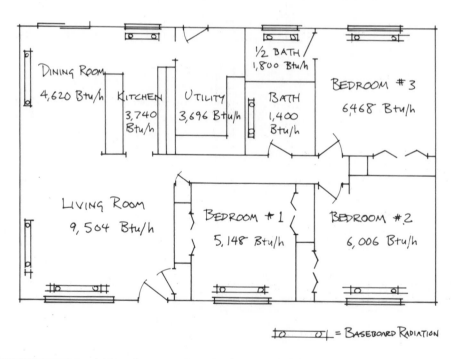

FIGURE 20-2
Floor plan from Figure 20-1 with the terminal devices shown for a simple hydronic layout

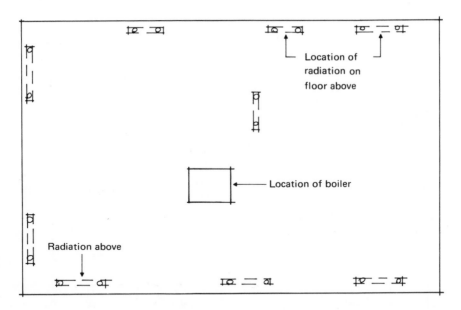

FIGURE 20-3

Boiler and terminal devices show on an overlay of the plan for Figure 20-1

shown on the same plan as the boiler. Figure 20–3 shows an overlay, made from the plan in Figure 20–1, with the boiler and the terminal devices located. In selecting the terminal devices and the boiler, as well as other mechanical parts of the system, the designer should refer to the manufacturer's specification sheets for the parts, which will give the necessary data for selection and sizing.

The location of the supply and return piping is next shown on the plan, in accordance with the system selected as best for the job. The pipe can also be sized at this time, using the gpm required for the terminal devices to determine the flow rate for each section of pipe. This step will be described in detail later in this chapter. Figure 20–4 shows the overlay with the piping layout on it.

After the piping layout is made and the piping is sized, all of the other components of the system can be selected and a detailed drawing made of the boiler piping and accessories. Figure 20–5 is a schematic drawing showing some of the details that should appear on this drawing. Details concerning the boiler are found

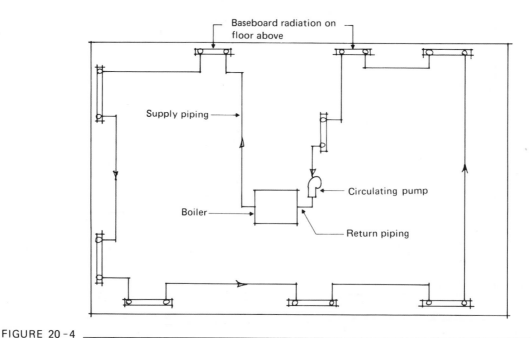

FIGURE 20-4

Supply and return pipe locations shown on overlay of the plan for Figure 20-1

in Chapter 10 in this text. The distribution systems were described in Chapter 19. Sizing of the piping and the other components of the system will be covered in the rest of this chapter.

## 20.4  TERMINAL DEVICES

The selection of the type of terminal devices to use is usually made when the distribution system is decided on. The two parts of the system are so dependent upon each other that they must be considered at the same time in the design process. The designer should select the terminal devices from those described in Chapter 19 and show the location of each one on the drawing.

After the type of terminal unit has been selected, it is necessary to select the proper size of unit. The heat load of each space should already have been figured, on a room-by-room basis, and the loads shown on the floor plan of the building. The location of the terminal devices should also be shown on the plan. This plan will be used during the rest of the design procedure and also later for balancing the system.

## gpm per Terminal Unit

The next step in sizing the terminal devices is to convert the Btu/h required per unit to gpm per unit. The equipment, piping, valves, and other components used in the system are all sized for handling water. The water in the system will be flowing at a rate measured in gpm, so the loads to be handled by each terminal device should be converted to gpm. The following formula can be used:

$$gpm = \frac{Btu/h}{60 \times 1 \times 8.345 \times TD}$$

where:

gpm = amount of water to be circulated
Btu/h = amount of heat to be delivered
60 = min in an hr
1 = specific heat of water
8.345 = lb of water per gal
TD = temperature drop of the water

In use, the divisor $60 \times 1 \times 8.345$ is constant. Its value of 500.7 is close enough to 500.0 so it can be rounded off and 500 used as a constant value. The formula then becomes:

$$gpm = \frac{Btu/h}{500 \times TD}$$

The use of this formula makes figuring the gpm required for any Btu/h fairly easy and is accurate enough for practical design use.

To make the conversion, it is necessary only to select the proper TD expected through a terminal unit and use the formula to determine the gpm for that unit. For design purposes, a TD of 20°F through any major part of the system can be used. This TD will have to be verified when the equipment is selected, but it is close enough for preliminary design purposes. Also for design purposes, it can be assumed that the initial supply water temperature will be 200°F and the leaving water temperature would be 180°F.

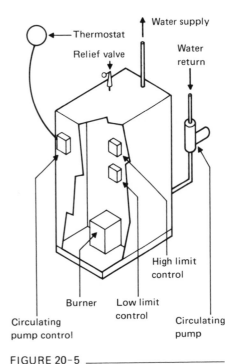

FIGURE 20-5 _____

Schematic drawing of a boiler and some of the parts and components related to it

### EXAMPLE 20–1

A zone in a building has a heat loss of 5568 Btu/h, and the TD through a terminal unit to serve that zone is 20°F. Find the gpm to be circulated.

**Friction loss**    Pressure drop caused by resistance to water flow in a pipe

**Static head**[*]

**Friction head**[*]

**Mils per foot (mils/ft)**[*]

**Feet per hundred feet (ft/100 ft)**[*]

## Solution

$$\text{gpm} = \frac{\text{Btu/h}}{500 \times 20°\text{F}}$$

$$= \frac{5568}{500 \times 20}$$

$$= 0.56$$

The quantity of water to be circulated is 0.56 gpm.

Some thought will have to be given to the type of piping system that will be used. If a series loop system is used, then the *TD* through each unit affects the entire system. If a primary-secondary system is used, the effect of each unit is much less.

## Selecting the Terminal Units

The selection of the terminal units must be made by type to fit the application and also by capacity for the heating load. Sizing selection charts for each of the different types of units are furnished by the manufacturers of the units. Designers should refer to these charts when making a selection. The selection chart shows the Btu/h output of each unit at a given flow rate (gpm) and for a variety of *TD*. If a unit is to be sized to heat a room requiring a given Btu/h, then reference should be made to the chart for a unit with that Btu/h capacity, and the gpm and *TD* for the unit will be shown. Conversely, if a flow rate has been established for a particular part of a system and the Btu/h required is known, then the designer can use the gpm and Btu/h to see if the *TD* is acceptable.

## 20.5    PIPE SIZING

With the location of the supply and return piping shown on the plan, the next step is to size the pipe for the amount of water that will have to flow through it. Since the water is the medium that carries the heat to the different parts of the system, proper sizing will ensure

[*]Term defined in text.

that the correct amount of heat is available where it is needed.

In sizing the pipe, the designer must consider the **friction loss** through the pipe, as well as the gpm required. The friction loss in piping is like static pressure drop in ductwork. Friction loss is an indication of the amount of resistance to the flow of the water in the pipe.

Water pressure can be measured in feet of head. This term indicates the amount of pressure exerted by a column of water of a given height, or conversely, the amount of pressure that would support a column of water of a given height. For instance, water pressure of 6 ft of head would raise a column of water 6 ft in vertical height. **Static head** is another term for this pressure. See Figure 20–6.

Friction in a water pipe is measured in feet of friction head. This term relates to the amount of resistance in a given length of pipe. **Friction head** is measured in feet of head and is equivalent to the pressure reduction caused by the friction in the pipe. Figure 20–7 illustrates friction head. Friction head is measured in **mils per foot (mils/ft)** or **feet per hundred feet (ft/100 ft)** of pipe. The term "mils/ft" is actually milinches (1/1000th of an in.) of head per ft of pipe; the term "ft/100 ft" is 1 ft of head per 100 ft of pipe.

Just as a duct system is designed for a given amount of friction per 100 ft of duct, so is pipe. An arbitrary selection of 300 mils/ft, or 2.5 ft/100 ft, will give a design that is right for most small- to medium-sized jobs. The resulting total friction loss will usually be within the range of a standard circulation pump.

A pipe sizing chart is used for the actual sizing of the pipe. The chart is similar to the equal friction chart used for ducts but has data for pipes. Different charts are used for different types of pipes because of the difference in the roughness of the inside surface. Table 20–1 shows a pipe sizing chart with pipe size as a function of the flow rate of water in gpm and the friction loss in mils/ft. To use the chart, the gpm required is plotted against the desired friction, and the pipe size is shown for the two. When they do not coincide on a pipe size exactly, then the next larger size pipe should be used. For example, if a section of pipe must carry 17.0 gpm of water at 300 milinches per foot, Table 20–1 shows that a 1.5 in. pipe should be used. A 1.5 in.

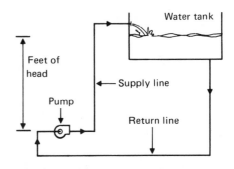

FIGURE 20-6 _____
Static head. The height in feet that the water has to be raised in a system is static head.

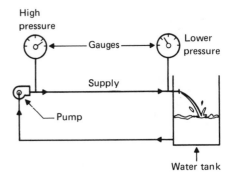

FIGURE 20 -7 _____
Friction head and pressure difference. Friction head is the friction in a piping system given in equivalent feet of height and pressure difference is an indication of friction in the piping system.

TABLE 20–1

Pipe Sizing Chart

| | ¼ | ⅜ | ½ | ¾ | 1 | 1¼ | 1½ | 2 | 2½ | 3 | 3½ | 4 | 5 | 6 | 8 | 10 | 12 |
|---|---|---|---|---|---|---|---|---|---|---|---|---|---|---|---|---|---|
| | | | | | | Nominal Pipe Size (in.) | | | | | | | | | | | |
| | | | | Carrying Capacity of Various Pipe Sizes in Gallons per Minute* | | | | | | | | | | | | |
| 700 | 0.53 | 1.2 | 2.2 | 4.8 | 9.3 | 19.0 | 29.0 | 56.0 | 90.5 | 170.0 | 245.0 | 350.0 | 650.0 | 1,100.0 | 2,300.0 | — | — |
| 650 | 0.50 | 1.13 | 2.1 | 4.6 | 9.0 | 18.0 | 28.0 | 54.0 | 88.0 | 160.0 | 235.0 | 330.0 | 620.0 | 1,040.0 | 2,200.0 | — | — |
| 600 | 0.48 | 1.1 | 2.0 | 4.4 | 8.5 | 17.3 | 26.8 | 51.0 | 85.0 | 150.0 | 225.0 | 320.0 | 590.0 | 1,000.0 | 2,100.0 | 4,000.0 | — |
| 550 | 0.46 | 1.03 | 1.95 | 4.2 | 8.1 | 16.5 | 25.3 | 49.0 | 80.0 | 145.0 | 215.0 | 300.0 | 560.0 | 940.0 | 2,000.0 | 3,800.0 | — |
| 500 | 0.44 | 1.0 | 1.8 | 3.9 | 7.7 | 16.0 | 24.0 | 47.0 | 76.0 | 140.0 | 210.0 | 290.0 | 530.0 | 900.0 | 1,900.0 | 3,600.0 | — |
| 450 | 0.41 | 0.94 | 1.7 | 3.7 | 7.3 | 15.0 | 22.7 | 44.0 | 73.0 | 131.0 | 190.0 | 270.0 | 505.0 | 840.0 | 1,800.0 | 3,400.0 | — |
| 400 | 0.39 | 0.87 | 1.6 | 3.5 | 6.9 | 14.0 | 21.2 | 41.0 | 68.0 | 123.0 | 180.0 | 250.0 | 480.0 | 800.0 | 1,700.0 | 3,200.0 | — |
| 375 | 0.37 | 0.85 | 1.55 | 3.4 | 6.6 | 13.5 | 20.6 | 40.0 | 65.0 | 120.0 | 175.0 | 245.0 | 460.0 | 760.0 | 1,600.0 | 3,100.0 | — |
| 350 | 0.36 | 0.81 | 1.5 | 3.2 | 6.3 | 13.0 | 20.0 | 38.5 | 63.0 | 115.0 | 165.0 | 235.0 | 440.0 | 730.0 | 1,550.0 | 2,950.0 | — |
| 325 | 0.35 | 0.80 | 1.45 | 3.1 | 6.1 | 12.5 | 19.0 | 37.0 | 60.0 | 110.0 | 160.0 | 225.0 | 420.0 | 700.0 | 1,500.0 | 2,850.0 | — |
| 300 | 0.33 | 0.75 | 1.38 | 3.0 | 5.8 | 11.8 | 18.0 | 35.0 | 57.0 | 104.0 | 150.0 | 215.0 | 400.0 | 670.0 | 1,400.0 | 2,700.0 | — |
| 275 | 0.31 | 0.72 | 1.3 | 2.9 | 5.6 | 11.3 | 17.2 | 33.5 | 55.0 | 100.0 | 145.0 | 205.0 | 380.0 | 640.0 | 1,350.0 | 2,600.0 | — |
| 250 | 0.30 | 0.68 | 1.25 | 2.7 | 5.3 | 10.9 | 16.5 | 32.0 | 52.3 | 95.0 | 138.0 | 195.0 | 360.0 | 600.0 | 1,300.0 | 2,450.0 | — |
| 225 | 0.28 | 0.65 | 1.18 | 2.6 | 5.0 | 10.2 | 15.5 | 30.0 | 49.0 | 90.0 | 130.0 | 182.0 | 340.0 | 570.0 | 1,200.0 | 2,300.0 | 4,000.0 |
| 200 | 0.27 | 0.60 | 1.1 | 2.4 | 4.7 | 9.5 | 14.5 | 28.0 | 46.0 | 83.0 | 122.0 | 170.0 | 320.0 | 530.0 | 1,100.0 | 2,150.0 | 3,800.0 |
| 175 | 0.25 | 0.57 | 1.01 | 2.25 | 4.4 | 8.9 | 13.5 | 26.0 | 43.0 | 78.0 | 113.0 | 160.0 | 300.0 | 500.0 | 1,050.0 | 2,000.0 | 3,500.0 |
| 150 | 0.23 | 0.52 | 0.96 | 2.07 | 4.0 | 8.2 | 12.5 | 24.0 | 38.5 | 71.0 | 104.0 | 145.0 | 275.0 | 450.0 | 950.0 | 1,800.0 | 3,250.0 |
| 125 | 0.21 | 0.48 | 0.85 | 1.85 | 3.6 | 7.4 | 11.2 | 22.0 | 36.0 | 65.0 | 95.0 | 132.0 | 247.0 | 410.0 | 850.0 | 1,650.0 | 3,000.0 |
| 100 | 0.18 | 0.42 | 0.77 | 1.64 | 3.2 | 6.5 | 10.0 | 19.0 | 31.0 | 56.0 | 83.0 | 117.0 | 217.0 | 360.0 | 750.0 | 1,450.0 | 2,700.0 |
| 75 | 0.16 | 0.36 | 0.64 | 1.4 | 2.7 | 5.5 | 8.5 | 16.4 | 28.0 | 48.0 | 72.0 | 100.0 | 182.0 | 310.0 | 650.0 | 1,220.0 | 2,350.0 |
| 50 | 0.13 | 0.28 | 0.51 | 1.13 | 2.2 | 4.5 | 6.7 | 13.0 | 21.5 | 39.0 | 56.0 | 80.0 | 152.0 | 245.0 | 500.0 | 980.0 | 2,000.0 |
| | | | | | | | | | | | | | | | | | 1,550.0 |

Note: The pipe is sized to carry water at a given flow rate and at a fixed friction loss.

Source: Bell & Gossett

pipe will actually carry 18.0 gpm at 300 milinches, but since it is the size closest to 17.0 gpm it should be used.

## Sizing Procedure

The gpm required to each terminal device should be shown on a heating unit selection chart and also on the layout of the building. A typical heating unit selection chart is illustrated in Figure 20–8.

If the chart was used during the selection process for the terminal units, the gpm will already be listed on it. If it was not, the data should be entered now. The other columns can be used for data relating to the pipe sizing.

The next step is to size the pipe. The following steps should be taken:

1. Figure the gpm needed in each section of pipe;
2. Select the friction head to use;
3. Find the pipe size on the pipe sizing chart;
4. Record the pipe size on the unit selection chart and on the plan;

5. Indicate on the plan where the pipe size transitions take place.

## Sizing a One-Pipe, Series Loop System

A one-pipe, series loop system is simple to size since the flow rate (gpm) is the same all around the loop. Using a friction head of 300 mils/ft of pipe, the designer uses the pipe sizing chart to find the pipe size for the desired gpm. The total gpm for the loop should have been determined when the terminal units were selected. If it was not, it can be found at this point.

### EXAMPLE 20–2

If a one-pipe, series loop is to be sized for a job that requires a total of 28,250 Btu/h of heating, with a 20°F *TD* through the system, regardless of the number of terminal units, the gpm for the Btu/h must first be found. This is done by using the following formula:

$$\text{gpm} = \frac{\text{Btu/h}}{500 \times TD}$$

Then the pipe is sized for the gpm from the pipe sizing table.

| Heating Unit Selection | | | | |
|---|---|---|---|---|
| Space | Btu/h | gpm | TD | Unit selected |
|  |  |  |  |  |
|  |  |  |  |  |
|  |  |  |  |  |
|  |  |  |  |  |
|  |  |  |  |  |
|  |  |  |  |  |
|  |  |  |  |  |
|  |  |  |  |  |

FIGURE 20-8

Heating unit selection chart used to record the data for proper selection of the terminal devices for a given job

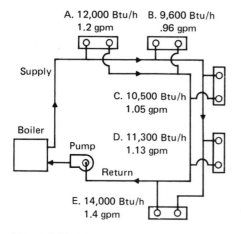

**FIGURE 20-9**

Diagram of a two-pipe reverse return system showing the Btu/h load and the gpm requirements for each terminal unit

## Solution

$$gpm = \frac{Btu/h}{500 \times TD}$$

$$= \frac{28,250}{500 \times 20}$$

$$= 2.825$$

The quantity of water is 2.825 gpm. The pipe is then sized by referring to the sizing chart, which shows that a 3/4 in. pipe will carry 3 gpm at 300 mils/ft. Since this is the next size pipe above that for 2.825 gpm, then that is the size pipe to be used.

Because all of the water in a one-pipe, series system flows through all sections of the pipe, the loop pipe will be the same size all around the loop. This size should be marked on the pipe layout on the floor plan.

## Sizing a Two-Pipe System

A better example of pipe sizing can be shown by using a two-pipe, reverse return system. In this system, the terminal units are in parallel with the supply piping. The return pipe is separate from the supply but also in parallel with it. See Figure 20–9 for a diagram of a two-pipe, reverse return system. The water quantity in the supply pipe diminishes as it flows through the system, but the flow rate increases in the return line as it goes around the loop.

To size the pipe, the initial flow rate is used for the first section of pipe. This rate should be the total required for all units in the system. The pipe sizing is taken from the sizing chart, as explained for a one-pipe system. After the first unit, the flow rate through the supply pipe will be less because of the water taken out at the first unit, so the flow rate between the first and second unit is now used for sizing the pipe in that section. The same procedure is followed around the loop, sizing each section of pipe for the flow rate in that section, until all of the supply pipe has been sized.

The return pipe is sized in the same way, except that the pipe will increase in size as it goes around the loop. A pipe sizing data chart, in which the sections of pipe are identified by letters or numbers and the gpm is listed for each section, is invaluable for sizing pipe in any system where different flow rates occur in the different sections of the system. Figure 20–10 shows such a chart with data on it for Example 20–3, which follows next.

## EXAMPLE 20–3

Find the pipe sizing for the job shown in Figure 20–9.

### Solution

First, find the gpm through each unit so the gpm for each section of pipe can be figured. To do this calculation, the total system *TD* should be decided upon

| Pipe Sizing Data Chart | | | |
|---|---|---|---|
| Pipe | Designation | gpm | Size at 300 mil/ft |
| SUPPLY | A-B | 5.74 | 1" ∅. |
| DO | B-C | 4.54 | 1" ∅. |
| DO | C-D | 3.58 | 1" ∅. |
| DO | D-E | 2.53 | 3/4" ∅. |
| DO | E-F | 1.40 | 1/2" ∅. |
| RETURN | G-H | 1.20 | 1/2" ∅. |
| DO | H-I | 2.16 | 3/4" ∅. |
| DO | I-J | 3.21 | 1" ∅. |
| DO | J-K | 4.34 | 1" ∅. |
| DO | K-L | 5.74 | 1" ∅. |
| | | | |
| | | | |

FIGURE 20-10

Pipe sizing data chart with data recorded for Example 20–3

| Heating Unit Selection | | | | |
|---|---|---|---|---|
| Space | Btu/h | gpm | TD | Unit selected |
| A | 12,000 | 1.20 | 4.18 | |
| B | 9,600 | .96 | 3.34 | |
| C | 10,500 | 1.05 | 3.66 | |
| D | 11,300 | 1.13 | 3.94 | |
| E | 14,000 | 1.40 | 4.88 | |
| | | | | |
| | | | | |
| TOTALS | 57,400 | 5.74 | 20°F | |

FIGURE 20-11 _____

Heating unit selection chart with data recorded for Example 20–3

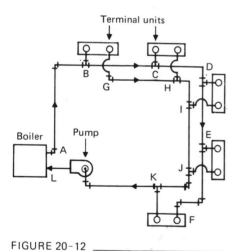

FIGURE 20-12 _____

Sections of pipe identified with letters to match those on the pipe sizing chart shown in Figure 20-10

and then the total system gpm found. If a total *TD* of 20°F is chosen, then the total flow rate is found by the following formula:

$$\text{gpm} = \frac{\text{Btu/h}}{500 \times TD}$$

$$= \frac{57,400}{500 \times 20}$$

$$= 5.74$$

The flow rate through each terminal unit can be found by proportioning the 5.74 gpm for the total heat load to each unit by unit heat load. The results are shown on the heating unit selection chart, Figure 20–11. The total *TD* can also be proportioned per each unit. The result should be shown on the chart.

Then the sections of pipe are identified by letter or number as on the plan. See Figure 20–12. The sections are also as listed on the pipe sizing data chart, Figure 20–10.

The gpm through each section of pipe is figured by starting with the full flow through the first section of pipe (A–B) and then deducting the gpm as the water is taken out at each terminal, as recorded on the pipe sizing data chart. The gpm of the return pipe is figured in the same way, but the water is added as it comes out of each terminal unit. This information is also shown on the pipe sizing data chart.

Finally, the pipe sizes are selected from the pipe sizing chart. The sizes are recorded on the pipe sizing data chart and also on the plan.

Next, the sections of pipe feeding each terminal unit are sized. Since the largest gpm is 1.40, and a 1/2 in. pipe will provide enough flow for that gpm, then 1/2 in. pipe can be used for all of the supply and return branches to the terminal units.

The final step is to show by use of a symbol (X) on the pipes where the transition in pipe sizes takes place. This marking is shown on the plan.

## Sizing Fittings

The selection of balancing valves, shutoff valves, and other fittings for the distribution system can usually be made on the basis of the pipe sizing to the units. In some cases where special flow valves are used for automatic balancing, however, the flow rate through the valves is also required to make the selection. See Figure 20–13 for an illustration of some of these special valves. The flow rate in this case would be that required for the unit being provided water by the valve. The selection data are found in the valve manufacturer's specification sheets.

Other fittings, such as elbows and tees, are normally sized the same as the pipe to which they are connected. The additional pressure drop caused by these fittings is added to the total pressure drop of the system when sizing the pumps.

## 20.6  TOTAL SYSTEM FRICTION

The total system friction must be determined for the selection of a circulating pump. The pump will have to move the amount of water required for heating the building against the friction in the piping.

Total system friction is simple to calculate after the pipe has been sized for a given friction head per given length of pipe. In Examples 20–2 and 20–3, the friction head was to be 300 mils/ft, which is the equiv-

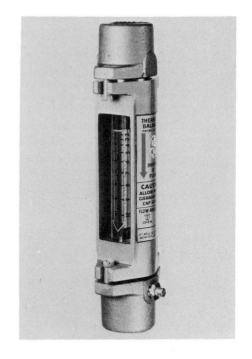

FIGURE 20–13 _____

Special valves used to balance the water flow in the different sections of the pipe

alent of 2.50 ft/100 ft. To find the total friction in the piping in the system, it is necessary only to find the length of the pipe in the system, multiply the length by 2.5 ft/100 ft, and divide by 100 ft. The following formula is used:

$$\text{friction loss} = \frac{\text{length of pipe} \times 2.5}{100}$$

To find the total friction head in the system, the friction loss through any other components in the system, such as terminal devices, valves, or fittings, must be added to that for the piping.

## Single-Pipe Systems

To size the piping for a single-pipe, series loop system, the length of the pipe loop should be scaled off the plan, or otherwise determined, and the length multiplied by 2.5 ft/100 ft. The result is divided by 100. This calculation gives the friction loss in the piping only. To find the total friction loss for the system, the pressure drop through each terminal device should be added to the piping loss. This figure is used to size the circulating pumps.

## Two-Pipe Systems

The total friction loss through a two-pipe system is figured similarly to that for a one-pipe system, with one difference: the friction loss through each unit is not added. The water flows through one device only in any one circuit around the system. The friction loss through the device having the greatest loss is added to the friction loss for half of the supply pipe length and half of the return pipe length. This calculation will give the total friction drop for the system, disregarding the loss for fittings. For small jobs and preliminary figures, the total is close enough to use for sizing pumps.

This method gives the proper steps to follow for sizing any piping system for a typical loop system. The method should be followed for each loop, whether the system has only one simple loop or multiple loops for different zones.

# 20.7  CIRCULATING PUMP

After the total system friction is determined, the next step is to select and size the circulating pump, or pumps, for the system. Pumps are selected and sized to circulate the amount of water required in the system, against the friction, or resistance to flow, in the pipes. The flow rate is that determined for the total system when the pipe was sized. The resistance is measured in feet of water per 100 ft of pipe. The pump is selected to provide the desired gpm at that friction.

The total amount of water to be circulated should already have been calculated and entered on the pipe sizing data chart and the plan. The amount is the sum of the gpm required through each terminal device in the system. This amount was calculated on the basis of the Btu/h required in each space when the terminal units were selected. The *TD* of the water and the design characteristics of the terminal units determine the total amount for each unit. Figure 20–14 is a photograph of an installation with two circulating pumps and some piping, valves, and accessories.

The total friction loss through the system has also been calculated. It is the friction loss through the pipe itself, plus the loss through the terminal units and the fittings in the system.

A circulating pump is selected by using a pump selection chart. This chart is furnished by the pump manufacturer. It shows the gpm delivery rate of the pump, as compared to the head in feet of water against which the pump must work. See Figure 20–15 for a pump selection chart.

The term **head in feet of water** is indicative of the resistance to water flow in the pipe, as measured by the feet of water in a vertical column. A designer should always use a chart that has been developed for the particular pump under consideration.

**Head in feet of water***

## EXAMPLE 20–4

Using the pump selection chart shown in Figure 20–15, select a pump for a job requiring a 20 gpm flow rate in a particular loop against a head of 8 ft.

FIGURE 20-14

Installation with two circulating pumps, piping, valves, and accessories

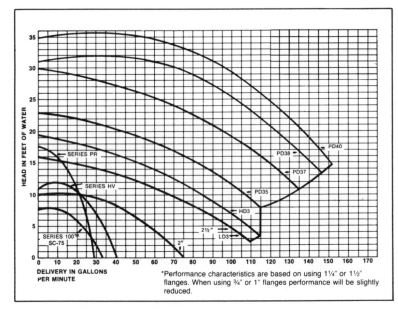

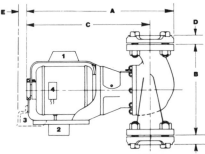

**How to select a B&G Booster Pump**
Required: 10 GPM at 6 ft. head. Look first at the bottom of the Booster Pump Capacity Chart where pump delivery is shown. Run a line straight upward from the 10 gallon point until it intersects a horizontal line from the 6 ft. head on scale at left. The nearest pump curve, or one slightly above this intersection is the proper selection. The Series 100 Booster pump should be used. It is not advisable to select a Booster pump with a head under 2½ ft.

FIGURE 20-15

Circulating pump selection chart from a manufacturer's catalog

## Solution

Locate 8 ft of head on the vertical scale on the left of the chart. Then move across to the right on the horizontal line. Where the horizontal line intersects a vertical line coming up from the 20 gpm point on the flow rate scale at the bottom of the chart, look for the curving line related to the capacity of the pump. In this example, using the chart in Figure 20–15, the pump called for is a series PR booster pump. This pump would be the proper selection for this particular job.

If this were a multiloop job, the same procedure would be followed for each loop in the system. Individual pumps can then be individually controlled to give the proper temperature control to each zone in the building.

## 20.8  ACCESSORIES

In addition to the major components of a system, certain other components are necessary to a system for

**Expansion tank**   Air tank located above the boiler in a hydronic system and plumbed into the system in such a way that the tank absorbs the extra volume of water when the water is heated

**Relief valve**   Valve plumbed into the piping system above the boiler in a hydronic system that is designed to open and relieve the pressure in the boiler if that pressure exceeds a particular set point

**Air vents**   Fittings installed at high points in the piping system of a hydronic system to vent air from the water circulating through the pipes

safe and proper operation. These components are called accessories to the other parts. The most important ones are: the **expansion tank**, the **relief valve, air vents,** and certain other valves, such as pressure-reducing and flow-control valves.

The expansion tank is used to accommodate expansion of the water in the system as the water is heated. The hot water has a greater volume than cold water. Because water in the liquid state is noncompressible, a great deal of pressure would be exerted if there was no allowance for it to expand as it was heated. The expansion, or compression, tank is filled with air and connected into the piping system in such a way that water will not flow into the tank naturally. Instead, the water will be forced into the tank as the water in the piping system expands. The air in the tank acts as a cushion. Since the air will compress, the water in the system can expand, but only at a controlled rate.

A relief valve is always used on a boiler. The valve is spring loaded or weighted; it will open if the pressure in the system exceeds what is considered to be safe. If any malfunction of the controls causes the burner to stay on too long, the relief valve will open and prevent dangerous pressure build-up in the system. Most relief valves open on pressure build-up. Some will also sense high water temperature and open if either pressure or temperature exceeds a set limit.

As water is heated, air is released from it. This air naturally collects at high points in the piping system and prevents the water from flowing. Air vents are used to allow the air to escape into the atmosphere. These vents are installed in the piping system at the high points. Since the system cannot be left open or water would flow out, the valves have to be a type that will open only when there is air present but not water. One of two types is used. One type is manual. This valve can be opened by an operator when system operation indicates that air is present. The other valve is automatic. It will open by itself when air is in the line but will close as soon as it senses water.

Various other valves are used to control pressure or water flow. Among these are pressure-reducing valves to control the upper limit of the water pressure in the system, check valves used to make sure the water flows in one direction only, and gate and globe valves for flow control or shutoff control.

Other special flow-control valves can also be used in a typical system. The designer should refer to a plumbing diagram of the specific type of system to determine what valves and controls are needed. Manufacturers' specification sheets are the best source of detailed information concerning valves or controls.

## 20.9  CONTROLS

A hydronic control system is made up of two parts. One part controls the water temperature in the boiler. The other controls the air temperature at the terminal device. This arrangement is necessary in a hydronic system because the water is heated at one place in a building and then used to heat air in other parts of the building.

The main control that regulates the temperature of the water in the boiler is an aquastat. This device is a thermostat that senses water temperature and opens or closes an electric switch depending upon the temperature. An aquastat has a temperature scale and an adjustable indicator that can be set for a desired temperature, the set point of the system. If the temperature of the water goes below the set point, the electric switch closes. If the temperature goes above the set point, the switch opens. The switch is wired into an electric circuit that controls the burner. If the switch closes, the burner is actuated; if the switch opens, the burner is turned off. The burner cycles off and on to maintain the water temperature at or near the set-point temperature.

To make the system better able to compensate for fluctuating loads, an outdoor reset is sometimes used in conjunction with the aquastat. This control will reset the set-point temperature to compensate for changing temperatures outside the building. If the outdoor temperature goes down, the water temperature is reset higher. If the outdoor temperature goes up, the water temperature is set lower.

The second part of the control system—the part that controls the terminal devices—can be as simple

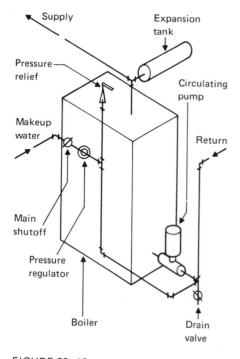

**FIGURE 20-16** _____

Typical piping layout for small system equipment room

as a balancing valve that is adjusted for a constant volume of water to flow. This type of control does not adjust to varying loads and so is practical only for the simple systems. If some control of the heat output of the device in relation to the load is needed, then a thermostatically controlled valve can be used to vary the flow rate according to the load. For better control of the heat output, air temperature controls are used with the water flow controls. This type of control will turn the blower off and on as the heating load varies.

In simple systems, the controls are mounted on the terminal devices themselves. For better control of the room air temperature, the thermostat is wall mounted. The thermostat is then connected to the motorized valves or blower control in the unit by an electric circuit.

## 20.10   EQUIPMENT ROOM LAYOUT

The final step in designing a hydronic heating system is to arrange the piping scheme that is to be used in the equipment room for connecting the boiler, expansion tank, control valves, and circulating pumps. This equipment is usually located in an equipment room in the building to be heated or adjacent to it in a special room for that purpose. The room itself should have been planned in the initial building plan. A typical piping layout for a small system equipment room is shown in Figure 20–16.

Consideration must be given to the location of many components necessary to complete the system. The designer should decide how best to fit these components into the space that has been provided in the building. Normally, the designer will find that the best method is to work on the floor plan of the equipment room first and then to use elevation and section drawings to show details.

Many different layouts are possible. The designer should select the one to be used for each job with care. The physical size of the equipment and components has to be considered, as well as the physical locations. Unfortunately, room is not often available to spread the equipment out too much.

# 20.11 SUMMARY

Designing the hydronic system and sizing the components is similar to designing a warm air system. The design steps are:

1. Obtain or draw the building plan;
2. Figure the heat load;
3. Locate the boiler and terminal devices on the building plan;
4. Locate the piping layout on the building plan;
5. Size all components;
6. Select the control system;
7. Show the details of the boiler room piping on the building plan.

At this stage in the design process, selection of the proper type of components for the hydronic system should have been completed, as covered in Chapter 19. Now the components must be properly sized physically in order to fit into the building and also according to capacity in order to meet the heating requirements of the building.

In order to size the terminal devices, the heat load for each device must be converted from Btu/h to gpm. When the gpm requirement for each terminal device has been determined, manufacturers' sizing selection charts are used to select the correct devices for the job.

Next, the return and supply pipes are sized so that they will carry the proper amount of water for each part of the system they service, taking into consideration the resistance to flow of water in each section of pipe. The total system friction, or pressure drop, is then calculated to determine the amount of resistance that the distributing pump has to overcome. The distributing pump is sized taking into consideration the total amount of water to be circulated and the total system friction. Then, the accessory items, such as expansion tank, relief valve, and shutoff and balancing valves, are selected and sized. Finally, the arrangement of all of these parts in the equipment room is decided upon.

Boilers, circulating pumps, accessories, and controls are all selected and sized by using manufacturers' data sheets that give the specific details for each one. The components are selected to match each other and to provide the proper flow rate and control to heat the building according to the heat load.

## 20.12 QUESTIONS

1. The mechanical plan should show which piping _____ to use, as well as the other major components of the system.

2. Two things must be considered when sizing pipe: _____ through the pipe and _____ of water required.

3. The term "static head" relates to the actual vertical _____ of the water in the system.

4. The term _____ head relates to the amount of resistance to flow in a pipe.

5. The term "mils/ft" relates to millimeters per foot. True or false?

6. _____ mils/ft is about the right figure to use for the design static head for a small system design.

7. A pipe sizing chart shows pipe sizes as a function of _____ and _____.

8. What would the flow rate in gpm be through a terminal unit with a 26,300 Btu/h capacity and a 10°F *TD*?

9. What would the flow rate be through a one-pipe, series loop system if the terminal units had a combined capacity of 63,200 Btu/h and the system *TD* was 20°F?

10. From the pipe sizing chart, find the size pipe required to carry 32 gpm of water at a friction head of 250 mils/ft.

11. Total system friction includes the friction head for piping, plus the friction through the _____ _____, valves, elbows, and fittings.

12. What would the piping friction loss be in a system designed for 2.25 ft/100 ft of pipe if the system had 632 ft of pipe?

13. A circulating pump is selected by comparing the _____ of water needed, with the _____ _____ of the system.

14. If a terminal device is to be used to provide 26,000 Btu/h, for heating a space, and the system is designed for a 6°F *TD* through the device, what will the flow rate be, in gpm?

15. Two terminal devices are used in a series loop system, and one has twice the Btu/h heating capacity of the other. If the smaller has a *TD* of 2.6°F and the larger has a heating output of 12,600 Btu/h, what is the Btu/h capacity of the two units combined, and what is the total *TD* in the system? Prove your answer by use of formulas.

## 21.1  INTRODUCTION

**Troubleshooting** is the act of locating and identifying the source of trouble when a unit is not working properly. The investigation is preceded by the observation that something is wrong with the system, and it is followed by **service** to correct the problem.

The first requirement for troubleshooting is knowledge of how the equipment operates normally. The normal function of gas-fired, oil-fired, and electric furnaces is covered in the preceding chapters. In this chapter, the troubleshooting procedure is described. The chapter covers analyzing the complaints, identifying system components related to the complaint, determining what part is faulty, and checking the part to verify the problem.

## 21.2  ANALYZING COMPLAINTS

When a heating system is not working properly, the building users usually complain. The first indication that something is wrong is that the building air temperature is different from that expected. The temperature may be too hot or too cold, but it will vary from the set-point temperature on the thermostat. Since the building user generally is not familiar enough with the system to identify the problem, he or she will make a complaint.

The complaint will be relayed as a general statement to the person who will do the troubleshooting, and this general statement will need some clarification as to specifics. The following is a list of some common complaints:

1. No heat,
2. Insufficient heat,
3. Too much heat,
4. Fluctuating heat,
5. Noise,
6. Odor,
7. Excessive cost of operation.

A general description of each of these complaints is

*Term defined in text.

**Troubleshooting***

**Service**   Corrective action taken to repair or adjust a malfunctioning unit so that it will work correctly

**Malfunction**    Not functioning correctly

given below. The troubleshooting steps to follow for each of them is given later in the chapter.

## No Heat

This complaint is self-explanatory. When a building does not warm up at all, generally the unit is not working. The troubleshooter must determine why it is not.

## Insufficient Heat

In some cases of unit malfunction, the heat output of the unit is reduced. The complaint will be that there is not enough heat. This complaint may not arise until the outside temperature of the air gets close to the design temperature, or the condition may be intermittent at other outside temperatures.

## Too Much Heat

This complaint is a cause for immediate concern. Too much heat may be caused by a unit that is not shutting off and can indicate a serious safety hazard. If a gas valve sticks open on a gas furnace or the relays fail to open on an electric furnace, overheating will occur. Ordinarily, the high temperature limit will shut the unit off. In some rare cases, the limit may fail, and a very dangerous fire hazard exists. Overheating can be a dangerous condition.

## Fluctuating Heat

In some cases of **malfunction,** the temperature in a building may fluctuate above or below the set-point temperature on the thermostat. This problem may not be the fault of the unit; other parts of the system may be at fault. A thermostat located on a cold outside wall or where cold drafts can affect it is one example. The troubleshooter should check the entire system to determine the actual cause.

## Noise

This complaint is usually created by mechanical failure of some part of the system or unit, but it may be caused by faulty ignition or operation of the burner. All noise complaints should be thoroughly investigated. They may indicate a failure that could lead to a more serious condition.

# Odor

In the case of a fuel-fired unit, this complaint can indicate a dangerous situation, such as leaking fuel or flue gas. The troubleshooter may have difficulty identifying the source of an unpleasant odor. Because of the danger created by escaping gas or incomplete combustion, the furnace should not be used until a complete check of its operation has been made.

# Excessive Cost of Operation

This complaint is difficult to deal with because the cause can be so elusive. Perhaps there are problems with the operation of the unit; more likely, there are problems within the total system. The cost of operating any system is important. Because fuel costs are so high, this complaint should be considered carefully.

# 21.3  TROUBLESHOOTING THE SYSTEM

The troubleshooter should always use a systematic approach to the problem. First, a complete understanding of the normal operation of the system is necessary. Second, the troubleshooter must obtain enough information from the person making the complaint and from first-hand observation to know what the unit is doing, especially what is different from normal operation. Third, the troubleshooter must identify the general family of components in which the problem seems to lie. And fourth, the troubleshooter must identify the specific part or parts that may be faulty. These four steps are explained in detail in the following sections.

# Understanding the Operation

The first step in the troubleshooting process is to understand how the system is supposed to work. The basic operation of a heating unit is easy to understand, but the operation of the total system is more complex. An example is two pieces of equipment using common controls, such as an electronic cleaner used with a heating system. A problem in the cleaner control system will affect the operation of the heating system. The operation of the heating system can also be affected by the type of control system used. Outdoor reset controls or damper controls may be tied in with the heating controls. Malfunctions in the controls themselves or in the damper motors will adversely affect the operation of the heating system.

The troubleshooter should inspect the unit and the other components of the total system and make sure that he or she understands the operation of the entire system before starting to look for problems.

## Recognizing the Abnormal Function

The second step in the troubleshooting process is to understand how the system is operating now. Before starting to look for problems, the troubleshooter should analyze the complaint and by observation of the unit, attempt to visualize what part of the system is functioning incorrectly. A few minutes spent in careful thought at the beginning of a troubleshooting call may save a lot of wasted effort. If the complaint is one of too much heat, for instance, then there is no use starting at the disconnect and fuse box. Because the unit is working part of the time, there must be electric power to it. If, however, the complaint is that there is no heat, then the place to start is at the disconnect.

## Identifying the Problem-Causing Components

The third step in the process is to identify the general components causing a problem. The major unit component categories are as follows:

1. Electric power components,
2. Controls,
3. Burner and heat-producing components,
4. Air-handling components.

These components and their parts are shown in Table 21–1 for the three main types of furnaces: gas-fired furnace, oil-fired furnace, and electric furnace. The let-

TABLE 21-1

**Major Components and Parts of Gas-Fired Furnaces, Oil-Fired Furnaces, and Electric Furnaces**

**Gas-fired furnace**
Electric power
    1. Disconnect
    2. Fuses (circuit breakers)
    3. Control transformer

Controls
    G-4.    Thermostat
    G-5.    Limit control
    G-6.    Blower control
    G-7.    Blower relay

Burner and heating components
    G-9.    Gas shut-off
    G-10.    Piping
    G-11.    Gas valve
    G-12.    Thermocouple
    G-13.    Pilot safety
    G-14.    Pilot burner
    G-15.    Burners
    G-16.    Heat exchanger
    G-17.    Diverter

Air-handling components
    19. Blower
    20. Blower motor
    21. Belt and drives
    22. Filter
    23. Outside air
    24. Ductwork

**Oil-fired furnace**
Electric power
    1. Disconnect
    2. Fuses (circuit breakers)
    3. Control transformer

Controls
    O-4.    Thermostat
    O-5.    Limit control
    O-6.    Blower control
    O-7.    Combustion safety
            controls
    O-8.    Primary control

Oil-fired furnace (continued)
Burner and heating components
    O-9.    Oil supply
    O-10.    Piping
    O-11.    Pump
    O-12.    Nozzle
    O-13.    Ignition transformer
    O-14.    Electrodes
    O-15.    Combustion blower
    O-16.    Fire pot
    O-17.    Heat exchanger
    O-18.    Barometric damper

Air-handling components
    19. Blower
    20. Blower motor
    21. Belt and drives
    22. Filter
    23. Outside air
    24. Ductwork

**Electric furnace**
Electric power
    1. Disconnect
    2. Fuses (circuit breakers)
    3. Control transformer

Controls
    E-4.    Thermostat
    E-5.    Limit control
    E-6.    Blower control
    E-7.    Blower relay
    E-8.    Sequencers

Heating components
    E-9.    Wiring
    E-10.    Elements

Air-handling components
    19. Blower
    20. Blower motor
    21. Belt and drives
    22. Filter
    23. Outside air
    24. Ductwork

ter and number designations of the parts relate to the troubleshooting chart (Table 21–2) and the procedure charts (Tables 21–3, 21–4, and 21–5). The letter designation *G* is for a gas-fired furnace, *O* is for an oil-fired furnace, and *E* is for an electric furnace.

TABLE 21–2

Troubleshooting Chart for Gas-Fired Furnaces, Oil-Fired Furnaces, and Electric Furnaces

| Components/Parts | | No heat | Insufficient heat | Too much heat | Fluctuating heat | Noise | Odor | High operation costs |
|---|---|---|---|---|---|---|---|---|
| **Gas-fired furnace** | | | | | | | | |
| Electric power | | | | | | | | |
| 1. Disconnect | | 2 | | | | | | |
| 2. Fuses (circuit breakers) | | 3 | | | | | | |
| 3. Control transformer | | 9 | 3 | | | | | |
| Controls | | | | | | | | |
| G–4 | Thermostat | 1 | 1 | 1 | 1 | | | |
| G–5 | Limit control | 10 | 2 | | | | | |
| G–6 | Blower control | | 4 | | 2 | | | |
| G–7 | Blower relay | | | | | 1 | | |
| Blower and heating components | | | | | | | | |
| G–9 | Gas shut-off | 4 | | | | | | |
| G–10 | Piping | | | | | | 1 | |
| G–11 | Gas valve | 11 | | 2 | | | | |
| G–12 | Thermocouple | 5 | | | | | | |
| G–13 | Pilot safety | 8 | | | | | | |
| G–14 | Pilot burner | 6 | | | | | | |
| G–15 | Burners | 7 | | | | 2 | 2 | |
| G–16 | Heat exchanger | | 8 | | | | 3 | |
| G–17 | Diverter | | | | | | 4 | |
| Air-handling components | | | | | | | | |
| 19. Blower | | 12 | 5 | | 3 | 3 | | |
| 20. Blower motor | | 13 | 6 | | | 4 | | |
| 21. Belt and drives | | | | | | 5 | | |
| 22. Filter | | | 7 | | 4 | | | 1 |
| 23. Outside air | | | | | 5 | | 5 | 2 |
| 24. Ductwork | | | | | | | | 3 |
| **Oil-fired furnace** | | | | | | | | |
| Electric power | | | | | | | | |
| 1. Disconnect | | 2 | | | | | | |
| 2. Fuses (circuit breakers) | | 3 | | | | | | |
| 3. Control transformer | | 7 | 3 | | 1 | | | |
| Controls | | | | | | | | |
| O–4 | Thermostat | 1 | 1 | 1 | 1 | | | |
| O–5 | Limit control | 8 | 2 | | | | | |
| O–6 | Blower control | | 4 | | 2 | | | |
| O–7 | Combustion safety controls | 5 | | | | | | |
| O–8 | Primary control | 6 | | 2 | | | | |

| Components/Parts | | Complaints | | | | | | |
|---|---|---|---|---|---|---|---|---|
| | | No heat | Insufficient heat | Too much heat | Fluctuating heat | Noise | Odor | High operation costs |
| **Burner and heating components** | | | | | | | | |
| O-9 | Oil supply | 4 | 5 | | | | 1 | |
| O-10 | Piping | 9 | 6 | | | | 2 | |
| O-11 | Pump | 10 | | | | 2 | 3 | |
| O-12 | Nozzle | 11 | | | | | | |
| O-13 | Ignition transformer | 12 | | | | | | |
| O-14 | Electrodes | 13 | | | | | | |
| O-15 | Combustion blower | | | | | 3 | | |
| O-16 | Fire pot | | | | | | 4 | |
| O-17 | Heat exchanger | | | | | | 5 | |
| O-18 | Barometric damper | | | | | | 6 | |
| **Air-handling components** | | | | | | | | |
| 19. | Blower | | 7 | | 3 | 4 | | |
| 20. | Blower motor | 14 | 8 | | 4 | 5 | | |
| 21. | Belt and drives | 15 | 9 | | 5 | | | |
| 22. | Filter | | | | 6 | | | 1 |
| 23. | Outside air | | | | | | 7 | 2 |
| 34. | Ductwork | | | | | | | 3 |
| **Electric furnace** | | | | | | | | |
| **Electric power** | | | | | | | | |
| 1. | Disconnect | 2 | | | | | | |
| 2. | Fuses (circuit breakers) | 3 | | | | | | |
| 3. | Control transformer | 4 | | | | | | |
| **Controls** | | | | | | | | |
| E-4 | Thermostat | 1 | 1 | 1 | 1 | | | |
| E-5 | Limit control | 6 | 2 | | | | | |
| E-6 | Blower control | | 3 | | 2 | | | |
| E-7 | Blower relay | | 4 | | | | | |
| E-8 | Sequencers | 5 | 5 | 2 | | 1 | | |
| **Heating components** | | | | | | | | |
| E-9 | Wiring | 7 | | | | | | |
| E-10 | Elements | 8 | | | | | | |
| **Air-handling components** | | | | | | | | |
| 19. | Blower | 9 | 6 | | 3 | 2 | | |
| 20. | Blower motor | 10 | 7 | | 4 | 3 | | |
| 21. | Belt and drives | 11 | 8 | | 5 | | | |
| 22. | Filter | | | | 6 | | | 1 |
| 23. | Outside air | | | | 7 | | | 2 |
| 24. | Ductwork | | | | | | | 3 |

TABLE 21-3 _____

Troubleshooting Procedures for Gas-Fired Furnaces

**Electric power components**
1. Disconnect
   a. Check main disconnect for power
   b. Check unit disconnect for power (may be circuit breakers)
2. Fuses (circuit breakers)
   a. Check main fuses for opens
   b. Check unit fuses for opens (may be circuit breakers)
3. Control transformer
   a. Check for voltage on primary side
   b. Check secondary voltage
   c. Check for 60 cycle hum

**Control components**
G-4 Thermostat
   a. Check setting
   b. Check power input from transformer
   c. Do contacts close on call?
   d. Check location (sun, drafts, etc.)
   e. Is thermostat level on wall?
   f. Check anticipation
G-5 Limit control
   a. Is limit control closed in normal operation, and does it open at proper temperature?
G-6 Blower control
   a. Is blower control open when unit is cold, and does it close at proper temperature?
G-7 Blower relay
   a. Does blower relay close when fan switch is closed?

**Burner and heating components**
G-9 Gas shut-off
   a. Is shut-off valve open?
G-10 Piping
   a. Check piping for leaks

**Burner and heating components** (continued)
G-11 Gas valve
   a. Is gas valve closed when deenergized?
   b. Does valve open when energized?
   c. Check gas pressure at manifold
G-12 Thermocouple
   a. Does thermocouple produce adequate voltage?
G-13 Pilot safety
   a. Does pilot safety hold open when energized and close when deenergized?
   b. Does safety close both pilot and main valves?
G-14 Pilot burner
   a. Does pilot flame impinge properly on the thermocouple?
   b. Is burner clean?
   c. Check gas adjustment at pilot valve
G-15 Burners
   a. Is burner flame correct?
   b. Are ports clean?
   c. Are orifices correct size?
   d. Is there sufficient combustion air?
G-16 Heat exchanger
   a. Check for cracks or burn-out
   b. Is there any flame impingement?
   c. Check for sooting
G-17 Diverter
   a. Check for spillage

**Air-handling components**
19. Blower
   a. Is blower wheel clean?
   b. Check for correct rotation
   c. Check speed of rpm
   d. Is wheel noisy?

Air-handling components (continued)

20. Blower motor
    a. Does motor operate when blower control closes?
    b. Is motor noisy?
    c. Does motor run hot?

21. Belt and drives
    a. Is belt broken or cracked?
    b. Check belt tension
    c. Are pulley sizes correct for cfm required?

22. Filter
    a. Is filter in place?

Air-handling components (continued)
    b. Does filter need cleaning or replacing?
    c. Is filter efficiency correct for job?

23. Outside air
    a. Check for proper quantity of outside air

24. Ductwork
    a. Is ductwork leaky?
    b. Is ductwork too small?
    c. Does duct run in uninsulated space?

---

**TABLE 21–4**

Troubleshooting Procedures for Oil-Fired Furnaces

---

**Electric power components**

1. Disconnect
    a. Check main disconnect for power
    b. Check unit disconnect for power (may be circuit breakers)

2. Fuses (circuit breakers)
    a. Check main fuses for opens
    b. Check unit fuses for opens (may be circuit breakers)

3. Control transformer
    a. Check for voltage on primary side
    b. Check secondary voltage
    c. Check for 60 cycle hum

**Control components**

O–4 Thermostat
    a. Check setting
    b. Is thermostat receiving voltage from transformer?
    c. Do contacts close on call?
    d. Check location (sun, drafts, etc.)
    e. Is thermostat mounted level on the wall?

Control Components (continued)

O–5 Limit control
    a. Are limit control contacts normally closed, and do they open at the proper temperature?

O–6 Blower control
    a. Are blower control contacts normally open, and do they close at proper temperature?
    b. Does sure-start fan switch close after proper time interval?

O–7 Combustion safety controls
    a. Does stack switch or cad cell bring on unit within proper time interval after call for heat?
    b. Does cad cell "see" flame properly during firing?

O–8 Primary control
    a. Is there power at T terminals?
    b. Does switch respond to signals from combustion safety device?

TABLE 21–4 (continued)

Control Components (continued)

    c. Does safety switch open on call from combustion safety?

**Burner and heating components**

O–9 Oil supply

    a. Is there oil in the tank?

    b. Has sludge plugged the intake?

    c. Has water gotten into lines?

    d. Are all line fittings tight?

O–10 Piping

    a. Check oil lines for oil or air leaks

O–11 Pump

    a. Is pump running on a call for heat?

    b. Is pump noisy?

    c. Is pump producing the correct oil pressure?

O–12 Nozzle

    a. Is nozzle proper size?

    b. Is nozzle clean?

    c. Is spray pattern correct?

O–13 Ignition transformer

    a. Does transformer produce proper voltage?

    b. Is electric connection good from transformer to electrodes?

O–14 Ignition electrodes

    a. Check insulators to make sure they are clean and not cracked

    b. Are electrodes properly located and spaced?

    c. Are points proper shape?

O–15 Combustion blower

    a. Does combustion blower come on on a call for heat?

    b. Check blower adjustment

O–16 Fire pot

    a. Is fire pot lining intact?

Burner and heating components (continued)

O–17 Heat exchanger

    a. Check for cracks or burn-out

    b. Is there any flame impingement?

    c. Is inside of exchanger clean?

O–18 Barometric damper

    a. Adjust damper for proper draft over fire

**Air-handling components**

19. Blower

    a. Is blower wheel clean?

    b. Check for correct rotation

    c. Check speed of rpm

    d. Is wheel noisy?

20. Blower motor

    a. Does motor operate when blower control closes?

    b. Is motor noisy?

    c. Does motor run hot?

21. Belt and drives

    a. Is belt broken or cracked?

    b. Check belt tension

    c. Are pulley sizes correct for cfm required?

22. Filter

    a. Is filter in place?

    b. Does filter need cleaning or replacing?

    c. Is filter efficiency proper for job?

23. Outside air

    a. Check for proper quantity of outside air

24. Ductwork

    a. Is ductwork leaky?

    b. Is ductwork too small?

    c. Does duct run in uninsulated space?

TABLE 21–5
Troubleshooting Procedures for Electric Furnaces

**Electric power components**
1. Disconnect
   a. Check main disconnect for power
   b. Check unit disconnect for power (may be circuit breakers)
2. Fuses (circuit breakers)
   a. Check main fuses for opens
   b. Check unit fuses for opens (may be circuit breakers)
3. Control transformer
   a. Check for voltage on primary side
   b. Check secondary voltage
   c. Check for 60 cycle hum

**Control components**
E–4 Thermostat
   a. Check setting
   b. Check for power to thermostat from transformer
   c. Do contacts close on call?
   d. Check location (sun, drafts, etc.)
   e. Is thermostat level?

E–5 Limit control
   a. Is limit control closed in normal operating conditions?
   b. Is thermal link intact in power circuit?

E–6 Blower control
   a. Is blower control normally open, and does it close on temperature rise?

E–7 Blower relay
   a. Are relay contacts open when coil is not energized?
   b. Do contacts close when coil is energized?

E–8 Sequencer
   a. Does sequencer start staging on a call for heat?
   b. Does each stage come on in proper sequence?

**Control components (continued)**
   c. Does sequencer step each stage off in proper sequence?
   d. Does sequencer signal blower relay?

**Heating components**
E–9 Wiring
   a. Are all connections tight?
   b. Is insulation intact?

E–10 Elements
   a. Do elements heat up when energized?
   b. Are all insulators in place?
   c. Are elements sagging excessively?

**Air-handling components**
19. Blower
   a. Is blower wheel clean?
   b. Check for correct rotation
   c. Check speed of rpm
   d. Is wheel noisy?
20. Blower motor
   a. Does motor operate when blower control closes?
   b. Is motor noisy?
   c. Does motor run hot?
21. Belt and drives
   a. Is belt broken or cracked?
   b. Check belt tension
   c. Are pulley sizes correct for cfm required?
22. Filter
   a. Is filter in place?
   b. Does filter need cleaning or replacing?
   c. Is filter efficiency proper for job?
23. Outside air
   a. Check for proper quantity of outside air
24. Ductwork
   a. Is ductwork leaky?
   b. Is ductwork too small?
   c. Does duct run in uninsulated space?

# Checking the Individual Parts

After determining which part of the system may be working incorrectly, the troubleshooter will start examining the individual parts for proper operation.

**Identifying Unit Parts.**    Table 21–2 is a troubleshooting chart for gas-fired furnaces, oil-fired furnaces, and electric furnaces. The major components are listed on the left side of the table, with the related parts listed below each component. The most common complaints are shown at the top right of the table. In each column below the listed complaint, there are numbers across from the various parts. These numbers start at 1 and continue in order. The numbers give the order in which the parts should be checked.

**Checking Unit Parts for Faults.**    The numbers in the columns under the complaints on the troubleshooting chart indicate the order in which the parts should be checked for faults. For instance, if it is reported that the temperature in a building using a gas-fired furnace fluctuates from the set-point temperature on the thermostat, then refer to that portion of Table 21–2 that shows the steps to take in troubleshooting a gas-fired unit.

In the column under the heading "heat fluctuates," the number 1 is found across from "thermostat" in the components column. The thermostat is the first part to check. If the thermostat is operating correctly, the troubleshooter goes to number 2 on the chart, found across from "blower control." The troubleshooter checks the operation of the blower control. The procedure to follow in checking the parts of the gas-fired furnace is shown in Table 21–3 and is also described later in this chapter. If the blower control is operating correctly, then the troubleshooter continues to number 3 and checks the part indicated there. This procedure is followed until the faulty part is found.

Tables 21–3, 21–4, and 21–5 are troubleshooting procedure charts for gas-fired furnaces, oil-fired furnaces, and electric furnaces. The tables contain the list of parts and descriptions of the procedures for checking each part.

Each troubleshooting procedure chart contains four lists. The four lists are for the main components: electric power, controls, burner and heating components, and air-handling components. The electric power component list and the air-handling list are basically the same for all three types of furnaces, but the controls and heating components lists are different for each.

On each chart, the parts related to each component are identified by number and listed below each component. One or more checks to make concerning each part are also listed, identified by letters. For example, in Table 21–3, Troubleshooting Procedures for Gas-Fired Furnaces, under electric power components, the disconnect is listed as one of three parts. Below the disconnect, two procedures are listed: (a) check the main disconnect for power and (b) check the unit disconnect for power. If the troubleshooter determines from the troubleshooting chart for a given unit that the problem is in the power supply, then these two checks should be made in the order shown.

## EXAMPLE 21–1

A complaint is received that a building heated by an oil-fired furnace is not heating properly (insufficient heat).

**Solution**

Under the section for oil-fired furnaces in the troubleshooting chart in Table 21–2, the troubleshooter locates the complaint heading "Insufficient heat." The first part of the unit to check is the thermostat. Previous investigation would have indicated this problem to be in the controls section. Using the troubleshooting procedure chart for an oil-fired furnace (Table 21–4), under the section for control components, the first check for the thermostat is to make sure that the thermostat is set higher than the room temperature—in other words, to see whether it is calling for heat. If the thermostat is calling for heat and the furnace still is not on, then the troubleshooter should check whether there is power to the thermostat from the transformer. If that check does not find the problem, the troubleshooter should see whether the thermostat contacts are closing when they should. The troubleshooter proceeds in this manner until the fault is found. If the fault is not found in the thermostat or its circuit, then the troubleshooter will see from Table 21–2 that the second part to check under "Insufficient heat" is the limit controls. The limit controls could be short-cycling the unit, preventing it from producing its full rated capacity. Using the troubleshooting procedure (Table 21–4), the troubleshooter locates the procedure for checking the limit control under the control components section. The troubleshooter follows this general procedure until he or she identifies the faulty part.

## 21.4    TROUBLESHOOTING INDIVIDUAL PARTS

The description of some of the main parts and controls in earlier chapters included some troubleshooting and service procedures. This section will cover some troubleshooting procedures that can be used for individual parts. Many of the procedures listed on the troubleshooting procedure charts are so obvious, however, that they do not need explanation and will not be covered here at all.

### Electric Power Components Common to All Units

The basic electric power components are common to all types of furnaces. These components are the disconnects, fuses, and, in some cases, transformers. Procedures for troubleshooting each of these follow.

**Disconnects.**    First, the troubleshooter should make a visual check of the main disconnect to see if the switch is closed. If it is closed and the unit does not come on, then a voltmeter should be used to check for power, first on the line side and then on the load side of the disconnect. The voltmeter is set at a higher voltage setting than the expected voltage for the system to protect the meter against damage due to overload. Then one probe is placed on each of the two terminals on the disconnect. For a three-wire system, the three legs of the system will have to be checked in combinations of two. The voltmeter will indicate if there is power to the disconnect and also power through the disconnect. If there is no power on the line side, the electric utility that provides the power should be notified. If there is power on the load side of the main disconnect but the unit still does not run, then the unit disconnect should be checked, using the same procedure as that outlined for the main disconnect.

**Fuses and Circuit Breakers.**    If there is power on the line side of a fused disconnect or a circuit breaker but not on the load side, the fuses should be checked. Two methods can be used: (1) the fuses can be removed and checked for continuity with an ohmmeter, or (2) they can be checked in place with a voltmeter.

To check with an ohmmeter, one of the meter probes is held to each end of the fuse, and the ohm value of the fuse is read on the meter. If the fuse is good, the meter will show very little resistance. If the fuse is burned out, the reading should show infinite resistance. To check fuses in place, the troubleshooter places the voltmeter probes on the line side of one fuse and the load side of another. A volt reading on the meter indicates that the fuse with the probe on the load side is good. No reading indicates that fuse is burned out. The procedure should be reversed for checking the other fuse(s) in the system.

A circuit breaker has an indicating switch built in with its disconnect switch. If the circuit breaker has opened, the toggle switch handle will have moved to the off position, and a red signal will indicate it is open.

**Transformer.**    If there is electric power on the load side of the unit disconnect and all fuses are good but the unit still will not run, the control transformer should be checked. A voltmeter can be used to take readings on the primary and the secondary sides. The voltage on the primary side should be the same as that on the load side of the unit disconnect. The voltage on the secondary side should be within plus or minus 10% of the rated secondary voltage of the transformer.

# Controls Common to All Units

The operation of most of the controls is covered in Chapters 11 and 12, but a general procedure for trouble-shooting some of the most common controls is included here. The controls examined are the thermostat, limit controls, blower control, and blower relay.

**Thermostat.** The first check to make of a thermostat is visual. Is the set-point temperature higher than the room temperature? If it is, check to see if the thermostat contacts are closing by placing a jumper wire across the terminals. If the unit comes on, the thermostat is faulty.

A thermostat should always be located on an inside wall where it will sense average room temperature. It should not be positioned where drafts from doors or windows will affect it or where direct sunshine will strike it. Any one of these locations can cause a reading by the thermostat that is very different from that of the room air. The thermostat should always be mounted level on the wall so the mercury bulb contacts work properly.

A thermostat anticipator has to be set to coincide with the amperage draw of the thermostat circuit. The amperage in the thermostat circuit is checked with an ammeter while the unit is in operation. The anticipator is set at the reading by adjusting the lever in the thermostat.

**Limit Control.** To check a limit control, the trouble-shooter disconnects the blower by removing one of the wires from the blower motor or blower control. A high temperature thermometer is placed in the supply plenum of the furnace. The unit is then started by turning up the thermostat. The limit control should open and turn the unit burner off when the thermometer reading indicates the temperature that the control is rated for, usually about 180°F to 200°F. The blower should then be started by reconnecting the wire. When the plenum temperature drops below the rated temperature for the control, the unit should come back on.

**Blower Control.** Proper operation of a blower control can be checked by observation. On a call for heat, the burner should come on; as soon as the air going through the heat exchanger reaches the set-point temperature of the blower control, the blower motor should come on. The temperature of the air can be checked by a thermometer in the supply plenum. On an off cycle, when the burner shuts off, the blower control should open and turn off the blower after a delay of approximately 30 seconds to 1 minute.

**Blower Relay.**    If the thermostat has a blower control switch on it, there will be a relay in the system operated by that switch for continuous blower operation. When the switch is set at "continuous," the relay brings the blower on. The contacts in this relay are wired in parallel with the contacts in the blower control, so either one can operate the blower motor. This relay should be checked for proper operation by checking the coil for continuity with an ohmmeter and the contacts are checked, in both the open and closed positions, the same as any other relay.

## Gas Furnace Parts

The two general types of gas furnace burners are described in Chapter 7. This section will describe troubleshooting procedures to use for some of the components used with the burners: the gas valve, thermocouple, pilot safety, and pilot burner.

**Gas Valve.**    The operation of the gas valve can be checked by using a voltmeter to see if there is power to the valve on a call for heat. The meter should be set to the appropriate voltage scale and one probe placed on each terminal of the solenoid on the valve. If there is power to the terminals, the valve should open. The opening of the valve can be checked by observing the burners. If the burner does not come on, the combustion safety controls should be checked before the valve is condemned. If the combustion safety controls are functioning properly and the burner still does not come on, the troubleshooter should disconnect the control wiring from the solenoid and check the solenoid for continuity with an ohmmeter. An ohmmeter reading indicating an open circuit on the coil shows that the coil is burned out. If the coil is all right and the valve still does not open, the problem is within the valve itself. The valve should then be replaced.

If a gas valve sticks open and the burner will not shut off after the thermostat is satisfied, the gas should be shut off at the unit shut-off valve, and the valve should be changed.

**Thermocouple.**    If on lighting a pilot light, the pilot stays lit while the pilot switch is held on the valve but goes out as soon as the switch is released, the thermocouple should be checked. To check this part, the connecting lead to the pilot safety valve, or the gas valve if the safety is in it, is disconnected. With the pilot flame burning, a reading is taken with a millivoltmeter between the terminal on the end of the lead and the outer part of the lead cable. The positive lead probe of the millivoltmeter is placed on the outside conductor and the other probe on the inside lead, or terminal at the end. The acceptable minimum voltage for most valves is approximately 18 millivolts. (Check the proper data in the valve manufacturer's data sheets for the valve used.) If the reading is less than the minimum allowed for the specific valve, the thermocouple should be replaced.

**Pilot Safety.**    If the thermocouple checks out all right but the pilot still goes out as soon as the pilot reset button is released, the pilot safety is faulty and should be replaced.

**Pilot Burner.**    The pilot burner is checked first by visual inspection of the flame. The flame should be steady, with about the last one-third of the flame surrounding the tip of the thermocouple. If the flame is too low, the pilot gas valve should be adjusted, or the screen in the pilot burner should be cleaned, or both. If the flame is too high or is lifting off the burner, the pilot gas input should be reduced by turning down the pilot gas valve or pilot adjustment screw.

# Oil Furnace Parts

Oil furnace parts are described in detail in Chapter 8. Procedures for checking the separate parts of a burner and the combustion safety control specifically related to oil furnaces are covered here.

**Oil Burner.**    If oil fails to get to the burner, the following should be checked: the oil supply in the tank, piping for leaks, fuel pump, and/or the fuel pump motor. A dip stick can be used to check the oil supply. If the tank has been allowed to run dry and then filled, sludge may be present in the lines or pump. If the pump whines when it comes on, the suction pressure should be checked with a gauge on the intake side of the pump. The suction pressure should be less than 5 in. if the system is gravity feed or not more than 10 in. if the tank is below the level of the burner.

If the vacuum is all right, the bleed valve should be opened or the oil line disconnected from the pump to the gun assembly. When the pump is running, oil should come out of the line. If oil does not, the line is plugged, a restriction is in the line from the tank to the pump, or the pump screen is plugged. To check for a plugged or restricted line, the troubleshooter should disconnect the line at the burner pump intake and at the oil tank. Air pressure should be used to force the oil out of the line and to clear any obstructions. If air will not go through the line, the line will have to be replaced. To check for a plugged screen, remove the head from the pump and check the screen.

A faulty burner motor is checked by following the same procedure as used for any other electric motor. If the fuel pump is faulty, the problem will usually show up by causing excessive amp draw on the pump motor. Mechanical failure of the pump will make the motor turn hard, overloading the motor.

**Oil Nozzle.**    A poor or unbalanced flame can indicate a faulty nozzle. To check for this problem, the troubleshooter can inspect the flame with an inspection mirror. An inspection mirror is a small metal mirror on a long handle. The mirror can be inserted into the fire pot through the inspection door. It is used to inspect the inside of the fire pot or the burner face. A visual inspection of the nozzle should be made for a plugged orifice or carbon build-up around the orifice. The strainer on the inlet side of the nozzle should also be checked for dirt. If any of these problems are found, the nozzle should be replaced with a new nozzle of the same type and size.

**Ignition Transformer.**    A faulty ignition transformer can cause a weak ignition spark, resulting in delayed ignition or intermittent ignition.

**Barometric damper***

To check the transformer, the troubleshooter should first check the primary voltage and make sure it is within 15% of the rated voltage for the unit. If it is not, excessive voltage drop to the unit should be considered. If the primary voltage is all right, then the secondary voltage output should be checked. It should be within the limits given by the manufacturer for the unit. Usually the secondary voltage is approximately 10,000 V from either terminal to ground. If the transformer is faulty, it must be replaced.

**Ignition Electrodes.**   Electrodes that are burned on the end or misaligned in the unit can be the cause of faulty ignition. The electrodes should be checked visually. If the ends are burned, they can be dressed with a file to a near point with a blunt end. They should then be replaced in the unit as the manufacturer of the unit recommends. Normally this recommendation is for the tips 1/8 in. apart and 1/8 in. beyond the nozzle. The tips are normally 1/2 in. above the nozzle.

**Fire Pot and Heat Exchanger.**   A visual check of the fire pot and heat exchanger should be made with an inspection mirror to determine if there are any breaks in the lining of the pot or cracks or burnouts in the exchanger. The burner can be removed from the burner port for examining. Some units have inspection plates that can be removed from the upper part of the exchanger to facilitate cleaning and inspection.

**Barometric Damper.**   A **barometric damper** is a tee installed in the vent of a furnace. The open side of the tee faces the furnace room. This side is covered with a damper mounted on pivots. The damper is balanced so it will open and allow room air to enter the vent if the draft in the vent exceeds a predetermined limit. The barometric damper is used on heating units with power burners.

The barometric damper should be adjusted to provide the proper draft over the fire. The draft is necessary for proper combustion and to prevent the flame from being pulled up out of the fire pot.

Draft over the fire is checked by using a draft gauge or an inclined manometer. The reading is taken with the unit operating. The gauge probe is inserted in the small hole in the inspection door in the front of the furnace. The draft over the fire should be between 0.02 in. wg and 0.04 in. wg. Another reading should be taken between the flue outlet and the barometric damper. This reading should be between 0.04 in. wg and 0.06 in. wg. There should be no more than 0.02 in. wg difference between the reading at the flue outlet and the inspection door. If the readings indicate more than 0.02 in. wg difference, the heat exchanger flue passages may be clogged with soot. Too low a difference indicates air may be leaking into the fire areas through cracks in the unit.

**Combustion Safety Control.**    The troubleshooter should check the following parts of the combustion safety control: stack switch, cad cell, and cad cell detector.

*Stack Switch.*    To check a stack switch for proper operation, the control should be placed "in step" by pushing the reset button or setting the lever. If the burner starts, the unit was just off on safety shut-off. If the burner fails to start, then place a jumper wire across the cold contact terminals on the detector. If the burner fails to start with the jumper wire in place, the voltage between terminals 1 and 2 on the control should be checked. If there is less than 105 V across them, the voltage across the burner terminals should be checked. These are the terminals to which the burner leads are attached. The troubleshooter should make sure the thermostat is calling for heat and that the cold contacts are made. If, when the reset button is pushed, there is no voltage, then the fault is in the primary control. The problem could be dirty burner relay contacts. Contacts can be cleaned with a contact cleaner. If the internal circuitry is faulty, the control should be replaced.

*Cad Cell.*    A troubleshooter checking a cad cell safety control on a safety shut-off should make sure the thermostat is calling for heat and then reset the safety switch level on the primary control. If the burner does not start, the flame detector wires on the primary control should be disconnected. If the burner starts, the fault is in the flame detector circuit. The problem could be shorted leads, a flame detector exposed to light, or a short circuit in the detector. Each of these possibilities needs checking. Faulty parts must be repaired or replaced. If the burner still does not start, the voltage between the black and white leads should be checked. No voltage or low voltage indicates no power to the control. The power should be checked. The limit control could be open or faulty.

*Cad Cell Detector.*    To check a cad cell detector for proper operation, the troubleshooter should disconnect the cad cell from the flame detector terminals on the primary control. A jumper wire should be placed across the terminals. The burner should be allowed to run with the jumper in place. With an ohmmeter, the troubleshooter should check the resistance across the detector terminals. If the resistance exceeds the recommended reading for the particular type of cell (approximately 1500 ohms), then the cell is dirty or faulty; the wiring to the cell is faulty; or the cell is crooked in its mounting and is not seeing the flame directly. If the burner locks out on safety while firing with the jumper in place, the fault is in the primary control.

## Electric Furnace Parts

The two parts of an electric furnace not covered elsewhere in this book are the elements and the sequencer. This section gives some procedures that can be used for checking these parts.

**Electric Heating Elements.**    A complaint of not enough heat can result if some of the elements in an electric furnace are burned out. Because there are multiple elements in an electric furnace and they are brought on in sequence, burned-out elements often are not discovered until the outside temperature drops low enough to need all of them for heat. To check for burned-out elements, the troubleshooter should disconnect each element from the power and use an ohmmeter for checking continuity. A burned-out element will give an infinite ohm reading.

**Sequencer.**    To check a sequencer for proper operation, the troubleshooter should make sure that the thermostat is calling for heat. Then she or he should use an ohmmeter to check the on time for each set of con-

tacts. There should be about a 30 to 50 sec delay between each set. In some relay systems, if any one set
of contacts fails to close, the others following it in the
normal sequence will also fail. If any contacts fail to
close or are welded shut, the sequencer should be
changed.

## Air Distribution Parts

Another section of a system that is generally the same
for all units is the air distribution system, which includes the blower, filter, and ductwork. Procedures for
checking each of these are discussed here.

**Blower.**    To check the blower for proper operation,
the troubleshooter should make a visual check of the
wheel to see if it is rotating in the right direction and
if the pulleys are aligned so the belt runs straight. To
check the alignment of the pulleys, a straight rod or a
dowel can be laid in the pulley grooves, from one pulley
to the other. The grooves in the pulleys should line up
with the rod and with each other. If the pulleys are not
in line with each other, one or the other should be
moved on its shaft so they do line up. The set screw
that holds the pulley in place on the shaft should be
loosened and the pulley moved so it will line up with
the other one.

Tension on the blower belt should be checked to
see that the belt is neither too tight nor too loose. If it
is too tight, the bearings will wear excessively. If it is
too loose, the belt will slip on start-up. The correct
tension should allow about 1 in. of play when the two
sides of the belt are pulled toward each other. If the
tension is not right, it can be adjusted by loosening the
motor on its mounts and tipping it either toward
the blower or away, as needed. The motor mount should
have provisions for this adjustment and an adjusting
screw for setting the motor to the correct position.

The blower bearings should be checked for wear at
the same time by manually moving the shaft in the
bearings and observing the play. If the play is excessive,
the bearing should be changed.

**Blower Motor.**    The blower motor is checked for worn
bearings in the same way that the blower was. The
blower motor should also be checked for overheating

by placing a hand on it while it is operating. If the motor is too hot to keep a hand on it for any length of time, it is overloaded. The amperage should be checked on the motor circuit and this compared to the rated amperage on the nameplate. If the operating amperage exceeds the rated amperage, then the motor should be replaced with one of a higher horsepower rating.

**Filters.**    Mechanical filters should be checked visually on a regular basis. If the surface coating of dust and dirt particles starts to close off the air passage, the filters need to be cleaned or changed. If the filters are the replaceable type, they can usually be cleaned carefully perhaps once; then they will need to be replaced. If the filters are permanent, they can be washed in water and detergent and replaced in the unit.

Electronic cleaners usually have a test button or meter on the front of them that shows whether they are working. They should be checked occasionally. The prefilter should be pulled out and cleaned regularly. The electronic cell will need cleaning only about twice a season in a normally dirty atmosphere. If the cell is used in a building where the air is very dirty, the cell should be cleaned oftener. Service on the unit should be performed according to the manufacturer's instructions.

**Ductwork.**    Problems with the ductwork may show up when the system is balanced. The procedure used for balancing a system is found in Chapter 18.

## 21.5  FOLLOW-UP

After identifying the faulty part(s) of a system, the next step is to correct the problems or replace faulty parts. Usually the troubleshooter does this work, but in some cases a service specialist may be needed, especially if a major part or control must be repaired or replaced. The repair or replacement of the parts should be made. Then the unit should be thoroughly tested to make sure that it operates properly.

In most cases, replacing the parts rather than repairing them is less expensive and better in the long run. Most service people are not equipped to repair parts and controls. Also, the cost of the parts is usually much less than the cost for labor to effect a repair. New parts normally carry a warranty, while repaired parts do not, and the charge for repeat failure is less when new parts are used.

## 21.6  SUMMARY

Like all other mechanical equipment, heating equipment needs maintenance to keep it working properly. Occasionally it needs service to adjust or repair malfunctioning parts. The process of investigating the problem when a unit stops working correctly is called troubleshooting. Troubleshooting takes place after the equipment is no longer doing what it is supposed to and before the actual service work is performed to correct the problems.

The usual result of a breakdown of the heating equipment is that the temperature in the building will no longer stay at the set point of the thermostat. Then usually the building users will complain. The complaints generally fall into one of the following categories: no heat, insufficient heat, too much heat, fluctuating heat, noise, odor, or excessive cost of operation.

A systematic investigation is made in troubleshooting a piece of equipment. First, the troubleshooter must understand the operation of the equipment and the system. Second, the troubleshooter must understand what the complaint is and its probable cause. Third, the troubleshooter must decide what general part of the unit is affected by the breakdown. Fourth, the troubleshooter must check that section of the system to identify the actual part (or parts) that is faulty.

After the part that is not functioning correctly is identified, it must be adjusted or replaced as necessary to make the unit functional again. After the repair, the unit should be operated through a normal cycle to make sure it is functioning properly.

## 21.7  QUESTIONS

1. Complaints about malfunctioning equipment usually originate with the building _____.

2. Match the complaints in the first column with the conditions in the second column.

1. No heat
2. Insufficient heat
3. Too much heat
4. Heat fluctuates
5. Noise
6. Odor
7. Cost of operation

a. Temperature above set point
b. Dangerous
c. Heat output reduced
d. Total system involved
e. Unit does not work
f. Unit will not shut off
g. Mechanical failure

3. A complaint of too much heat is not one that needs immediate attention. True or false?

4. When a complaint of fluctuating heat is received, a troubleshooter knows that only one part of the system must be at fault. True or false?

5. Four steps should be taken when troubleshooting a heating unit. They are: (a) _____, (b) _____, (c) _____, and (d) _____.

6. The actual operation of a heating unit can be affected by other components in the system, such as _____, _____, or _____.

7. List the four major unit component categories.

8. If a system that is heated by an electric furnace overheats, what major family of components should be suspected? What is the first part to check in that family?

9. When troubleshooting an oil furnace, if a complaint of odor has been made, what part should be checked after the piping has been inspected?

10. If a gas furnace is being checked on a complaint of insufficient heat, why aren't the electric disconnect and fuses checked first?

11. When checking the electric power components on a system, is it necessary only to check the power at the main disconnect? Why?

12. There are two methods used to check fuses to see if they are faulty. One method uses an ohmmeter. What meter is used for the other?

13. The location of a thermostat in a building has nothing to do with its operation of the system. True or false?

14. Match each of the controls with the proper description
    of its action in the unit.

    1. Limit control
    2. Blower control
    3. Blower relay

    a. Starts blower on
       temperature rise
    b. Controlled by switch
       on thermostat
    c. Turns off furnace on
       abnormal
       temperature rise

15. Air distribution is a common function in any of the
    systems described in this text. Troubleshooting of the
    air distribution system is nearly the same as doing an
    air balance. True or false?

# CHAPTER

# 22

# Heat Pumps

## 22.1 INTRODUCTION

So far, this book has covered conventional types of heating systems. Heat pumps are discussed here because their use is becoming more common and because their high operating efficiency will certainly cause an increase in that use.

A heat pump is a combination cooling-heating unit. Until recently, heat pumps were used mainly in applications where the primary need was for cooling, and heating was of secondary concern. The fact that a heat pump provided heat so efficiently was considered a good reason to use one instead of a cooling unit with a furnace. Because of the realization that known sources of energy are fast being depleted and because these sources are becoming so expensive, the use of heat pumps as a primary source of heat has become important.

This chapter explains how a heat pump works and describes the parts and controls. The chapter also covers the types of units that are available and the application and sizing of units.

## 22.2 HEAT SOURCE

A heat pump, unlike a typical heating unit, does not produce heat. It simply transfers heat from one place to another. For a heat pump to be practical, heat must be available for transfer. Any material with a temperature above absolute zero has heat in it however, so sources are common. The soil, **groundwater,** and the air in our atmosphere are all good potential sources of heat.

A typical heat pump used for heating buildings operates on the same principle as does a household refrigerator. A refrigerator extracts heat from the food that is placed in it and expels that heat outside the cabinet; a heat pump extracts heat from a source, called a **heat sink,** and expels the heat where it can be used for heating a building.

The two most common sources of heat for a heat pump system are outside air and groundwater. In the winter, the outside air temperature will range from below zero in some localities to as high as the inside design temperature. Groundwater tends to remain about the same temperature all year, usually around 45°F. A typical heat pump will work well with a heat sink temperature down to about 10°F. In almost all habitable areas, the average air temperature is higher than 10°F at any time of the year. Groundwater is always a good source of heat for a heat pump.

In a typical heating system, the temperature of the air inside the building is maintained at about 75°F. The temperature of outside air and groundwater is too cold to heat this air directly. But a heat pump not only extracts heat from the air or water; it also raises the temperature. The higher-temperature air from the heat pump is a good source of heat for the building.

**Groundwater**  Water from an underground source, such as a spring or a well

**Heat sink**  Source of heat for a heat pump system, usually outside air or groundwater

**Refrigerant***

## 22.3  SAVING ENERGY WITH A HEAT PUMP

Because picking up heat from a heat sink and moving the heat to a location where it is wanted is less expensive than generating that heat by a combustion process or with electricity, a heat pump is a very efficient device for heating. It saves both energy and money.

Since the heat in outside air or groundwater is available without direct cost, transferring this heat is less expensive than generating heat by the combustion process or by electric heating elements. The only cost is the cost for the energy to operate the heat pump. Electricity is usually used for this purpose. One watt of electricity converts directly to 3.41 Btu of heat in a resistance heating element. Much more heat than this can be transferred by the use of 1 W when it is used to operate a heat pump. This savings in energy to produce heat can be as high as 3.5 to 1. In other words, 1 W of energy will produce the same amount of heat as 3.5 W would in a conventional system. Most of this heat is extracted from the heat sink.

## 22.4  HOW A HEAT PUMP WORKS

An understanding of how a heat pump works requires an understanding of the refrigeration cycle. In a refrigeration system, a chemical compound called a **refrigerant** is used to transfer the heat. A refrigerant is a liquid

*Term defined in text.

**Ambient**    "Surrounding," such as the temperature in the surrounding air

**Expansion device**    Device used in a refrigeration system to meter the flow of refrigerant in such a way that a high pressure is maintained on one side of the device and a low pressure on the other

**Evaporator**    Coil in an air-conditioner or a heat pump in which the refrigerant evaporates to produce cooling

**Condenser***

**Air-to-air system***

**Reversing valve***

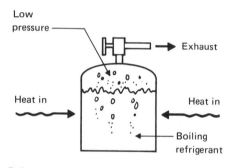

FIGURE 22-1 _____

Boiling refrigerant. When a refrigerant boils, or changes state, heat is absorbed.

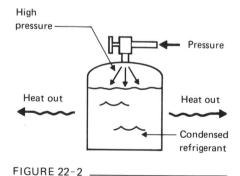

FIGURE 22-2 _____

Condensed refrigerant. When a refrigerant condenses, heat is given up.

with a boiling point temperature below that of normal **ambient** temperatures at atmospheric pressure. The refrigerant would boil, turning into a gas, at normal room temperature if it were not kept under pressure.

The boiling point temperature of any liquid is directly related to the pressure exerted on it. If the pressure on a liquid is increased, the boiling point temperature of that liquid goes up; if the pressure is reduced, the temperature goes down. By controlling the pressure on the refrigerant in a refrigeration system, the boiling point temperature is controlled.

If the boiling point temperature of a refrigerant in a container is lowered by reducing the pressure, then the temperature of the refrigerant will be lower than the air around the container. Because of the temperature difference, heat will flow from the air around the container into the refrigerant, as shown in Figure 22-1. The refrigerant will boil, or change to a gas, and as a result will cool the surrounding air. Conversely, if the pressure on the refrigerant is increased, so the boiling point temperature is higher than the surrounding air, then the refrigerant will give up heat to the air. See Figure 22-2. At this point, the refrigerant is in the vapor state. As the refrigerant gives up heat, it is condensed back into a liquid. Because the refrigerant gives up heat, the medium surrounding the container is heated.

In an air-conditioning system, the pressure is reduced on the refrigerant in that part of the system where it is desired to pick up heat. The boiling point temperature is lowered below that of the air or water, and heat is picked up by the refrigerant. The pressure on the refrigerant is raised in that part of the system where the heat is to be rejected. In this case, the temperature of the refrigerant is raised above that of the air or water in that part of the system, and heat flows from the refrigerant to that air or water. This process is the typical refrigeration cycle.

An operating refrigeration system has both a high pressure side and a low pressure side. Coils are used for heat exchangers. One coil has refrigerant at a high pressure flowing through it, and another has refrigerant at a low pressure. See Figure 22-3.

The pressure is maintained on the high side of the system by a compressor. It is controlled on the low side by an **expansion device.** The coil on the low pressure side of the system is called an **evaporator** and is used to cool the air or water that is used to cool a building. The coil on the high side of the system is called the **condenser** and has air or water flowing over it to carry away the heat. Figure 22-4 shows a diagram of a typical air-to-air system. An **air-to-air system** is one in which the air is used as both a heat source and as a medium to carry the cool air.

When the system is used as a heat pump, the relative position of the two coils is reversed by changing the pressures in the two sides of the system. The coil that is rejecting heat in the air-conditioning cycle is now collecting it, and the other coil is giving up the heat. On an air-to-air system, the outside coil is picking up heat from the outside air and rejecting it inside, where the heat can be used to heat a building. A refrigerant control device called a **reversing valve** is used to reverse the direction of flow of the refrigerant. Figure 22-5 is a diagram of a heat pump system showing the relative position of the indoor and outdoor coils and the location of the reversing valve.

Heat pumps usually are built so they can be used for heating or cooling. They have controls and devices that make it possible to reverse the direction of the refrigerant flow, which changes the pressure in the two sides of the system. The heat pump can operate as an air-conditioner in one mode of operation and as a heater in the other. A heat pump can be called a reverse-cycle air-conditioner.

## 22.5  OPERATING DEVICES AND CONTROLS

A refrigeration system, or air-conditioning system, has six major components. They are:

1. Evaporator coil,
2. Condenser coil,
3. Compressor,
4. Expansion device,
5. Refrigeration lines,
6. Controls.

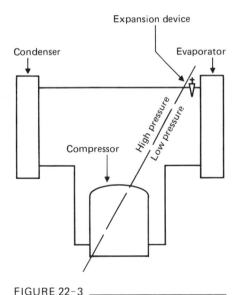

FIGURE 22-3 _____
Diagram of a refrigeration system showing the high pressure side and the low pressure side

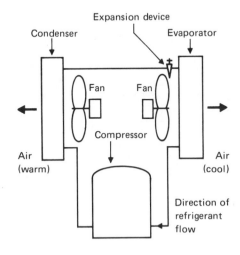

FIGURE 22-4 _____
Diagram of a typical air-to-air air-conditioning system

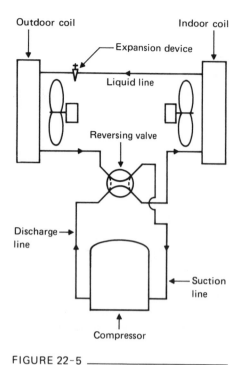

FIGURE 22-5 _____

Diagram of heat pump system showing reversing valve

These major components are shown on the diagram in Figure 22–6.

A heat pump has these same major components, but some additional parts are used to make it operable in the reverse cycle. These additional parts are:

1. Reversing valve,
2. Additional expansion device and check valves,
3. Heat-cool thermostat,
4. Defrost controls.

These parts are shown on the diagram in Figure 22–7. Each will be discussed in the sections that follow.

## Operating Devices

The reversing valve and the additional expansion device with the **check valves** are needed to make a heat pump operate in the heating mode. The reversing valve is the device that actually reverses the direction of flow of the refrigerant in the system. The added expansion device and the check valves are used to change the

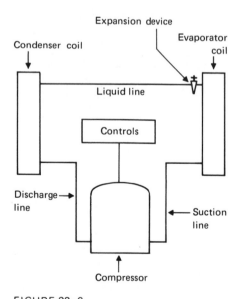

FIGURE 22-6 _____

Major components of an air-conditioning system

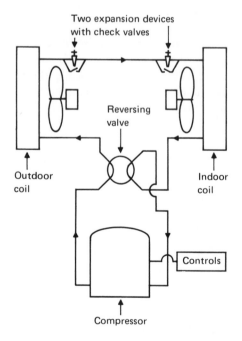

FIGURE 22-7 _____

Heat pump system showing the major parts

point in the system where the pressure changes from high to low.

**Reversing Valve.**    To change a heat pump operation from cooling to heating, it is necessary to reverse the direction of flow of the refrigerant. The compressors used in refrigeration systems, however, are nonreversible; that is, the refrigerant goes through them in only one way. It always goes in one port, called the **suction port,** and comes out the other, called the **discharge port.** Another device is needed to reverse the flow of the refrigerant. This part is the reversing valve. The reversing valve takes the discharge gas from the compressor and routes it to one or the other of the coils, depending upon whether cooling or heating is wanted. Figure 22–8 shows a reversing valve. The valve also takes the return vapor from the opposite coil and routes it to the compressor suction port.

Most reversing valves are piston type. A piston is moved back and forth inside a cylinder, and it opens or closes ports to the lines according to its position in the cylinder. The piston and the line ports can be seen in the photograph in Figure 22–9.

**Check valve**    Mechanical device placed in a pipe that will allow liquids or gases to flow in only one direction in the pipe

**Suction port***

**Discharge port***

FIGURE 22-8 _____

Typical reversing valve used in a residential heat pump

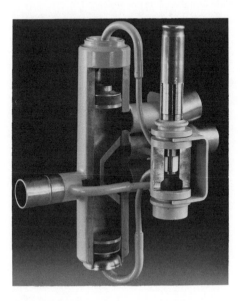

FIGURE 22-9 _____

Cutaway photograph of a reversing valve showing the operating parts and the pilot-operated valve

**Pilot-operated valve**   Valve operated by system pressure that is controlled by a smaller valve

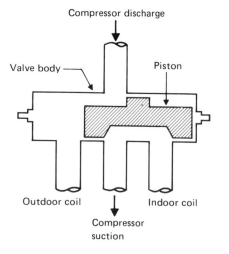

FIGURE 22-10 _____

Diagram of a reversing valve showing the slide in the cooling position

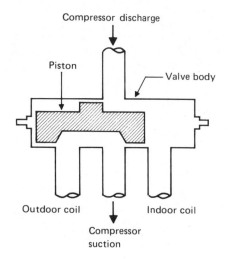

FIGURE 22-11 _____

Diagram of a reversing valve showing the slide in the heating position

The reversing valve is a **pilot-operated valve.** An electric solenoid valve controls pressure on the ends of the cylinder in the main valve, and the piston is actually moved by that pressure, which is exerted by the refrigerant from the operating system. Figure 22–9 also shows the pilot valve.

The solenoid of the pilot valve is energized by the thermostat. On a call for cooling, a small piston in the pilot valve opens a port so that one end of the master cylinder is connected to the suction line of the refrigeration system. The reduction of pressure in the end of the cylinder causes the master piston to move in that direction. The ports in the master valve are lined up so the suction line from the compressor is connected to the line from the inside coil, and the discharge line from the compressor is connected to the line to the outside coil. When the piston is in this position, the unit is in the cooling cycle. Figure 22–10 shows a reversing valve in this position.

On a call for heat, the pilot valve moves the master piston the other way. The ports in the valve are then lined up so the suction line from the compressor is connected to the line from the outside coil, and the discharge line from the compressor is connected to the line going to the inside coil. This is the heating cycle. Figure 22–11 shows a reversing valve in this position.

**Expansion Device.**   The expansion device on an air-conditioner is located on the liquid line where the line enters the evaporator coil; but on a heat pump, the function of the two coils is reversed when the system is operating in the heating mode. The expansion device used for the cooling cycle will not work on the heating cycle. So that the system will be workable for both cooling and heating, another expansion device is located on the liquid line at the entrance to the other (outside) coil.

**Check Valves.**   Refrigerant will flow through a valve in only one direction, so a bypass line is installed around each valve. A check valve in each bypass makes sure refrigerant goes through the valve in the proper direction. The check valve allows the refrigerant to pass through on one cycle but checks the flow so the refrigerant will go through the other valve on the other cycle.

See Figure 22–12 for the location of the bypass and the extra expansion device.

# Controls

In addition to the normal controls used on air-conditioning units, a heat pump must have a heat-cool thermostat and controls to defrost the outside coil when frost or ice builds up on it.

**Heat-Cool Thermostat.**    Because a heat pump is a unit that can provide both heating and cooling, a thermostat must be used that will call for either heat or cooling as needed. This purpose can be accomplished in one of several ways. The most common way is to use a four-bulb thermostat, with each bulb controlling one part of the operation. Two bulbs control the cooling and two bulbs the heating. A heat pump thermostat is shown in Figure 22–13.

With this type of thermostat, the system is normally in the heating mode. When the first stage of cooling is called for, the reversing valve switches to the cooling position. The second-stage cooling bulb then brings on the compressor. When the thermostat is in the normal heating position, the first-stage heat bulb energizes the compressor, and if the second-stage heat bulb calls for more heat, the thermostat brings on the

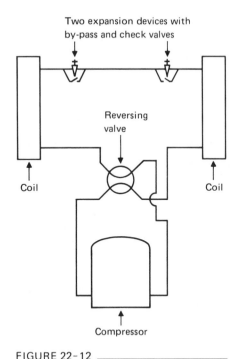

FIGURE 22-12 _____

Diagram showing the heat pump system with the extra expansion device, the refrigerant by passes, and the check valves

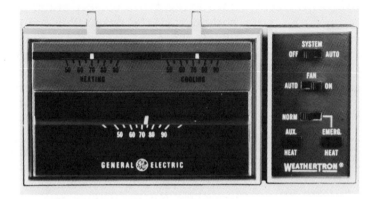

FIGURE 22-13 _____

Thermostat used on a heat pump system

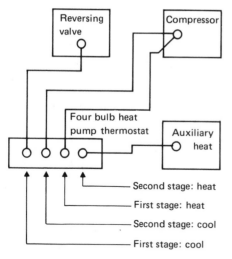

FIGURE 22-14
Block diagram showing the heat pump control components in relation to the thermostat

auxiliary heat. This sequence is shown in the block diagram in Figure 22–14.

**Defrost Controls.**    On a heating cycle, when the outside coil is acting as an evaporator, the temperature of the refrigerant in the coil will often be below 32°F. When the temperature falls below 32°F, frost or ice can form on the coil from the moisture in the air that condenses on it. If this frost or ice starts to build up, the amount of air across the coil is reduced, and the refrigerant temperature drops still more. Eventually, the unit will go off on low pressure, or damage will occur to the compressor.

To prevent this buildup of ice, controls are used to monitor the condition. If frosting or icing starts, the unit operation is reversed. Hot refrigerant gas is cycled through the coil so defrosting takes place. These controls are called the **defrost controls.** When the system is reversed, it is actually operating in the cooling mode. To prevent cold air from being blown into the building, the defrost controls also bring on the **auxiliary heat.** The defrost cycle lasts only a few minutes. As soon as the controls sense that the coil is free of ice or frost, they put the unit back into the heating mode. This sequence is shown on the block diagram in Figure 22–15.

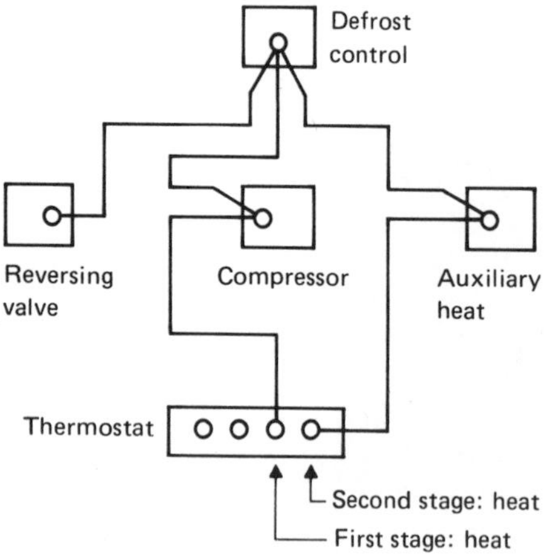

## 22.6   TYPES OF UNITS AVAILABLE

Two major types of heat pump systems are used on buildings: (1) the **compatible-component type unit,** in which the parts are manufactured so they work together, and (2) the **built-up type unit,** in which the components are selected and the system designed by an engineer. The first type is almost always used on residential buildings and smaller commercial buildings, and the capacities are seldom very large. The second type is most often used on larger buildings, and the capacity of the system is usually much greater. Within these two major types, other divisions are found concerning operating characteristics, physical characteristics, and heating-cooling capacity of the units.

There are four different possibilities for operating characteristics. These characteristics can be listed by the heat source and the heating-cooling medium:

FIGURE 22-15
Block diagram showing the components of the defrost system in relation to each other

1. A unit that uses water as a heat source and water as the heating-cooling medium,
2. A unit that uses water as a heat source but air as the heating-cooling medium,
3. A unit that uses air as the heat source and water as the heating-cooling medium,
4. A unit that uses air as the heat source and also as the heating-cooling medium.

The division by physical characteristics includes the **package units,** which have all of the components in one sheet metal box, and the **split systems,** which have the components located in different parts of the building or even outside the building.

The division by equipment capacity will be discussed in sections by categories.

# Residential and Small Commercial Units

Heat pumps for small commercial and residential buildings are nearly all the compatible-component type unit—that is, they have components that are meant to work together. Some are split systems, and some are packaged units. Figure 22–16 shows an outdoor unit used with a split system heat pump. The indoor unit is a coil mounted on the furnace or a blower coil unit in the building. The two parts of the system are connected by refrigerant lines.

The capacity of package units and compatible-component units is somewhat limited. The units are rated from 1 ton to about 15 to 20 tons capacity. This measurement is cooling capacity. One **ton of cooling** is equivalent to 12,000 Btu/h. The heat capacity of a heat pump is more difficult to state because the output varies as the temperature of the heat source medium varies. Generally, the heating capacity of an air source unit when the outside air temperature is 45°F is about the same as the cooling capacity of the unit at design conditions.

**Package Heat Pumps.**    Package heat pumps are completely self-contained units. All of the components are contained in one sheet metal cabinet. The pumps are built so they need only to be connected to a source of water, with a drain, if they are water source units; or,

**Defrost controls***

**Auxiliary heat**    Heat that is added in a heat pump system by means other than the heat pump

**Compatible-component type unit***

**Built-up type unit***

**Package unit***

**Split system***

**Ton of cooling***

FIGURE 22-16

Outdoor unit of a split-system heat pump

if they are air source units, they need to be installed outside the building. Figure 22–17 illustrates a package unit.

Some smaller package units are built to be installed in windows, with part of the unit outside and part inside, or through an outside wall with the same arrangement. These units are quite small. Their maximum capacity is about 2 tons of air-conditioning and proportionate heating ability. These units are meant to be used for conditioning one room at a time. Figure 22–18 illustrates a typical window unit.

Larger package heat pumps—from about 2 tons to 15 tons of capacity—are designed to be used with duct systems.

**Split Systems.**    Nearly all split-system heat pump units use outside air as a source of heat. The outside part of the system includes the outside coil, a blower to move the air across the coil, and the compressor with the controls for these components. The inside part of the system includes the inside coil and whatever

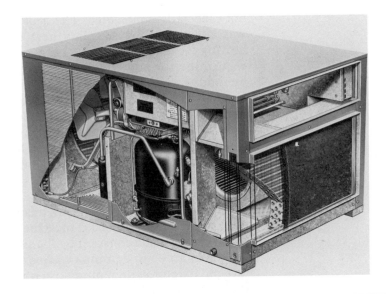

FIGURE 22-17

Package heat pump

blower is used to move the conditioned air. Split systems are available with indoor coil units to be used with remote blowers and also with blower coil units. Split systems are usually used in applications where it is not convenient to duct the air into the system from outside.

## Units for Larger Buildings

Heat pump systems for larger buildings that need more capacity are often the built-up type of unit, that is, the components are selected separately. Water is most often the source of heat and also the medium used to carry the heat to the point of use. Large multistory buildings and buildings heated and cooled from a central station system are included in this category. The capacities of these units may be in the hundreds of tons.

## 22.7  SIZING AND APPLICATION

In the past, when heat pumps were used mainly as air-conditioning units, they were sized to take care of the cooling load on a building. Their heating capacity was

FIGURE 22-18

Window-model heat pump

**Balance point***

not especially considered. The heat output of the unit was augmented by auxiliary heat to provide enough heating for the building. In climates with high cooling requirements and low heating requirements, this method of sizing is fine. In areas where the cooling load is small and the heating load is great, however, this method of sizing is not adequate. In such climates— or in situations when a heat pump is used primarily as a heating unit—the selection should be made to provide as much heat as possible, compared to the total load. The cooling capacity is not as important.

The heat output of a heat pump is dependent upon the temperature of the heat sink medium and the quantity of the medium used. Since the quantity of air or water can be made constant by the sizing of the outdoor fan on an air source pump or adjusting the water flow rate on a water source pump, then the temperature of the air or water is the only variable. The temperature of groundwater is quite stable, so the heat output of a water source pump does not vary too much. The outside air temperature, when used as a source, will vary greatly during the heating season.

Since both the output of the heat pump and the heat loss of a building vary with the outside temperature, a correlation can be drawn between the two. Unfortunately, the heat output of the unit goes down as the air temperature goes down.

The graph in Figure 22–19 shows the heating capacity of various heat pumps at varying outside temperatures. The Btu/h output of the pumps is shown in Btu/h on the scale on the left of the graph. Outside design temperatures are shown on the scale across the bottom. The heat output of the pumps is shown on the diagonal lines coming down from the right upper corner. Where any heat pump line crosses a vertical temperature line, the Btu/h output of the unit at that temperature is read on the Btu/h scale at the left. Each of the heat pump output curves is for a different capacity unit.

## Sizing the Unit

To size a heat pump for a specific job, the heat output of the unit should be compared to the heat loss of the building. Because the output varies with the outside temperature and the heat loss also varies with the tem-

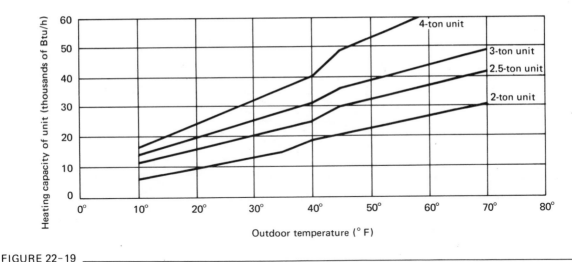

FIGURE 22-19

Typical capacity curves for four sizes of heat pumps

perature, the two can be compared on one graph. Using the heat output graph shown in Figure 22–19 and adding a heat loss curve to it for the particular building under consideration, the loss can be visually compared to the output and the temperature at which the output offsets the loss can be seen, as in Figure 22–20. The heat loss curve is drawn from 65°F outside temperature at 0 Btu/h, to the Btu/h loss figured for the building at the outside design temperature at which the loss was figured. The point where the heat pump output curve crosses the heat loss curve is called the **balance point** of the system. The heat pump for a particular job can be selected by picking the unit that will give the most desirable balance point.

**Balance Point.**    The balance point for any job is related to the heat loss curve for that job only and the heat output curve for a specific heat pump. Since the heat output of a heat pump goes down as the outside temperature goes down, the efficiency of the unit also goes down with the temperature. It is important to select a unit that will operate as efficiently as possible yet is not too large because the cost of a heat pump is relative to its size and capacity. A balance point of about 35°F is right for most jobs. When the heat loss curve is drawn on the graph, a unit should be selected that gives a balance point temperature of 35°F.

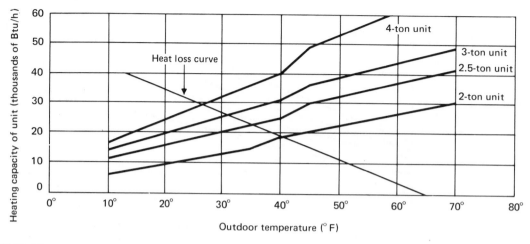

FIGURE 22-20

Heat loss for a given building plotted against the capacity curves for a selection of heat pumps

## EXAMPLE 22–1

A building to be constructed in Seattle, Washington, where the outside design temperature is 15°F, has a total heat loss of 38,600 Btu/h. Select a heat pump for that building.

### Solution

Draw the heat loss curve on the balance point chart, as shown in Figure 22–21. Look for the heat pump output curve that crosses the heat loss line closest to 35°F outside temperature. For this job, a 2.5-ton heat pump should be used.

**Coefficient of Performance.** Since the heating output of a heat pump varies with the temperature of the heat source but the amount of energy required to operate the unit does not vary significantly, then the efficiency of the unit varies with the temperature also. The efficiency of the unit is found by dividing the heat output by the heat equivalent of the energy used to operate the unit. The resulting figure is called the coefficient of performance (COP) of the unit:

$$COP = \frac{\text{Btu/h output of unit}}{\text{Btu/h input of energy}}$$

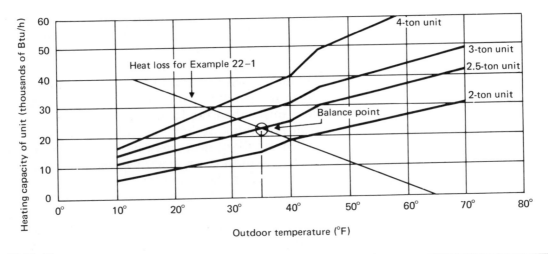

FIGURE 22-21

Balance point chart for the building described in the text

This COP figure will vary from about 1.5 at the lowest outside operating temperatures to a high of about 3.5 at the highest temperatures.

For an electric heat pump, the heat equivalent of the energy used to operate the unit is provided in the manufacturer's specification sheets. The wattage to operate the unit at various outside temperatures is shown there with the heating capacity for the same conditions. The wattage is multiplied by the factor of 3.41 Btu/W to find the Btu in heat for the electric energy.

## EXAMPLE 22–2

A heat pump produces 34,500 Btu/h of heating at a given set of conditions and requires 3679 W to operate at that set of conditions. Find the COP.

### Solution

$$COP = \frac{\text{Btu/h produced}}{\text{Btu/h used}}$$

$$= \frac{35,400}{3679 \times 3.41}$$

$$= 2.75$$

The COP of the unit at the operating conditions is 2.75.

The COP of a family of typical heat pumps at various outside temperatures is shown in Figure 22–22.

**Auxiliary Heat.**    The heating capacity of a heat pump is enough to heat the building only down to the balance point temperature. Anytime the outside temperature goes lower than the balance point temperature, extra heat is required to heat the building. This extra heat, called auxiliary heat, may be supplied by any other source. In most cases, electric resistance units are used as an auxiliary heat source.

Most heat pumps have a cut-out point at about 10°F to 15°F. The low pressure cut-out on the unit will shut off the operation if the outside temperature gets below that point. Thus, if the design temperature used for the heat loss is higher than 10°F, theoretically only enough heat needs to be supplied by the auxiliary heat to provide enough to make up the difference between the output of the pump and the heat loss of the building at the design temperature. See Figure 22–23.

If the design temperature used in the heat loss figures is less than 10°F, enough auxiliary heat should be furnished to provide the total amount of heat for the building. See Figure 22–24.

In practice, it is better to provide enough auxiliary heat to carry the entire heating load at design temperature in any case. Enough heat must be available during defrost cycles and in case the heat pump is shut down for service.

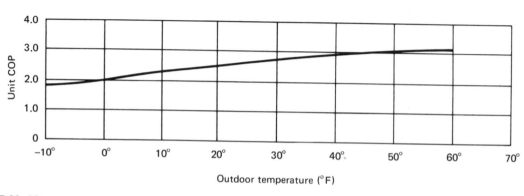

FIGURE 22-22

Typical coefficient of performance for air-to-air heat pumps

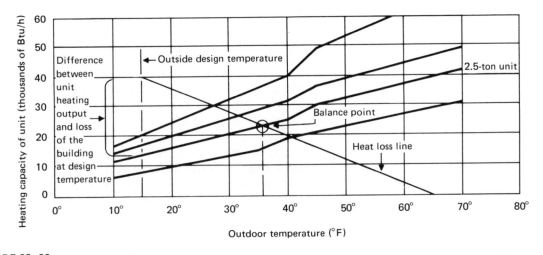

FIGURE 22-23

Amount of heat required as auxiliary heat for the example job

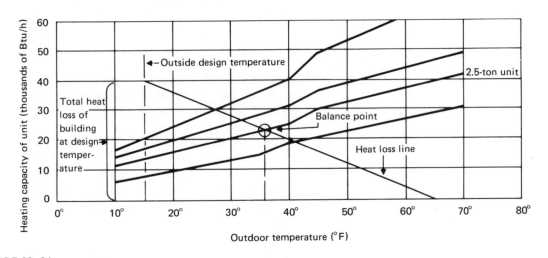

FIGURE 22-24

Amount of heat required as auxiliary heat if the design temperature of the job
is less than the cut-out temperature for the heat pump

# Application

The design of a heat pump system requires more attention to certain application details than does the design of a conventional heating system. Because a heat
pump is a refrigeration unit, the quantity of air or water
used as the heat source is critical. If the cfm of air or

the gpm of water on either side of the system drops too low, the temperature of the refrigerant in the system will increase, and the unit will either cycle off on high pressure, or damage to the compressor will be sustained. The designer must make sure that there is an adequate supply of air or water to each side of the system and that the ducts or piping, blowers, and pumps are sized correctly.

On the heat sink side of the system, an adequate source of air or water must be available at a high enough temperature to allow the extraction of heat at a constant rate. The source must be such that the air or water can be used with the equipment that the designer chooses for the job.

On the heat distribution side of the system, the ducts or piping and the blowers or pumps must be designed so the quantity of air or water to carry the heat through the building will not vary.

The distribution system that carries the air or water must be designed so the building will be heated properly.

**Heat Source.**    The heat sink, or source of heat for the heat pump, can be air or water. If it is air, the usual practice is to use the outside air. If it is water, the usual sources are public system, stream or lake, or wells.

A heat pump system that uses air as its heat source is relatively simple to design. The part of the system containing the outside coil needs to be located outside the building so it has a source of clean air. This part of the system contains a blower to draw the air across the coil. In some cases, a designer may be tempted to install the outside part of a unit inside the building and duct the air to it from the outside. Unless the unit is especially designed for this application, this placement should not be done. The air mover in a typical unit does not have a blower that will move air against the resistance of the ductwork, and the unit will be under aired if it is installed in that way.

Two precautions should be taken in locating the outside unit: (1) the discharge air from one unit should not enter the intake of another, and (2) the discharge air from any unit should not recycle back into its own entering air side. In either case, the temperature of the entering air would be too low, and the heating capacity of the unit would be reduced.

Water source heat pumps usually are used in one of two situations: when there is a plentiful, cheap supply of water available or when the unit or units need to be installed inside a building where it is difficult to get air to them from outside. Water is a good choice in both cases because it is a good source of heat for a heat pump. The temperature of the water is normally quite stable, and the output of the unit does not vary greatly.

When public utility water is used as the heat source, the cost of the water can be an important factor. In some cases, the waste of the water after it is used is more important. In most large U.S. cities, the waste of large quantities of water is not acceptable.

Lake or stream water that is used must be from a location where the quantity does not vary. In most cases, the water from either source should be filtered and possibly may have to be treated for contaminants before it is used. The water will normally have to be pumped from the source to the point of use.

If the size of the installation warrants the cost, the use of well water may be the best answer to the source for water for a heat pump system. When wells are used, usually at least two are drilled. One well is used for the source of the water, and the other is for the waste of the used water. The wells are spaced far enough apart so the waste water will mix with groundwater before entering the supply well again. Usually the system is designed so the use of the two wells can be alternated to control the water level that supplies the wells.

**Heat Distribution.**     On the distribution side of the system, either air or water can be used as the medium to carry the heat. When air is used, it is normally the same air as is circulated through the building for heating. When water is used, it is distributed to terminal units where a heat exchanger is used to heat air, which is then used to heat the building.

When air is the medium, the temperature rise through the inside coil is much lower than that for any other type of heating system, probably in the range of 15°F to 20°F. The air introduced into the building through the registers will be much cooler than that from a fuel-fired or electric heating unit. Therefore, the registers must be located very carefully to prevent the relatively cool air from blowing on occupants of the building. Air at a discharge temperature normal with heat pumps blowing directly on a person will feel like a cold draft.

When a heat pump is used on a hydronic heating system, the main concern is that the entire system be properly sized for the quantity of water required through the heat pump and for the temperature of water available from the heat pump.

## 22.8 SUMMARY

A heat pump is a refrigeration unit that collects heat and moves it from one place where it is not needed to a place where it is needed. The pump works on the same principle as any refrigeration unit. When the pump is operating in the heating mode, it can be called a reverse-cycle air-conditioning unit.

A heat pump is a more sophisticated piece of equipment than a gas, oil, or electric furnace. It also makes more heat available per unit of energy used than any of the other furnaces. Because of its high efficiency, it is fast becoming one of the most popular units used for heating buildings.

Manufacturers of heat pumps build units in a wide variety of types and capacities. Self-contained units and compatible component units, to be used as split systems, are available from 1 ton of air-conditioning capacity up to 15- or 20-ton capacity. The components to build up larger systems are available for almost any capacity desired.

Some special consideration has to be given to the application of heat pumps. They need a constant supply of air or water as a source of heat and an adequate distribution system to ensure the proper air or water to be used as the heating medium.

## 22.9 QUESTIONS

1. A heat pump, unlike more conventional units, does not produce heat; it simply _____ it from one place to another.

2. A heat pump extracts _____ from a source and raises the _____ of the medium high enough so it is usable in a heating system.

3. The two most common sources of heat for a heat pump to transfer are _____ and _____ _____.

4. The source of heat utilized by a heat pump for transfer is called the: (a) heat tank, (b) heat reservoir, (c) heat sink, or (d) none of the above.

5. A heat sink has to have a temperature greater than that in the building to be heated. True or false?

6. A heat pump transfers heat so efficiently that there is no cost involved in its operation. True or false?

7. In some cases, the efficiency of a heat pump is as high as: (a) 2 to 1, (b) 3.5 to 1, (c) 4 to 1, or (d) 1.5 to 1.

8. If the pressure is increased on the liquid in a closed container, the boiling point temperature of that liquid will: (a) increase, (b) decrease, or (c) stay the same.

9. In an air-conditioning system, the low pressure side of the system is _____ the building, and the high pressure side is _____ the building.

10. When a heat pump is operating in the heating mode, the inside coil is on the _____ pressure side of the system, and the outside coil is on the _____ pressure side.

11. A heat pump can be called a _____ _____ air-conditioning unit.

12. Name the six major divisions of parts found in an air-conditioning unit.

13. Name the four additional parts a heat pump has besides those found in an air-conditioning unit.

14. A reversing valve is called a pilot-operated valve because a _____ operates a small valve, to operate the large valve.

15. The check valves on a heat pump are placed in a _____ line around the expansion valve.

16. A heat-cool thermostat is necessary on a heat pump system because the unit can be used for either _____ or _____ the building.

17. A defrost system is necessary on a heat pump to prevent frost or ice from building up on the _____ coil.

18. Residential-sized heat pumps are available from about _____ to _____ tons of air-conditioning.

19. When a building heat loss curve is graphed against a heat pump output curve, the place where the two cross is called the _____ _____.

20. _____ _____ is used to supply heat to take care of heat loss below the balance point temperature.

21. Water is usually considered to be better for a heat sink than air because _____.

# Appendix Tables

## TABLE A-1
Weight and Heat Value for Some Fuel Woods

| Species | Btu/lb[a] | Weight/cord[b] | Btu/cord[c] |
|---|---|---|---|
| Alder | 6720 | 2540 | 17,068,800 |
| Apple | 7890 | 4400 | 34,716,000 |
| Ash | 8300 | 3440 | 28,552,000 |
| Aspen | 5980 | 2160 | 12,916,800 |
| Birch, white | 7470 | 3040 | 22,708,800 |
| Cedar, western | 6720 | 2060 | 13,843,200 |
| Cherry | 7470 | 3200 | 23,904,000 |
| Cottonwood | 5980 | 2160 | 12,916,800 |
| Dogwood | 8300 | 4230 | 35,109,000 |
| Elm | 7470 | 2260 | 16,882,200 |
| Fir, Douglas | 7470 | 2970 | 22,185,900 |
| Fir, grand | 5980 | 2160 | 12,916,800 |
| Hemlock | 6720 | 2700 | 18,144,000 |
| Hickory | 9500 | 4240 | 40,280,000 |
| Juniper | 7470 | 3150 | 23,530,500 |
| Larch, western | 7890 | 3330 | 26,273,700 |
| Madrone | 8300 | 4320 | 35,856,000 |
| Maple, red | 7885 | 3200 | 25,232,000 |
| Maple, sugar | 7890 | 3680 | 29,035,200 |
| Oak, red | 8300 | 3680 | 30,544,000 |
| Oak, white | 9500 | 4200 | 38,900,000 |
| Pine, lodgepole | 5980 | 2610 | 15,607,800 |
| Pine, ponderosa | 6720 | 2240 | 15,052,800 |
| Pine, white | 6720 | 2250 | 15,120,000 |
| Pine, yellow | 7890 | 2610 | 20,592,900 |
| Poplar | 5980 | 2080 | 12,438,400 |
| Redwood | 7470 | 2400 | 17,928,000 |
| Spruce | 5980 | 2070 | 15,462,900 |
| Spruce, Norway | 5980 | 2240 | 13,395,200 |

[a] Approximate heat value, Btu/lb, for air-seasoned (20% moisture) wood

[b] Approximate weight, lb/cord, for average density of air-seasoned wood

[c] Approximate Btu/cord for air-seasoned (20% moisture) wood and for average density

TABLE A–2

Orifice Sizing Chart for Utility Gas (Cubic Feet per Hour at Sea Level)

Specific Gravity = 0.60
Orifice Coefficient = 0.9
For utility gases of another specific gravity, select factor from Table 5,
For altitudes above 2,000 feet, first select the equivalent orifice size at sea level
from Table 6.

| Orifice Size (Decimal or DMS) | Gas Pressure at Orifice - Inches Water Column | | | | | | | | |
|---|---|---|---|---|---|---|---|---|---|
| | 3 | 3.5 | 4 | 5 | 6 | 7 | 8 | 9 | 10 |
| .008 | .17 | .18 | .19 | .23 | .24 | .26 | .28 | .29 | .30 |
| .009 | .21 | .23 | .25 | .28 | .30 | .33 | .35 | .37 | .39 |
| .010 | .27 | .29 | .30 | .35 | .37 | .41 | .43 | .46 | .48 |
| .011 | .33 | .35 | .37 | .42 | .45 | .48 | .52 | .55 | .59 |
| .012 | .38 | .41 | .44 | .50 | .54 | .57 | .62 | .65 | .70 |
| 80 | .48 | .52 | .55 | .63 | .69 | .73 | .79 | .83 | .88 |
| 79 | .55 | .59 | .64 | .72 | .80 | .84 | .90 | .97 | 1.01 |
| 78 | .70 | .76 | .78 | .88 | .97 | 1.04 | 1.10 | 1.17 | 1.24 |
| 77 | .88 | .95 | .99 | 1.11 | 1.23 | 1.31 | 1.38 | 1.47 | 1.55 |
| 76 | 1.05 | 1.13 | 1.21 | 1.37 | 1.52 | 1.61 | 1.72 | 1.83 | 1.92 |
| 75 | 1.16 | 1.25 | 1.34 | 1.52 | 1.64 | 1.79 | 1.91 | 2.04 | 2.14 |
| 74 | 1.33 | 1.44 | 1.55 | 1.74 | 1.91 | 2.05 | 2.18 | 2.32 | 2.44 |
| 73 | 1.51 | 1.63 | 1.76 | 1.99 | 2.17 | 2.32 | 2.48 | 2.64 | 2.78 |
| 72 | 1.64 | 1.77 | 1.90 | 2.15 | 2.40 | 2.52 | 2.69 | 2.86 | 3.00 |
| 71 | 1.82 | 1.97 | 2.06 | 2.33 | 2.54 | 2.73 | 2.91 | 3.11 | 3.26 |
| 70 | 2.06 | 2.22 | 2.39 | 2.70 | 2.97 | 3.16 | 3.38 | 3.59 | 3.78 |
| 69 | 2.25 | 2.43 | 2.61 | 2.96 | 3.23 | 3.47 | 3.68 | 3.94 | 4.14 |
| 68 | 2.52 | 2.72 | 2.93 | 3.26 | 3.58 | 3.88 | 4.14 | 4.41 | 4.64 |
| 67 | 2.69 | 2.91 | 3.12 | 3.52 | 3.87 | 4.13 | 4.41 | 4.69 | 4.94 |
| 66 | 2.86 | 3.09 | 3.32 | 3.75 | 4.11 | 4.39 | 4.68 | 4.98 | 5.24 |
| 65 | 3.14 | 3.39 | 3.72 | 4.28 | 4.62 | 4.84 | 5.16 | 5.50 | 5.78 |
| 64 | 3.41 | 3.68 | 4.14 | 4.48 | 4.91 | 5.23 | 5.59 | 5.95 | 6.26 |
| 63 | 3.63 | 3.92 | 4.19 | 4.75 | 5.19 | 5.55 | 5.92 | 6.30 | 6.63 |
| 62 | 3.78 | 4.08 | 4.39 | 4.96 | 5.42 | 5.81 | 6.20 | 6.59 | 6.94 |
| 61 | 4.02 | 4.34 | 4.66 | 5.27 | 5.77 | 6.15 | 6.57 | 7.00 | 7.37 |
| 60 | 4.21 | 4.55 | 4.89 | 5.52 | 5.95 | 6.47 | 6.91 | 7.35 | 7.74 |
| 59 | 4.41 | 4.76 | 5.11 | 5.78 | 6.35 | 6.78 | 7.25 | 7.71 | 8.11 |
| 58 | 4.66 | 5.03 | 5.39 | 6.10 | 6.68 | 7.13 | 7.62 | 8.11 | 8.53 |
| 57 | 4.84 | 5.23 | 5.63 | 6.36 | 6.96 | 7.44 | 7.94 | 8.46 | 8.90 |
| 56 | 5.68 | 6.13 | 6.58 | 7.35 | 8.03 | 8.73 | 9.32 | 9.92 | 10.44 |
| 55 | 7.11 | 7.68 | 8.22 | 9.30 | 10.18 | 10.85 | 11.59 | 12.34 | 12.98 |
| 54 | 7.95 | 8.59 | 9.23 | 10.45 | 11.39 | 12.25 | 13.08 | 13.93 | 14.65 |
| 53 | 9.30 | 10.04 | 10.80 | 12.20 | 13.32 | 14.29 | 15.27 | 16.25 | 17.09 |
| 52 | 10.61 | 11.46 | 12.31 | 13.86 | 15.26 | 16.34 | 17.44 | 18.57 | 19.53 |
| 51 | 11.82 | 12.77 | 13.69 | 15.47 | 16.97 | 18.16 | 19.40 | 20.64 | 21.71 |
| 50 | 12.89 | 13.92 | 14.94 | 16.86 | 18.48 | 19.77 | 21.12 | 22.48 | 23.65 |
| 49 | 14.07 | 15.20 | 16.28 | 18.37 | 20.20 | 21.60 | 23.06 | 24.56 | 25.83 |
| 48 | 15.15 | 16.36 | 17.62 | 19.88 | 21.81 | 23.31 | 24.90 | 26.51 | 27.89 |
| 47 | 16.22 | 17.52 | 18.80 | 21.27 | 23.21 | 24.93 | 26.62 | 28.34 | 29.81 |
| 46 | 17.19 | 18.57 | 19.98 | 22.57 | 24.72 | 26.43 | 28.23 | 30.05 | 31.61 |
| 45 | 17.73 | 19.15 | 20.52 | 23.10 | 25.36 | 27.18 | 29.03 | 30.90 | 32.51 |
| 44 | 19.45 | 21.01 | 22.57 | 25.57 | 27.93 | 29.87 | 31.89 | 33.96 | 35.72 |

TABLE A–2 *continued*

| Orifice Size (Decimal or DMS) | 3 | 3.5 | 4 | 5 | 6 | 7 | 8 | 9 | 10 |
|---|---|---|---|---|---|---|---|---|---|
| 43 | 20.73 | 22.39 | 24.18 | 27.29 | 29.87 | 32.02 | 34.19 | 36.41 | 38.30 |
| 42 | 23.10 | 24.95 | 26.50 | 29.50 | 32.50 | 35.24 | 37.63 | 40.07 | 42.14 |
| 41 | 24.06 | 25.98 | 28.15 | 31.69 | 34.81 | 37.17 | 39.70 | 42.27 | 44.46 |
| 40 | 25.03 | 27.03 | 29.23 | 33.09 | 36.20 | 38.79 | 41.42 | 44.10 | 46.38 |
| 39 | 26.11 | 28.20 | 30.20 | 34.05 | 37.38 | 39.97 | 42.68 | 45.44 | 47.80 |
| 38 | 27.08 | 29.25 | 31.38 | 35.46 | 38.89 | 41.58 | 44.40 | 47.27 | 49.73 |
| 37 | 28.36 | 30.63 | 32.99 | 37.07 | 40.83 | 43.62 | 46.59 | 49.60 | 52.17 |
| 36 | 29.76 | 32.14 | 34.59 | 39.11 | 42.76 | 45.77 | 48.88 | 52.04 | 54.74 |
| 35 | 32.36 | 34.95 | 36.86 | 41.68 | 45.66 | 48.78 | 52.10 | 55.46 | 58.34 |
| 34 | 32.45 | 35.05 | 37.50 | 42.44 | 46.52 | 49.75 | 53.12 | 56.55 | 59.49 |
| 33 | 33.41 | 36.08 | 38.79 | 43.83 | 48.03 | 51.46 | 54.96 | 58.62 | 61.55 |
| 32 | 35.46 | 38.30 | 40.94 | 46.52 | 50.82 | 54.26 | 57.95 | 61.70 | 64.89 |
| 31 | 37.82 | 40.85 | 43.83 | 49.64 | 54.36 | 58.01 | 61.96 | 65.97 | 69.39 |
| 30 | 43.40 | 46.87 | 50.39 | 57.05 | 62.09 | 66.72 | 71.22 | 75.86 | 79.80 |
| 29 | 48.45 | 52.33 | 56.19 | 63.61 | 69.62 | 74.45 | 79.52 | 84.66 | 89.04 |
| 28 | 51.78 | 55.92 | 59.50 | 67.00 | 73.50 | 79.50 | 84.92 | 90.39 | 95.09 |
| 27 | 54.47 | 58.83 | 63.17 | 71.55 | 78.32 | 83.59 | 89.27 | 95.04 | 99.97 |
| 26 | 56.73 | 61.27 | 65.86 | 74.57 | 81.65 | 87.24 | 93.17 | 99.19 | 104.57 |
| 25 | 58.87 | 63.58 | 68.22 | 77.14 | 84.67 | 90.36 | 96.50 | 102.74 | 108.07 |
| 24 | 60.81 | 65.67 | 70.58 | 79.83 | 87.56 | 93.47 | 99.83 | 106.28 | 111.79 |
| 23 | 62.10 | 67.07 | 72.20 | 81.65 | 89.39 | 94.55 | 100.98 | 107.49 | 113.07 |
| 22 | 64.89 | 70.08 | 75.21 | 85.10 | 93.25 | 99.60 | 106.39 | 113.24 | 119.12 |
| 21 | 66.51 | 71.83 | 77.14 | 87.35 | 95.63 | 102.29 | 109.24 | 116.29 | 122.33 |
| 20 | 68.22 | 73.68 | 79.08 | 89.49 | 97.99 | 104.75 | 111.87 | 119.10 | 125.28 |
| 19 | 72.20 | 77.98 | 83.69 | 94.76 | 103.89 | 110.67 | 118.55 | 125.82 | 132.36 |
| 18 | 75.53 | 81.57 | 87.56 | 97.50 | 108.52 | 116.03 | 123.92 | 131.93 | 138.78 |
| 17 | 78.54 | 84.82 | 91.10 | 103.14 | 112.81 | 120.33 | 128.52 | 136.82 | 143.91 |
| 16 | 82.19 | 88.77 | 95.40 | 107.98 | 118.18 | 126.78 | 135.39 | 144.15 | 151.63 |
| 15 | 85.20 | 92.02 | 98.84 | 111.74 | 122.48 | 131.07 | 139.98 | 149.03 | 156.77 |
| 14 | 87.10 | 94.40 | 100.78 | 114.21 | 124.44 | 133.22 | 142.28 | 151.47 | 159.33 |
| 13 | 89.92 | 97.11 | 104.32 | 118.18 | 128.93 | 138.60 | 148.02 | 157.58 | 165.76 |
| 12 | 93.90 | 101.41 | 108.52 | 123.56 | 135.37 | 143.97 | 153.75 | 163.69 | 172.13 |
| 11 | .95.94 | 103.62 | 111.31 | 126.02 | 137.52 | 147.20 | 157.20 | 167.36 | 176.03 |
| 10 | 98.30 | 106.16 | 114.21 | 129.25 | 141.82 | 151.50 | 161.81 | 172.26 | 181.13 |
| 9 | 100.99 | 109.07 | 117.11 | 132.58 | 145.05 | 154.71 | 165.23 | 175.91 | 185.03 |
| 8 | 103.89 | 112.20 | 120.65 | 136.44 | 149.33 | 160.08 | 170.96 | 182.00 | 191.44 |
| 7 | 105.93 | 114.40 | 123.01 | 139.23 | 152.56 | 163.31 | 174.38 | 185.68 | 195.30 |
| 6 | 109.15 | 117.88 | 126.78 | 142.88 | 156.83 | 167.51 | 178.88 | 190.46 | 200.36 |
| 5 | 111.08 | 119.97 | 128.93 | 145.79 | 160.08 | 170.82 | 182.48 | 194.22 | 204.30 |
| 4 | 114.75 | 123.93 | 133.22 | 150.41 | 164.36 | 176.18 | 188.16 | 200.25 | 210.71 |
| 3 | 119.25 | 128.79 | 137.52 | 156.26 | 170.78 | 182.64 | 195.08 | 207.66 | 218.44 |
| 2 | 128.48 | 138.76 | 148.61 | 168.64 | 184.79 | 197.66 | 211.05 | 224.74 | 235.58 |
| 1 | 136.35 | 147.26 | 158.25 | 179.33 | 194.63 | 209.48 | 223.65 | 238.16 | 250.54 |
| A | 145.34 | 155.48 | 165.62 | 189.28 | 206.18 | 219.70 | 236.60 | 250.12 | 263.63 |
| B | 150.36 | 160.85 | 171.33 | 195.81 | 213.29 | 227.25 | 244.76 | 258.74 | 272.73 |
| C | 155.45 | 166.30 | 177.14 | 202.44 | 220.52 | 234.98 | 253.06 | 267.51 | 281.97 |
| D | 160.63 | 171.84 | 183.04 | 209.19 | 227.87 | 242.81 | 261.48 | 276.44 | 291.38 |
| E | 165.89 | 177.47 | 189.05 | 216.05 | 235.34 | 250.77 | 270.06 | 285.49 | 300.93 |
| F | 175.43 | 187.61 | 199.78 | 227.75 | 248.70 | 265.02 | 285.40 | 301.70 | 318.02 |

TABLE A-2 *continued*

| Orifice Size (Decimal or DMS) | 3 | 3.5 | 4 | 5 | 6 | 7 | 8 | 9 | 10 |
|---|---|---|---|---|---|---|---|---|---|
| G | 180.81 | 193.43 | 206.04 | 235.49 | 256.50 | 273.33 | 294.35 | 311.18 | 327.99 |
| H | 187.81 | 200.91 | 214.01 | 244.59 | 266.42 | 283.89 | 305.74 | 323.20 | 340.67 |
| I | 196.38 | 210.08 | 223.77 | 255.75 | 278.58 | 296.85 | 319.68 | 337.95 | 356.20 |
| J | 203.66 | 217.87 | 232.08 | 265.24 | 288.92 | 307.87 | 331.55 | 350.49 | 396.44 |
| K | 209.59 | 224.22 | 238.84 | 272.95 | 297.33 | 316.82 | 341.19 | 360.69 | 380.18 |
| L | 223.23 | 240.49 | 257.75 | 290.71 | 316.68 | 337.44 | 363.40 | 384.17 | 404.92 |
| M | 231.00 | 247.12 | 263.23 | 300.83 | 327.69 | 349.18 | 376.03 | 397.52 | 419.01 |
| N | 242.09 | 258.98 | 275.06 | 315.27 | 343.42 | 365.94 | 394.09 | 416.61 | 439.13 |
| O | 265.05 | 283.54 | 302.03 | 345.18 | 376.00 | 400.66 | 431.47 | 456.13 | 480.79 |
| P | 276.92 | 296.24 | 315.56 | 360.64 | 392.84 | 418.60 | 450.80 | 476.56 | 502.32 |
| Q | 292.57 | 312.98 | 333.39 | 381.03 | 415.05 | 442.26 | 476.28 | 503.49 | 530.71 |
| R | 305.03 | 326.33 | 347.62 | 397.26 | 432.73 | 461.10 | 496.59 | 524.95 | 553.32 |
| S | 321.45 | 344.01 | 366.57 | 418.62 | 456.01 | 485.91 | 523.28 | 553.19 | 583.09 |
| T | 340.19 | 363.92 | 387.65 | 443.04 | 482.59 | 514.24 | 553.79 | 585.44 | 617.09 |
| U | 359.46 | 383.54 | 409.61 | 468.14 | 509.93 | 543.36 | 585.17 | 618.60 | 652.04 |
| V | 377.26 | 403.58 | 429.90 | 491.31 | 535.17 | 570.27 | 614.14 | 649.23 | 684.33 |
| W | 395.48 | 423.07 | 450.66 | 515.05 | 561.04 | 597.83 | 643.82 | 680.60 | 717.39 |
| X | 418.34 | 447.53 | 476.72 | 544.82 | 593.47 | 632.39 | 681.03 | 719.94 | 758.86 |
| Y | 433.23 | 463.45 | 493.67 | 564.20 | 614.58 | 654.87 | 705.25 | 745.56 | 785.86 |
| Z | 452.75 | 484.33 | 515.91 | 589.64 | 642.26 | 684.38 | 737.02 | 779.14 | 821.26 |

Source: American Gas Association

# TABLE A-3

## Outside Design Temperatures for the United States

| State & City | Winter DB | Summer DB | Daily Range | Summer WB | Latitude Deg. |
|---|---|---|---|---|---|
| **ALABAMA** | | | | | |
| Anniston | 12 | 94 | M | 78 | 35 |
| Birmingham | 14 | 94 | M | 78 | 35 |
| Gadsden | 11 | 94 | M | 78 | 35 |
| Mobile | 21 | 93 | L | 80 | 30 |
| Montgomery | 18 | 95 | M | 78 | 30 |
| Tuscaloosa | 14 | 96 | M | 78 | 35 |
| **ALASKA** | | | | | |
| Anchorage | -29 | 70 | M | 61 | 60 |
| Barrow | -49 | 54 | — | 51 | 70 |
| Bethel | -43 | — | — | — | 60 |
| Cordova | -13 | — | — | — | 60 |
| Fairbanks | -59 | 78 | M | 63 | 65 |
| Juneau | -11 | 71 | — | 64 | 60 |
| Ketchikan | 4 | — | — | — | 55 |
| Kodiak | 4 | 66 | — | 60 | 55 |
| Kotzebue | -46 | — | — | — | 65 |
| Nome | -37 | 62 | — | 56 | 60 |
| Seward | -4 | — | — | — | 60 |
| Sitka | 2 | — | — | — | 60 |
| **ARIZONA** | | | | | |
| Bisbee | 30 | 100 | H | 72 | 30 |
| Flagstaff | -10 | 82 | H | 60 | 35 |
| Globe | 30 | 105 | H | 70 | 30 |
| Nogales | 15 | 98 | H | 71 | 30 |
| Phoenix | 25 | 106 | H | 76 | 35 |
| Tucson | 23 | 102 | H | 73 | 30 |
| Winslow | 2 | 95 | H | 65 | 35 |
| Yuma | 32 | 109 | H | 78 | 35 |
| **ARKANSAS** | | | | | |
| Bentonville | 0 | 95 | M | 76 | 35 |
| Fort Smith | 9 | 99 | M | 78 | 35 |
| Hot Springs | 12 | 97 | M | 78 | 35 |
| Little Rock | 13 | 96 | M | 79 | 35 |
| Pine Bluff | 14 | 96 | M | 80 | 35 |
| Texarkana | 16 | 97 | M | 79 | 35 |
| **CALIFORNIA** | | | | | |
| Bakersfield | 26 | 101 | H | 71 | 35 |
| El Centro | 26 | 109 | H | 80 | 35 |
| Eureka | 27 | 65 | M | 59 | 40 |
| Fresno | 25 | 99 | H | 72 | 35 |
| Long Beach | 31 | 84 | M | 70 | 35 |
| Los Angeles | 36 | 83 | M | 70 | 35 |
| Montague | 15 | 95 | M | 70 | 40 |
| Needles | 27 | 110 | H | 75 | 35 |
| Oakland | 30 | 81 | M | 63 | 40 |
| Pasadena | 31 | 93 | M | 70 | 35 |
| Red Bluff | 15 | 100 | H | 70 | 40 |
| Sacramento | 24 | 97 | H | 70 | 40 |
| San Bernardino | 26 | 98 | H | 73 | 35 |
| San Diego | 38 | 83 | L | 70 | 35 |
| San Francisco | 32 | 79 | M | 63 | 40 |
| San Jose | 30 | 88 | M | 67 | 35 |
| **COLORADO** | | | | | |
| Boulder | -5 | 90 | M | 63 | 40 |
| Colorado Springs | -9 | 88 | H | 62 | 40 |
| Denver | -9 | 90 | H | 64 | 40 |
| Durango | -10 | 86 | H | 63 | 35 |
| Fort Collins | -18 | 89 | M | 62 | 40 |
| Grand Junction | -2 | 94 | H | 63 | 40 |
| Leadville | -18 | 73 | M | 55 | 40 |
| Pueblo | -14 | 94 | H | 67 | 40 |
| **CONNECTICUT** | | | | | |
| Bridgeport | -1 | 88 | L | 76 | 40 |
| Hartford | -4 | 88 | M | 76 | 40 |
| New Haven | 0 | 86 | M | 76 | 40 |
| New London | 0 | 86 | L | 75 | 40 |
| Norwalk | -5 | 89 | L | 76 | 40 |
| Torrington | 0 | 90 | M | 75 | 40 |
| Waterbury | -5 | 88 | M | 76 | 40 |
| **DELAWARE** | | | | | |
| Dover | 8 | 90 | M | 78 | 40 |
| Milford | 10 | 90 | M | 78 | 40 |
| Wilmington | 6 | 90 | M | 77 | 40 |
| **DIST. OF COLUMBIA** | | | | | |
| Washington | 12 | 92 | M | 77 | 40 |
| **FLORIDA** | | | | | |
| Apalachicola | 25 | 95 | L | 80 | 30 |
| Fort Myers | 34 | 92 | M | 80 | 25 |
| Gainesville | 24 | 94 | M | 79 | 30 |
| Jacksonville | 26 | 94 | M | 79 | 30 |
| Key West | 50 | 89 | L | 79 | 25 |
| Miami | 39 | 90 | L | 79 | 25 |
| Orlando | 29 | 94 | M | 79 | 30 |
| Pensacola | 25 | 90 | L | 81 | 30 |
| Tallahassee | 21 | 94 | M | 79 | 30 |
| Tampa | 32 | 91 | M | 80 | 30 |
| **GEORGIA** | | | | | |
| Athens | 12 | 94 | M | 77 | 35 |
| Atlanta | 14 | 92 | M | 77 | 35 |
| Augusta | 17 | 95 | M | 79 | 35 |
| Brunswick | 24 | 95 | L | 80 | 30 |
| Columbus | 19 | 96 | M | 79 | 35 |
| Macon | 18 | 96 | M | 79 | 35 |
| Rome | 11 | 95 | M | 77 | 35 |
| Savannah | 21 | 94 | M | 80 | 30 |
| Waycross | 20 | 95 | M | 79 | 30 |
| **IDAHO** | | | | | |
| Boise | 0 | 93 | H | 66 | 45 |
| Idaho Falls | -17 | 88 | H | 64 | 45 |
| Lewiston | 1 | 96 | H | 66 | 45 |
| Pocatello | -12 | 91 | H | 63 | 45 |
| Twin Falls | -5 | 94 | H | 64 | 40 |
| **ILLINOIS** | | | | | |
| Aurora | -13 | 91 | M | 77 | 40 |
| Bloomington | -7 | 92 | M | 78 | 40 |
| Cairo | 0 | 100 | M | 78 | 35 |
| Champaign | -6 | 94 | M | 78 | 40 |
| Chicago | -9 | 90 | M | 75 | 40 |
| Danville | -6 | 94 | M | 78 | 40 |
| Decatur | -6 | 93 | M | 78 | 40 |
| Elgin | -14 | 90 | M | 76 | 40 |
| Joliet | -11 | 92 | M | 77 | 40 |
| Moline | -12 | 91 | M | 77 | 40 |
| Peoria | -8 | 92 | M | 77 | 40 |
| Rockford | -13 | 90 | M | 76 | 40 |
| Rock Island | -10 | 95 | M | 76 | 40 |
| Springfield | -7 | 92 | M | 78 | 40 |
| Urbana | -10 | 95 | M | 77 | 40 |
| **INDIANA** | | | | | |
| Elkhart | -10 | 95 | M | 75 | 40 |
| Evansville | 1 | 94 | M | 78 | 40 |
| Fort Wayne | -5 | 91 | M | 76 | 40 |
| Indianapolis | -5 | 91 | M | 77 | 40 |
| Lafayette | -7 | 92 | M | 77 | 40 |
| South Bend | -6 | 89 | M | 76 | 40 |
| Terre Haute | -3 | 93 | M | 78 | 40 |
| **IOWA** | | | | | |
| Burlington | -10 | 92 | M | 78 | 40 |
| Cedar Rapids | -14 | 90 | M | 76 | 40 |
| Charles City | -20 | 95 | M | 75 | 45 |
| Clinton | -13 | 90 | M | 77 | 40 |
| Council Bluffs | -14 | 94 | M | 78 | 40 |
| Davenport | -10 | 95 | M | 78 | 40 |
| Des Moines | -13 | 92 | M | 77 | 40 |
| Dubuque | -17 | 90 | M | 76 | 40 |
| Fort Dodge | -18 | 92 | M | 77 | 40 |
| Keokuk | -9 | 93 | M | 78 | 40 |
| Marshalltown | -16 | 91 | M | 77 | 40 |
| Sioux City | -17 | 93 | M | 77 | 40 |
| Waterloo | -18 | 89 | M | 76 | 40 |
| **KANSAS** | | | | | |
| Atchison | -9 | 95 | M | 78 | 40 |
| Concordia | -10 | 97 | M | 78 | 40 |
| Dodge City | -5 | 97 | H | 73 | 40 |
| Iola | -5 | 100 | M | 75 | 40 |
| Leavenworth | -10 | 100 | M | 76 | 40 |
| Salina | -4 | 99 | M | 76 | 40 |
| Topeka | -4 | 96 | M | 78 | 40 |
| Wichita | -1 | 99 | M | 76 | 40 |
| **KENTUCKY** | | | | | |
| Bowling Green | 1 | 95 | M | 78 | 35 |
| Frankfort | 0 | 95 | M | 78 | 40 |
| Hopkinsville | 4 | 95 | M | 78 | 35 |
| Lexington | 0 | 92 | M | 77 | 40 |
| Louisville | 1 | 93 | M | 78 | 40 |
| Owensboro | 0 | 94 | M | 78 | 40 |
| Shelbyville | 0 | 95 | M | 78 | 40 |
| **LOUISIANA** | | | | | |
| Alexandria | 20 | 95 | M | 80 | 30 |
| Baton Rouge | 22 | 94 | M | 80 | 30 |
| New Orleans | 29 | 91 | L | 80 | 30 |
| Shreveport | 18 | 96 | M | 80 | 30 |
| **MAINE** | | | | | |
| Augusta | -13 | 86 | L | 73 | 45 |
| Bangor | -14 | 85 | L | 73 | 45 |
| Bar Harbor | -10 | 85 | L | 73 | 45 |
| Belfast | -10 | 85 | L | 73 | 45 |
| Eastport | -10 | 85 | L | 70 | 45 |
| Lewiston | -14 | 86 | L | 73 | 45 |
| Millinocket | -22 | 85 | M | 72 | 45 |
| Orono | -20 | 85 | M | 70 | 45 |
| Portland | -14 | 86 | M | 73 | 45 |
| Presque Isle | -20 | 85 | L | 73 | 45 |
| Rumford | -15 | 85 | L | 73 | 45 |
| **MARYLAND** | | | | | |
| Annapolis | 10 | 90 | M | 78 | 40 |
| Baltimore | 8 | 91 | M | 78 | 40 |
| Cambridge | 10 | 90 | L | 78 | 40 |
| Cumberland | 0 | 92 | M | 75 | 40 |
| Frederick | 2 | 92 | M | 77 | 40 |
| Frostburg | -5 | 90 | M | 75 | 40 |
| Salisbury | 10 | 90 | M | 78 | 40 |
| **MASSACHUSETTS** | | | | | |
| Amherst | -5 | 90 | M | 75 | 40 |
| Boston | -1 | 88 | M | 74 | 40 |
| Fall River | -1 | 86 | L | 74 | 40 |
| Fitchburg | -5 | 90 | M | 75 | 45 |
| Framingham | -7 | 89 | L | 74 | 40 |
| Lawrence | -9 | 88 | M | 74 | 40 |
| Lowell | -7 | 89 | M | 74 | 40 |
| Nantucket | 0 | 85 | L | 73 | 40 |
| New Bedford | 3 | 84 | L | 73 | 40 |
| Pittsfield | -11 | 84 | M | 72 | 40 |
| Plymouth | 0 | 85 | L | 74 | 40 |
| Springfield | -8 | 88 | M | 74 | 40 |
| Worcester | -8 | 87 | M | 73 | 40 |
| **MICHIGAN** | | | | | |
| Alpena | -11 | 85 | M | 73 | 45 |
| Ann Arbor | -5 | 99 | M | 75 | 40 |
| Big Rapids | -5 | 90 | M | 75 | 45 |
| Cadillac | -10 | 90 | M | 75 | 45 |
| Calumet | -20 | 80 | M | 73 | 45 |
| Detroit | 0 | 88 | M | 75 | 40 |
| Escanaba | -13 | 80 | M | 71 | 45 |
| Flint | -7 | 87 | M | 75 | 45 |
| Grand Haven | -5 | 90 | M | 75 | 45 |
| Grand Rapids | -3 | 89 | M | 74 | 45 |
| Houghton | -20 | 80 | M | 73 | 45 |
| Kalamazoo | -5 | 89 | M | 75 | 40 |
| Lansing | -4 | 87 | M | 75 | 45 |
| Ludington | -5 | 90 | M | 75 | 45 |
| Marquette | -14 | 86 | M | 71 | 45 |
| Muskegon | -2 | 85 | M | 74 | 45 |
| Port Huron | -6 | 88 | M | 74 | 45 |
| Saginaw | -7 | 86 | M | 75 | 45 |
| Sault Ste. Marie | -18 | 81 | M | 71 | 45 |
| **MINNESOTA** | | | | | |
| Alexandria | -26 | 88 | M | 74 | 45 |
| Duluth | -25 | 82 | M | 71 | 45 |
| Minneapolis | -19 | 89 | M | 75 | 45 |
| Moorhead | -30 | 95 | M | 75 | 45 |
| St. Cloud | -26 | 88 | M | 75 | 45 |
| St. Paul | -19 | 89 | M | 75 | 45 |
| **MISSISSIPPI** | | | | | |
| Biloxi | 26 | 92 | L | 81 | 30 |
| Columbus | 13 | 95 | M | 79 | 35 |
| Corinth | 5 | 95 | M | 78 | 35 |
| Hattiesburg | 18 | 95 | M | 79 | 30 |
| Jackson | 17 | 96 | M | 78 | 30 |
| Meridian | 15 | 95 | M | 79 | 30 |
| Natchez | 18 | 94 | L | 80 | 30 |
| Vicksburg | 13 | 96 | L | 79 | 30 |
| **MISSOURI** | | | | | |
| Columbia | -4 | 95 | M | 78 | 40 |
| Hannibal | -7 | 94 | M | 78 | 40 |
| Kansas City | -2 | 97 | M | 77 | 40 |
| Kirksville | -13 | 94 | M | 78 | 40 |
| St. Joseph | -8 | 95 | M | 78 | 40 |
| St. Louis | -2 | 95 | M | 78 | 40 |
| Springfield | 0 | 100 | M | 77 | 40 |
| **MONTANA** | | | | | |
| Anaconda | -30 | 85 | H | 59 | 45 |
| Billings | -19 | 91 | H | 66 | 45 |
| Butte | -34 | 83 | H | 59 | 45 |
| Great Falls | -29 | 88 | H | 63 | 50 |
| Havre | -32 | 87 | H | 64 | 50 |
| Helena | -27 | 87 | H | 63 | 45 |
| Kalispell | -17 | 84 | H | 63 | 50 |
| Miles City | -27 | 94 | H | 69 | 45 |
| Missoula | -16 | 89 | H | 63 | 45 |
| **NEBRASKA** | | | | | |
| Grand Island | -14 | 95 | M | 75 | 40 |
| Hastings | -11 | 96 | M | 75 | 40 |
| Lincoln | -10 | 96 | M | 77 | 40 |
| Norfolk | -18 | 95 | M | 76 | 40 |
| North Platte | -13 | 94 | H | 73 | 40 |
| Omaha | -12 | 94 | M | 78 | 40 |
| Valentine | -20 | 95 | M | 78 | 45 |
| York | -15 | 95 | M | 78 | 40 |
| **NEVADA** | | | | | |
| Elko | -21 | 92 | H | 62 | 40 |
| Las Vegas | 18 | 106 | H | 71 | 35 |
| Reno | -2 | 92 | H | 62 | 40 |
| Tonopah | 2 | 92 | M | 63 | 40 |
| Winnemucca | -8 | 95 | H | 62 | 40 |
| **NEW HAMPSHIRE** | | | | | |
| Berlin | -25 | 85 | H | 71 | 45 |
| Claremont | -19 | 87 | M | 73 | 45 |
| Concord | -17 | 88 | H | 73 | 45 |
| Franklin | -15 | 85 | M | 73 | 45 |
| Hanover | -15 | 85 | M | 73 | 45 |
| Keene | -17 | 88 | M | 73 | 45 |
| Manchester | -11 | 89 | M | 74 | 45 |
| Nashua | -10 | 85 | L | 74 | 45 |
| Portsmouth | -8 | 86 | L | 73 | 45 |
| **NEW JERSEY** | | | | | |
| Asbury Park | 5 | 90 | L | 78 | 40 |
| Atlantic City | 10 | 88 | L | 77 | 40 |
| Bayonne | 0 | 90 | L | 75 | 40 |
| Belvidere | 0 | 90 | M | 75 | 40 |
| Bloomfield | 0 | 90 | L | 75 | 40 |
| Bridgeton | 5 | 90 | L | 78 | 40 |
| Camden | 5 | 90 | L | 75 | 40 |
| East Orange | 0 | 90 | L | 75 | 40 |
| Elizabeth | 0 | 90 | L | 75 | 40 |
| Jersey City | 0 | 91 | M | 76 | 40 |
| Newark | 6 | 91 | M | 76 | 40 |
| New Brunswick | 3 | 89 | L | 76 | 40 |
| Paterson | 3 | 91 | L | 76 | 40 |
| Phillipsburg | 1 | 91 | M | 76 | 40 |
| Trenton | 7 | 90 | L | 77 | 40 |

## TABLE A–3 *continued*

| State & City | Winter DB | Summer DB | Daily Range | Summer WB | Latitude Deg. |
|---|---|---|---|---|---|
| **NEW MEXICO** | | | | | |
| Albuquerque | 6 | 94 | M | 65 | 35 |
| El Morro | 0 | 85 | H | 65 | 35 |
| Raton | –11 | 90 | H | 65 | 35 |
| Roswell | 5 | 99 | H | 70 | 35 |
| Santa Fe | –2 | 88 | M | 63 | 35 |
| Tucumcari | 1 | 97 | H | 70 | 35 |
| **NEW YORK** | | | | | |
| Albany | –14 | 88 | M | 74 | 45 |
| Auburn | –10 | 87 | M | 73 | 45 |
| Binghamton | –8 | 89 | M | 72 | 40 |
| Buffalo | –3 | 86 | M | 73 | 45 |
| Canton | –20 | 85 | M | 73 | 45 |
| Cortland | –11 | 88 | M | 73 | 45 |
| Elmira | –5 | 90 | M | 73 | 40 |
| Glens Falls | –17 | 86 | M | 72 | 45 |
| Ithaca | –10 | 88 | M | 73 | 40 |
| Jamestown | –5 | 86 | M | 73 | 40 |
| Lake Placid | –15 | 90 | M | 73 | 45 |
| New York | 6 | 91 | M | 76 | 40 |
| Niagara Falls | –2 | 86 | M | 74 | 45 |
| Ogdensburg | –20 | 85 | M | 73 | 45 |
| Oneonta | –13 | 87 | M | 72 | 45 |
| Oswego | 4 | 84 | M | 74 | 45 |
| Port Jervis | 0 | 90 | L | 75 | 40 |
| Rochester | –5 | 88 | M | 74 | 45 |
| Schenectady | –11 | 88 | M | 73 | 45 |
| Syracuse | –10 | 87 | M | 74 | 45 |
| Watertown | –20 | 84 | M | 74 | 45 |
| **NORTH CAROLINA** | | | | | |
| Asheville | 8 | 88 | M | 74 | 35 |
| Charlotte | 13 | 94 | M | 77 | 35 |
| Greensboro | 9 | 91 | M | 76 | 35 |
| Hatteras | 20 | 90 | L | 80 | 35 |
| New Bern | 14 | 92 | L | 80 | 35 |
| Raleigh | 13 | 92 | M | 78 | 35 |
| Salisbury | 10 | 90 | M | 78 | 35 |
| Wilmington | 19 | 91 | M | 81 | 35 |
| Winston-Salem | 9 | 91 | M | 76 | 35 |
| **NORTH DAKOTA** | | | | | |
| Bismarck | –31 | 91 | H | 72 | 45 |
| Devils Lake | –30 | 89 | M | 71 | 50 |
| Dickinson | –31 | 93 | H | 70 | 45 |
| Fargo | –28 | 88 | H | 74 | 45 |
| Grand Forks | –30 | 87 | M | 72 | 50 |
| Jamestown | –29 | 91 | M | 73 | 45 |
| Minot | –31 | 88 | M | 70 | 50 |
| Pembina | –35 | 90 | M | 73 | 50 |
| Williston | –28 | 90 | M | 69 | 50 |
| **OHIO** | | | | | |
| Akron | –5 | 87 | M | 73 | 40 |
| Cincinnati | 2 | 92 | M | 77 | 40 |
| Cleveland | –2 | 89 | M | 75 | 40 |
| Columbus | –1 | 88 | M | 76 | 40 |
| Dayton | –2 | 90 | M | 75 | 40 |
| Lima | –6 | 91 | M | 76 | 40 |
| Marion | –5 | 91 | M | 76 | 40 |
| Sandusky | –2 | 90 | M | 75 | 40 |
| Toledo | –5 | 90 | M | 75 | 40 |
| Warren | –6 | 88 | M | 74 | 40 |
| Youngstown | –5 | 86 | M | 74 | 40 |
| **OKLAHOMA** | | | | | |
| Ardmore | 9 | 101 | M | 78 | 35 |
| Bartlesville | –1 | 99 | M | 78 | 35 |
| Guthrie | 0 | 100 | M | 77 | 35 |
| Muskogee | 6 | 99 | M | 78 | 35 |
| Oklahoma City | 4 | 97 | M | 77 | 35 |
| Tulsa | 4 | 99 | M | 78 | 35 |
| Waynoka | –5 | 105 | M | 75 | 35 |

| State & City | Winter DB | Summer DB | Daily Range | Summer WB | Latitude Deg. |
|---|---|---|---|---|---|
| **OREGON** | | | | | |
| Arlington | 5 | 95 | M | 68 | 45 |
| Baker | –10 | 92 | M | 65 | 45 |
| Eugene | 16 | 88 | H | 67 | 45 |
| Medford | 15 | 94 | H | 68 | 40 |
| Pendleton | –2 | 94 | H | 65 | 45 |
| Port'and | 17 | 85 | M | 67 | 45 |
| Roseburg | 19 | 91 | H | 67 | 45 |
| Salem | 15 | 88 | H | 67 | 45 |
| Wamic | 0 | 90 | H | 66 | 45 |
| **PENNSYLVANIA** | | | | | |
| Altoona | –4 | 87 | M | 73 | 40 |
| Bethlehem | 0 | 90 | M | 75 | 40 |
| Coatesville | 5 | 90 | M | 75 | 40 |
| Erie | 1 | 85 | M | 74 | 40 |
| Harrisburg | 4 | 89 | M | 75 | 40 |
| New Castle | –7 | 89 | M | 74 | 40 |
| Oil City | –5 | 90 | M | 75 | 40 |
| Philadelphia | 7 | 90 | M | 77 | 40 |
| Pittsburgh | –1 | 87 | M | 74 | 40 |
| Reading | 1 | 90 | M | 76 | 40 |
| Scranton | –3 | 87 | M | 74 | 40 |
| Warren | –8 | 87 | M | 73 | 40 |
| Williamsport | 4 | 90 | M | 76 | 40 |
| York | –5 | 89 | M | 75 | 40 |
| **RHODE ISLAND** | | | | | |
| Block Island | 5 | 85 | L | 75 | 40 |
| Bristol | 0 | 90 | L | 75 | 40 |
| Kingston | 0 | 85 | L | 75 | 40 |
| Pawtucket | 0 | 90 | M | 75 | 40 |
| Providence | 0 | 86 | M | 75 | 40 |
| **SOUTH CAROLINA** | | | | | |
| Charleston | 19 | 92 | L | 80 | 35 |
| Columbia | 16 | 96 | M | 79 | 35 |
| Florence | 16 | 94 | M | 79 | 35 |
| Greenville | 14 | 93 | M | 76 | 35 |
| Spartanburg | 13 | 93 | M | 76 | 35 |
| **SOUTH DAKOTA** | | | | | |
| Aberdeen | –29 | 92 | M | 75 | 45 |
| Huron | –24 | 93 | H | 75 | 45 |
| Pierre | –21 | 96 | M | 74 | 45 |
| Rapid City | –17 | 94 | H | 71 | 45 |
| Sioux Falls | –21 | 92 | H | 75 | 45 |
| Watertown | –27 | 90 | M | 74 | 45 |
| **TENNESSEE** | | | | | |
| Chattanooga | 11 | 94 | M | 78 | 35 |
| Jackson | 8 | 95 | M | 79 | 35 |
| Johnson City | 0 | 95 | M | 78 | 35 |
| Knoxville | 9 | 92 | M | 76 | 35 |
| Memphis | 11 | 96 | M | 79 | 35 |
| Nashville | 6 | 95 | M | 78 | 35 |
| **TEXAS** | | | | | |
| Abilene | 12 | 99 | M | 75 | 30 |
| Amarillo | 2 | 96 | H | 71 | 35 |
| Austin | 19 | 98 | M | 78 | 30 |
| Brownsville | 32 | 92 | M | 80 | 25 |
| Corpus Christi | 28 | 93 | M | 80 | 30 |
| Dallas | 14 | 99 | M | 78 | 30 |
| Del Rio | 24 | 99 | H | 77 | 30 |
| El Paso | 16 | 98 | M | 69 | 30 |
| Fort Worth | 14 | 100 | M | 78 | 35 |
| Galveston | 28 | 89 | L | 81 | 30 |
| Houston | 23 | 94 | M | 80 | 30 |
| Palestine | 16 | 97 | M | 79 | 30 |
| Port Arthur | 25 | 92 | M | 80 | 30 |
| San Antonio | 22 | 97 | M | 77 | 30 |
| Waco | 16 | 99 | M | 78 | 30 |

| State & City | Winter DB | Summer DB | Daily Range | Summer WB | Latitude Deg. |
|---|---|---|---|---|---|
| **UTAH** | | | | | |
| Logan | –7 | 91 | H | 65 | 40 |
| Milford | –5 | 95 | H | 66 | 40 |
| Ogden | –3 | 92 | H | 65 | 40 |
| Salt Lake City | –2 | 94 | H | 66 | 40 |
| **VERMONT** | | | | | |
| Bennington | –10 | 90 | M | 73 | 45 |
| Burlington | –18 | 85 | M | 73 | 45 |
| Montpelier | –20 | 90 | M | 73 | 45 |
| Newport | –20 | 85 | M | 73 | 45 |
| Northfield | –20 | 90 | M | 73 | 45 |
| Rutland | –18 | 85 | M | 73 | 45 |
| **VIRGINIA** | | | | | |
| Cape Henry | 15 | 90 | L | 78 | 35 |
| Charlottesville | 7 | 90 | M | 77 | 40 |
| Danville | 9 | 92 | M | 78 | 35 |
| Lynchburg | 10 | 92 | M | 76 | 35 |
| Norfolk | 18 | 91 | L | 78 | 35 |
| Petersburg | 10 | 94 | M | 78 | 35 |
| Richmond | 10 | 93 | M | 78 | 40 |
| Roanoke | 9 | 91 | M | 76 | 35 |
| Wytheville | 5 | 90 | M | 76 | 35 |
| **WASHINGTON** | | | | | |
| Aberdeen | 19 | 80 | L | 60 | 45 |
| Bellingham | 8 | 82 | L | 65 | 50 |
| Everett | 13 | 78 | L | 65 | 50 |
| North Head | 20 | 80 | L | 65 | 50 |
| Olympia | 15 | 83 | H | 65 | 50 |
| Seattle | 17 | 80 | M | 65 | 50 |
| Spokane | –5 | 90 | H | 64 | 50 |
| Tacoma | 14 | 81 | M | 66 | 45 |
| Tatoosh Island | 20 | 80 | L | 65 | 50 |
| Walla Walla | 5 | 96 | H | 68 | 45 |
| Wenatchee | –2 | 92 | M | 66 | 50 |
| Yakima | –1 | 92 | H | 67 | 45 |
| **WEST VIRGINIA** | | | | | |
| Bluefield | 1 | 86 | M | 73 | 35 |
| Charleston | 1 | 90 | M | 75 | 40 |
| Elkins | –4 | 84 | M | 73 | 40 |
| Fairmont | 0 | 90 | M | 75 | 40 |
| Huntington | 4 | 93 | M | 76 | 40 |
| Martinsburg | 1 | 94 | M | 77 | 40 |
| Parkersburg | 2 | 91 | M | 76 | 40 |
| Wheeling | 0 | 89 | M | 75 | 40 |
| **WISCONSIN** | | | | | |
| Ashland | –27 | 83 | M | 71 | 45 |
| Beloit | –13 | 90 | M | 76 | 45 |
| Eau Claire | –21 | 88 | M | 74 | 45 |
| Green Bay | –16 | 85 | M | 73 | 45 |
| La Crosse | –18 | 88 | M | 76 | 45 |
| Madison | –13 | 88 | M | 75 | 45 |
| Milwaukee | –11 | 87 | M | 75 | 45 |
| Oshkosh | –20 | 90 | M | 75 | 45 |
| Sheboygan | –20 | 90 | M | 75 | 45 |
| **WYOMING** | | | | | |
| Casper | –20 | 90 | H | 62 | 45 |
| Cheyenne | –15 | 86 | H | 62 | 40 |
| Lander | –26 | 90 | H | 63 | 45 |
| Sheridan | –21 | 92 | H | 65 | 45 |
| Yellowstone Park | –35 | 85 | H | 62 | 45 |

The U.S. summer design temperatures represent the maximum wet and dry bulb temperatures recorded for each area reduced by 2½%. It is unusual for both wet and dry bulb temperatures to exceed the figures shown here simultaneously. The 2½% reduction is applied to the U.S. data for the months of June through September only.

The U.S. winter design temperatures shown represent the average of the coldest temperatures recorded for that area each year for approximately a twenty-five year period, often referred to as the Median of Extremes.

Source: ARI, Air Conditioning and Refrigeration Institute

## TABLE A-4
## Heat Transfer Factors Table

| Item | Design Temperature Difference, Degrees | | | | | | | | | | | | | | |
|---|---|---|---|---|---|---|---|---|---|---|---|---|---|---|---|
| | 30 | 35 | 40 | 45 | 50 | 55 | 60 | 65 | 70 | 75 | 80 | 85 | 90 | 95 | 100 |
| **WINDOWS, WOOD OR METAL FRAME (Btuh per sq ft)** Factors include heat loss for transmission and infiltration @ 15 mph | | | | | | | | | | | | | | | |
| **No. 1 Double-Hung, Horizontal-Slide, Casement or Awning (Infiltration less than 0.50 cfm/ft @ 25 mph certified by test)** | | | | | | | | | | | | | | | |
| (a) Single glass | 40 | 50 | 55 | 60 | 70 | 75 | 80 | 90 | 95 | 105 | 110 | 115 | 125 | 130 | 135 |
| (b) With double glass or insulating glass | 25 | 30 | 35 | 40 | 45 | 45 | 50 | 55 | 60 | 65 | 70 | 75 | 75 | 80 | 85 |
| (c) With storm sash | 25 | 25 | 30 | 35 | 40 | 45 | 45 | 50 | 55 | 60 | 60 | 65 | 70 | 75 | 80 |
| **No. 2 Double-Hung, Horizontal-Slide, Casement or Awning (Infiltration less than 0.75 cfm/ft @ 25 mph certified by test)** | | | | | | | | | | | | | | | |
| (a) Single glass | 45 | 50 | 60 | 65 | 75 | 80 | 90 | 95 | 105 | 110 | 120 | 125 | 135 | 140 | 150 |
| (b) With double glass or insulating glass | 30 | 35 | 40 | 45 | 50 | 55 | 60 | 65 | 70 | 75 | 80 | 80 | 85 | 90 | 95 |
| (c) With storm sash | 25 | 30 | 35 | 40 | 45 | 50 | 55 | 60 | 60 | 65 | 70 | 75 | 80 | 85 | 90 |
| **No. 3 Other Double-Hung, Horizontal-Slide, Casement or Awning** | | | | | | | | | | | | | | | |
| (a) Single glass | 75 | 90 | 105 | 115 | 130 | 140 | 155 | 165 | 180 | 195 | 205 | 220 | 230 | 245 | 255 |
| (b) With double glass or insulating glass | 60 | 70 | 80 | 90 | 105 | 115 | 125 | 135 | 145 | 155 | 165 | 175 | 185 | 195 | 205 |
| (c) With storm sash | 35 | 40 | 45 | 55 | 60 | 65 | 70 | 75 | 85 | 90 | 95 | 100 | 105 | 110 | 120 |
| **No. 4 Fixed or Picture** | | | | | | | | | | | | | | | |
| (a) Single glass | 40 | 50 | 55 | 60 | 70 | 75 | 85 | 90 | 95 | 105 | 110 | 115 | 125 | 130 | 140 |
| (b) Double glass or with storm sash | 25 | 30 | 35 | 40 | 45 | 45 | 50 | 55 | 60 | 65 | 70 | 75 | 75 | 80 | 85 |
| **No. 5 Jalousie (Infiltration less than 1.5 cfm/sq ft @ 25 mph certified by test)** | | | | | | | | | | | | | | | |
| (a) Single glass | 60 | 70 | 80 | 90 | 100 | 105 | 115 | 125 | 135 | 145 | 155 | 165 | 175 | 185 | 195 |
| (b) With storm sash | 35 | 40 | 50 | 55 | 60 | 65 | 70 | 75 | 80 | 85 | 90 | 100 | 105 | 110 | 115 |
| **No. 6 Other Jalousie** | | | | | | | | | | | | | | | |
| (a) Single glass | 225 | 265 | 300 | 340 | 375 | 415 | 450 | 490 | 525 | 565 | 600 | 640 | 675 | 715 | 750 |
| (b) With storm sash | 65 | 75 | 90 | 100 | 110 | 120 | 135 | 145 | 155 | 165 | 175 | 190 | 200 | 210 | 220 |
| **DOORS (Btuh per sq ft)** Factors include heat loss for transmission and infiltration @ 15 mph | | | | | | | | | | | | | | | |
| **No. 7 Sliding Glass Doors (Infiltration less than 1.0 cfm/sq ft @ 25 mph certified by test)** | | | | | | | | | | | | | | | |
| (a) Single glass | 50 | 60 | 70 | 75 | 85 | 95 | 100 | 110 | 115 | 125 | 135 | 140 | 150 | 160 | 165 |
| (b) Double glass | 40 | 45 | 50 | 55 | 60 | 65 | 70 | 80 | 90 | 95 | 100 | 105 | 115 | 120 | 125 |
| **No. 8 Other Sliding Glass Doors** | | | | | | | | | | | | | | | |
| (a) Single glass | 75 | 85 | 100 | 115 | 125 | 140 | 150 | 165 | 175 | 190 | 200 | 210 | 225 | 240 | 250 |
| (b) Double glass | 60 | 70 | 80 | 90 | 100 | 110 | 120 | 130 | 140 | 150 | 160 | 170 | 180 | 190 | 200 |
| **No. 9 Other Doors** | | | | | | | | | | | | | | | |
| (a) Weatherstripped and with storm door | 40 | 45 | 55 | 60 | 65 | 75 | 80 | 85 | 90 | 100 | 105 | 110 | 120 | 125 | 130 |
| (b) Weatherstripped or with storm door | 70 | 85 | 95 | 110 | 120 | 135 | 145 | 155 | 170 | 180 | 195 | 205 | 215 | 230 | 240 |
| (c) No weatherstripping or storm door | 135 | 160 | 180 | 200 | 225 | 250 | 270 | 290 | 315 | 340 | 360 | 380 | 405 | 430 | 450 |
| **WALLS AND PARTITIONS (Btuh per sq ft)** **No. 10 Wood Frame with Sheathing and Siding, Veneer or Other Exterior Finish** | | | | | | | | | | | | | | | |
| (a) No insulation | 8 | 9 | 10 | 11 | 13 | 14 | 15 | 16 | 18 | 19 | 20 | 21 | 23 | 24 | 25 |
| (b) Expanded polystyrene extruded board sheathing (R-5) | 3 | 4 | 4 | 5 | 6 | 6 | 7 | 7 | 8 | 8 | 9 | 9 | 10 | 10 | 11 |
| (c) R-7 batt insulation (2"-2¾") | 3 | 4 | 4 | 5 | 5 | 6 | 6 | 7 | 7 | 8 | 8 | 9 | 9 | 10 | 10 |
| (d) R-11 batt insulation (3"-3½") | 2 | 2 | 3 | 3 | 4 | 4 | 4 | 5 | 5 | 5 | 6 | 6 | 6 | 7 | 7 |
| (e) R-13 batt insulation (3½"-3⅝") | 2 | 2 | 2 | 3 | 3 | 3 | 4 | 4 | 4 | 5 | 5 | 5 | 5 | 6 | 6 |
| **No. 11 Partition between conditioned and unconditioned spaces** | | | | | | | | | | | | | | | |
| (a) Finished one side only, no insulation | 17 | 19 | 22 | 25 | 28 | 30 | 33 | 36 | 39 | 41 | 44 | 47 | 49 | 52 | 55 |
| (b) Finished both sides, no insulation | 9 | 11 | 12 | 14 | 16 | 17 | 19 | 20 | 22 | 23 | 25 | 26 | 28 | 29 | 31 |

* Insulation is produced in different densities and materials such as mineral fibers, glass fibers, etc.; therefore, there is some variation in thickness for the same R-value among various manufacturers' products.

## TABLE A–4 *continued*

| Item No. 11 Cont'd | Design Temperature Difference, Degrees | | | | | | | | | | | | | | |
|---|---|---|---|---|---|---|---|---|---|---|---|---|---|---|---|
| | 30 | 35 | 40 | 45 | 50 | 55 | 60 | 65 | 70 | 75 | 80 | 85 | 90 | 95 | 100 |
| (c) Partition with 1" expanded polystyrene extruded board, R-5, finished both sides | 4 | 4 | 5 | 5 | 6 | 7 | 7 | 8 | 8 | 9 | 10 | 10 | 11 | 11 | 12 |
| (d) R-7 insulation (2"-2¾") finished both sides | 3 | 4 | 4 | 5 | 5 | 6 | 6 | 7 | 7 | 8 | 8 | 9 | 9 | 10 | 10 |
| (e) R-11 insulation (3"-3½") finished both sides | 2 | 3 | 3 | 4 | 4 | 4 | 5 | 5 | 6 | 6 | 6 | 7 | 7 | 8 | 8 |
| (f) R-13 insulation (3½"-3⅝") finished both sides | 2 | 2 | 3 | 3 | 4 | 4 | 4 | 5 | 5 | 5 | 6 | 6 | 6 | 7 | 7 |
| **No. 12 Solid Masonry, Block or Brick** | | | | | | | | | | | | | | | |
| (a) Plastered or plain | 14 | 16 | 18 | 20 | 22 | 25 | 27 | 29 | 32 | 34 | 36 | 38 | 40 | 43 | 45 |
| (b) Furred, no insulation | 9 | 10 | 12 | 13 | 14 | 16 | 17 | 19 | 20 | 22 | 23 | 25 | 26 | 28 | 29 |
| (c) Furred, with R-5 insulation (Nominal 1½") | 4 | 5 | 5 | 6 | 6 | 7 | 8 | 8 | 9 | 10 | 10 | 11 | 12 | 12 | 13 |
| **No. 13 Basement or Crawl Space** | | | | | | | | | | | | | | | |
| (a) Above grade, no insulation | 15 | 18 | 20 | 23 | 26 | 28 | 31 | 33 | 36 | 38 | 41 | 43 | 46 | 48 | 51 |
| (b) Wall of conditioned crawl space, R-3.57 insulation (molded bead bd.) | 5 | 6 | 7 | 8 | 9 | 10 | 11 | 12 | 13 | 14 | 14 | 15 | 16 | 17 | 18 |
| (c) Wall of conditioned crawl space, R-5 insulation (ext. polystyrene bd.) | 4 | 5 | 6 | 6 | 7 | 8 | 9 | 9 | 10 | 11 | 12 | 12 | 13 | 14 | 14 |
| (d) Wall of crawl space used as supply plenum, R-3.57 insulation (molded bead bd.) | 11 | 12 | 13 | 14 | 15 | 16 | 16 | 17 | 18 | 19 | 20 | 21 | 22 | 23 | 24 |
| (e) Wall of crawl space used as supply plenum, R-5 insulation (ext. polystyrene bd.) | 9 | 10 | 10 | 11 | 12 | 12 | 13 | 14 | 15 | 15 | 16 | 17 | 17 | 18 | 19 |
| (f) Below grade wall | 2 | 2 | 2 | 3 | 3 | 3 | 4 | 4 | 4 | 5 | 5 | 5 | 5 | 6 | 6 |
| **CEILINGS AND ROOFS (Btuh per sq ft)** | | | | | | | | | | | | | | | |
| **No. 14 Ceiling Under Unconditioned Space or Vented Roof** | | | | | | | | | | | | | | | |
| (a) No insulation | 18 | 21 | 24 | 27 | 30 | 33 | 36 | 39 | 42 | 45 | 48 | 51 | 54 | 57 | 60 |
| (b) R-11 insulation (3"-3¼") | 2 | 3 | 3 | 4 | 4 | 4 | 5 | 5 | 6 | 6 | 6 | 7 | 7 | 8 | 8 |
| (c) R-19 insulation (5¼"-6½") | 2 | 2 | 2 | 2 | 3 | 3 | 3 | 4 | 4 | 4 | 4 | 4 | 5 | 5 | 5 |
| (d) R-22 insulation (6"-7") | 1 | 1 | 2 | 2 | 2 | 2 | 2 | 3 | 3 | 3 | 3 | 3 | 4 | 4 | 4 |
| (e) Ceiling under unconditioned room | 9 | 11 | 12 | 14 | 15 | 17 | 18 | 20 | 21 | 23 | 24 | 26 | 27 | 29 | 30 |
| **No. 15 Roof on Exposed Beams or Rafters** | | | | | | | | | | | | | | | |
| (a) Roofing on 1½ in. wood decking, no insulation | 10 | 12 | 14 | 15 | 17 | 19 | 20 | 22 | 24 | 26 | 27 | 29 | 31 | 32 | 34 |
| (b) Roofing on 1½ in. wood decking, 1 in. insulation between roofing and decking | 5 | 6 | 7 | 8 | 8 | 9 | 10 | 11 | 12 | 13 | 13 | 14 | 15 | 16 | 17 |
| (c) Roofing on 1½ in. wood decking, 1½ in. insulation between roofing and decking | 4 | 5 | 5 | 6 | 7 | 7 | 8 | 9 | 10 | 10 | 11 | 12 | 12 | 13 | 14 |
| (d) Roofing on 2 in. coarse, shredded wood plank | 6 | 7 | 8 | 9 | 10 | 11 | 12 | 14 | 15 | 16 | 17 | 18 | 19 | 20 | 21 |
| (e) Roofing on 3 in. coarse, shredded wood plank | 5 | 5 | 6 | 7 | 8 | 8 | 9 | 10 | 11 | 11 | 12 | 13 | 14 | 14 | 15 |
| (f) Roofing on 1½ in. insulating fiberboard decking | 6 | 7 | 8 | 9 | 10 | 10 | 11 | 12 | 13 | 14 | 15 | 16 | 17 | 18 | 19 |
| (g) Roofing on 2 in. insulating fiberboard decking | 4 | 5 | 6 | 7 | 8 | 8 | 9 | 10 | 11 | 11 | 12 | 13 | 14 | 14 | 15 |
| (h) Roofing on 3 in. insulating fiberboard decking | 3 | 4 | 4 | 5 | 6 | 6 | 7 | 7 | 8 | 8 | 9 | 9 | 10 | 10 | 11 |
| **No. 16 Roof-Ceiling Combination** | | | | | | | | | | | | | | | |
| (a) No insulation | 9 | 11 | 12 | 14 | 16 | 17 | 19 | 20 | 22 | 23 | 25 | 26 | 28 | 29 | 31 |
| (b) R-11 insulation (3"-3½") between ceiling and decking | 2 | 2 | 3 | 3 | 4 | 4 | 4 | 5 | 5 | 5 | 6 | 6 | 6 | 7 | 7 |
| (c) R-19 insulation (5¼"-6½") between ceiling and decking | 1 | 2 | 2 | 2 | 2 | 2 | 3 | 3 | 3 | 3 | 4 | 4 | 4 | 4 | 5 |
| (d) R-22 insulation (6"-7") | 1 | 1 | 2 | 2 | 2 | 2 | 2 | 3 | 3 | 3 | 3 | 3 | 4 | 4 | 4 |
| **FLOORS (Btuh per sq ft)** | | | | | | | | | | | | | | | |
| **No. 17 Floors Over Unconditioned Space** | | | | | | | | | | | | | | | |
| (a) Over unconditioned room | 4 | 5 | 6 | 6 | 7 | 8 | 8 | 9 | 10 | 11 | 11 | 12 | 13 | 13 | 14 |
| (b) Over open or vented space or garage, no insulation | 8 | 10 | 11 | 13 | 14 | 15 | 17 | 18 | 20 | 21 | 22 | 24 | 25 | 27 | 28 |
| (c) Over open or vented space or garage, R-7 insulation (2"-2¾") | 3 | 3 | 4 | 4 | 5 | 5 | 6 | 6 | 7 | 7 | 8 | 8 | 9 | 9 | 9 |
| (d) Over open or vented space or garage, R-11 insulation (3"-3½") | 2 | 2 | 3 | 3 | 4 | 4 | 4 | 5 | 5 | 5 | 6 | 6 | 6 | 7 | 7 |
| (e) Over open or vented space or garage, R-19 insulation (5¼"-6½") | 1 | 2 | 2 | 2 | 2 | 2 | 3 | 3 | 3 | 3 | 4 | 4 | 4 | 4 | 4 |
| **No. 18 Floor of Room Over Heated Basement or Crawl Space** | 0 | 0 | 0 | 0 | 0 | 0 | 0 | 0 | 0 | 0 | 0 | 0 | 0 | 0 | 0 |
| **No. 19 Basement Floor** | 1 | 1 | 1 | 1 | 1 | 2 | 2 | 2 | 2 | 2 | 3 | 3 | 3 | 3 | 3 |
| **FLOORS (Btuh per ft of Perimeter)** | | | | | | | | | | | | | | | |
| **No. 20 Concrete Slab Floor, Unheated** | | | | | | | | | | | | | | | |
| (a) No edge insulation | 35 | 40 | 40 | 45 | 45 | 50 | 50 | 55 | 55 | 60 | 60 | 65 | 65 | 70 | 75 |
| (b) 1 in. edge insulation | 25 | 30 | 30 | 35 | 35 | 40 | 40 | 45 | 45 | 50 | 50 | 55 | 55 | 60 | 60 |

TABLE A–4 *continued*

| Item | Design Temperature Difference, Degrees | | | | | | | | | | | | | | |
|---|---|---|---|---|---|---|---|---|---|---|---|---|---|---|---|
| No. 20 cont'd | 30 | 35 | 40 | 45 | 50 | 55 | 60 | 65 | 70 | 75 | 80 | 85 | 90 | 95 | 100 |
| (c)  2 in. edge insulation | 15 | 20 | 20 | 25 | 25 | 30 | 30 | 35 | 35 | 40 | 40 | 45 | 45 | 50 | 50 |
| **No. 21 Concrete Slab Floor with Perimeter System in Slab** | | | | | | | | | | | | | | | |
| (a)  No edge insulation | 60 | 70 | 80 | 90 | 95 | 105 | 115 | 125 | 135 | 145 | 150 | 160 | 170 | 180 | 190 |
| (b)  1 in. edge insulation | 45 | 50 | 55 | 60 | 65 | 70 | 75 | 80 | 85 | 90 | 95 | 100 | 105 | 110 | 115 |
| (c)  2 in. edge insulation | 25 | 30 | 35 | 40 | 45 | 50 | 55 | 60 | 65 | 70 | 75 | 80 | 85 | 90 | 95 |
| **No. 22 Floor of Heated Crawl Space** | | | | | | | | | | | | | | | |
| (a)  Less than 18 in. below grade | 35 | 40 | 40 | 45 | 45 | 50 | 50 | 55 | 55 | 60 | 60 | 65 | 65 | 70 | 75 |
| (b)  18 in. or more below grade | 15 | 20 | 20 | 25 | 25 | 30 | 30 | 35 | 35 | 40 | 40 | 45 | 45 | 50 | 50 |
| **No. 23 Floor of Crawl Space Used as Supply Plenum** | | | | | | | | | | | | | | | |
| (a)  Less than 18 in. below grade | 45 | 50 | 50 | 60 | 60 | 65 | 65 | 70 | 70 | 80 | 80 | 85 | 85 | 90 | 95 |
| (b)  18 in. or more below grade | 20 | 25 | 25 | 30 | 30 | 40 | 40 | 45 | 45 | 50 | 50 | 60 | 60 | 65 | 65 |
| **OUTSIDE AIR (Btuh per cfm)** | | | | | | | | | | | | | | | |
| No. 24 Ventilation or Makeup Air | 30 | 40 | 45 | 50 | 55 | 60 | 65 | 70 | 75 | 80 | 85 | 90 | 95 | 105 | 110 |

Source: Reprinted from *Manual J* with permission of ACCA, Air Conditioning Contractors of America

TABLE A–5

Abbreviated Heat Transfer Factors Table

| DESIGN TEMPERATURE DIFFERENCE, DEGREES | 30 | 40 | 50 | 60 | 70 | 80 | 90 | 100 | U |
|---|---|---|---|---|---|---|---|---|---|
| **WINDOWS** | | | | | | | | | |
| **GLASS** | | | | | | | | | |
| a) no storm sash | 34 | 45 | 57 | 68 | 79 | 90 | 102 | 113 | 1.13 |
| b) w/storm sash | 23 | 30 | 38 | 45 | 53 | 60 | 68 | 75 | 0.75 |
| c) double glass | 18 | 24 | 31 | 37 | 43 | 49 | 55 | 61 | 0.61 |
| **DOORS** | | | | | | | | | |
| a) no storm door | 15 | 20 | 25 | 30 | 35 | 40 | 45 | 50 | 0.50 |
| b) w/storm door | 10 | 13 | 16 | 19 | 22 | 26 | 29 | 32 | 0.32 |
| **WALLS** | | | | | | | | | |
| **FRAME OR MASONRY** | | | | | | | | | |
| a) no insulation | 9 | 12 | 15 | 18 | 21 | 24 | 27 | 30 | 0.30 |
| b) 2" insulation | 4 | 6 | 7 | 8 | 10 | 11 | 13 | 14 | 0.14 |
| c) more than 2" insulation | 3 | 4 | 5 | 5 | 6 | 7 | 8 | 9 | 0.09 |
| **CEILINGS & ROOFS** | | | | | | | | | |
| **CEILINGS** | | | | | | | | | |
| a) uninsulated | 10 | 13 | 16 | 19 | 22 | 26 | 29 | 32 | 0.32 |
| b) 2" insulation | 4 | 5 | 6 | 7 | 8 | 10 | 11 | 12 | 0.12 |
| c) 4" insulation | 3 | 4 | 5 | 6 | 7 | 8 | 9 | 10 | 0.10 |
| d) more than 4" insulation | 2 | 3 | 4 | 5 | 6 | 6 | 7 | 8 | 0.08 |
| **FLOORS** | | | | | | | | | |
| UNCONDITIONED ROOMS | 4 | 6 | 7 | 8 | 10 | 11 | 13 | 14 | 0.14 |
| **CONCRETE SLAB, UNHEATED** | | | | | | | | | |
| a) no insulation | 5 | 7 | 9 | 10 | 12 | 14 | 15 | 17 | 0.17 |
| **CONCRETE, USING PERIMETER SYSTEM IN SLAB** | | | | | | | | | |
| a) no insulation | 57 | 76 | 95 | 114 | 133 | 152 | 181 | 190 | 1.90 |
| **INFILTRATION** | | | | | | | | | |
| **WINDOWS** | | | | | | | | | |
| a) average fit, weatherstripped or w/storm sash | 18 | 24 | 30 | 36 | 42 | 48 | 54 | 60 | 0.60 |
| **DOORS** | | | | | | | | | |
| a) average fit | 60 | 80 | 100 | 120 | 140 | 160 | 180 | 200 | 2.00 |
| b) average fit, weatherstripped or w/storm door | 30 | 40 | 50 | 60 | 70 | 80 | 90 | 100 | 1.00 |

Source: Lennox Industries Inc.

TABLE A–6

Short Form Calculation Sheet Used for Figuring Shell Load

JOB NAME _____  DATE _____

ADDRESS _____

DEALER _____

*O.S. Design Temp* _____  *I.S. Design Temp* _____
*Design Temp Diff* _____
*Wall Construction* _____ *Insulation* _____ " *Window Type* _____
*Roof Construction* _____ *Insulation* _____ " *Floor* _____

|  | AREA | HTF | BTUH LOSS |  |
|---|---|---|---|---|
| GROSS WALL |  |  |  |  |
| WINDOWS |  |  |  |  |
| DOORS |  |  |  |  |
| NET WALL |  |  |  |  |
| CEILING |  |  |  |  |
| FLOOR |  |  |  |  |
| CRACKAGE |  |  |  |  |
| INFILTRATION W |  |  |  |  |
| D |  |  |  |  |
| TOTAL LOSS |  |  |  |  |

Source: Lennox Industries Inc.

TABLE A–7
Long Form Calculation Sheet Used for Figuring Room-by-Room Load

HEATING CONTRACTOR ........................................    ADDRESS ........................................

NAME OF JOB ........................................    ADDRESS ........................................

JOB NO. ........................................
DATE ........................................    BY ........................................
OUTSIDE DESIGN TEMP. ........................................
INSIDE DESIGN TEMP. ........................................
DESIGN TEMP. DIFF. ........................................

| | Table No. | "U" Factor | Multi-plier | 1. Area Crack Cfm | 1. Btuh Loss | 2. Area Crack Cfm | 2. Btuh Loss | 3. Area Crack Cfm | 3. Btuh Loss | 4. Area Crack Cfm | 4. Btuh Loss | 5. Area Crack Cfm | 5. Btuh Loss | 6. Area Crack Cfm | 6. Btuh Loss | 7. Area Crack Cfm | 7. Btuh Loss | 8. Area Crack Cfm | 8. Btuh Loss | 9. Area Crack Cfm | 9. Btuh Loss |
|---|---|---|---|---|---|---|---|---|---|---|---|---|---|---|---|---|---|---|---|---|---|
| 1. Room Use and Floor Level | | | | | | | | | | | | | | | | | | | | | |
| 2. Room Length and Width | | | | | | | | | | | | | | | | | | | | | |
| 3. Running Ft of Exposed Wall | | | | | | | | | | | | | | | | | | | | | |
| 4. Ceiling Height | | | | | | | | | | | | | | | | | | | | | |
| 5. Exposure Type | | | | | | | | | | | | | | | | | | | | | |
| 6. Gross Exposed Wall | | | | | | | | | | | | | | | | | | | | | |
| 7. Windows and Doors | | | | | | | | | | | | | | | | | | | | | |
| 8. Net Exposed Wall | | | | | | | | | | | | | | | | | | | | | |
| 9. Cold Partition | | | | | | | | | | | | | | | | | | | | | |
| 10. Cold Ceiling | | | | | | | | | | | | | | | | | | | | | |
| 11. Cold Floor | | | | | | | | | | | | | | | | | | | | | |
| 12. Infiltration, (Leakage) or Air Change | | | | | | | | | | | | | | | | | | | | | |
| 13. Room Sub-Total Btuh | | | | | | | | | | | | | | | | | | | | | |
| 14. Btuh Adjustment | | | | | | | | | | | | | | | | | | | | | |
| 15. Room Total Btuh | | | | | | | | | | | | | | | | | | | | | |
| 16. Cfm @ ____ Temp. Rise | | | | | | | | | | | | | | | | | | | | | |
| 17. Adjusted cfm | | | | | | | | | | | | | | | | | | | | | |

Source: Lennox Industries Inc.

## TABLE A–8
### Equal Friction Chart

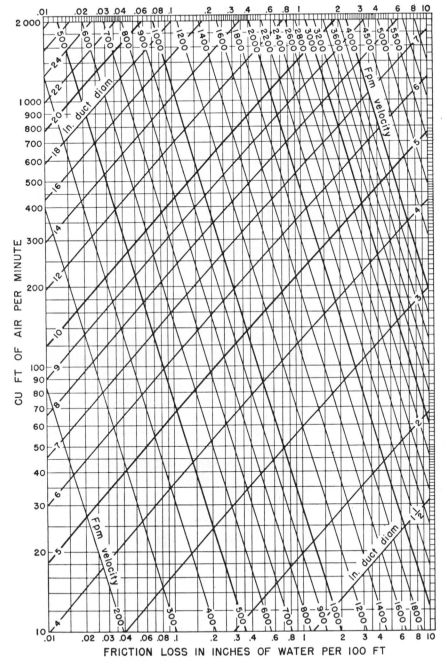

(Based on Standard Air of 0.075 lb per cu ft density flowing through average, clean, round, galvanized metal ducts having approximately 40 joints per 100 ft.) Caution: Do not extrapolate below chart.

Source: Reprinted with permission from American Society of Heating, Refrigerating and Air-Conditioning Engineers, Inc., *ASHRAE Handbook, 1977 Fundamentals* (New York, 1977)

# TABLE A-9
## Rectangular Duct Equivalent

| Duct Diameter, in. | Equivalent Rectangular Stack and Duct Sizes, in. | | | | | | | | | | | | |
|---|---|---|---|---|---|---|---|---|---|---|---|---|---|
| 4.0 | | 4.5×3 | | | | | | | | | | | |
| 4.5 | 8×2.25 | 6×3 | | | | | | | | | | | |
| 5.0 | 10×2.25 | | 8×3.25 | 5×4 | 4×5 | | | 3×8 | | | | | |
| 5.5 | 12×2.25 | | 10×3.25 | 7×4 | 5×5 | | | 4×8 | | | | | |
| 6.0 | 14×2.25 | | 10×3.25 | 8×4 | 6×5 | | | 4×8 | | | | | |
| 6.5 | | | 12×3.25 | 9×4 | 7×5 | 6×6 | | 5×8 | | | | | |
| 7.0 | | | 14×3.25 | 11×4 | 8×5 | 7×6 | | 5×8 | | | | | |
| 7.5 | | | | 13×4 | 10×5 | 8×6 | 7×7 | 6×8 | | | | | |
| 8.0 | | | | 15×4 | 11×5 | 9×6 | 8×7 | 7×8 | | | | | |
| 8.5 | | | | 17×4 | 13×5 | 10×6 | 9×7 | 8×8 | | | | | |
| 9.0 | | | | 20×4 | 15×5 | 12×6 | 10×7 | 8×8 | | | | | |
| 9.5 | | | | 22×4 | 17×5 | 13×6 | 11×7 | 9×8 | 8×9 | | | | |
| 10.0 | | | | 25×4 | 19×5 | 15×6 | 12×7 | 11×8 | 9×9 | | | | |
| 10.5 | | | | | 21×5 | 16×6 | 14×7 | 12×8 | 10×9 | 9×10 | | | |
| 11.0 | | | | | 23×5 | 18×6 | 15×7 | 13×8 | 11×9 | 10×10 | | | |
| 11.5 | | | | | 26×5 | 20×6 | 17×7 | 14×8 | 12×9 | 11×10 | | | |
| 12.0 | | | | | 29×5 | 22×6 | 18×7 | 16×8 | 14×9 | 12×10 | | | |
| 12.5 | | | | | 32×5 | 24×6 | 20×7 | 17×8 | 15×9 | 13×10 | 11×12 | | |
| 13.0 | | | | | 35×5 | 27×6 | 22×7 | 18×8 | 16×9 | 14×10 | 12×12 | | |
| 13.5 | | | | | | 30×6 | 24×7 | 20×8 | 17×9 | 16×10 | 13×12 | | |
| 14.0 | | | | | | 32×6 | 26×7 | 22×8 | 19×9 | 17×10 | 14×12 | | |
| 14.5 | | | | | | 35×6 | 28×7 | 24×8 | 20×9 | 18×10 | 15×12 | | |
| 15.0 | | | | | | 38×6 | 31×7 | 26×8 | 22×9 | 19×10 | 16×12 | 14×14 | |
| 15.5 | | | | | | 41×6 | 33×7 | 28×8 | 24×9 | 21×10 | 17×12 | 14×14 | |
| 16.0 | | | | | | 45×6 | 36×7 | 30×8 | 25×9 | 22×10 | 18×12 | 15×14 | |
| 16.5 | | | | | | | 38×7 | 32×8 | 27×9 | 24×10 | 19×12 | 16×14 | |
| 17.0 | | | | | | | 41×7 | 34×8 | 29×9 | 25×10 | 21×12 | 17×14 | 15×16 |
| 17.5 | | | | | | | 44×7 | 37×8 | 31×9 | 27×10 | 22×12 | 18×14 | 16×16 |
| 18.0 | | | | | | | | 39×8 | 33×9 | 29×10 | 23×12 | 20×14 | 17×16 |
| 18.5 | | | | | | | | 42×8 | 36×9 | 31×10 | 25×12 | 21×14 | 18×16 |
| 19.0 | | | | | | | | 45×8 | 38×9 | 33×10 | 26×12 | 22×14 | 19×16 |
| 19.5 | | | | | | | | 47×8 | 41×9 | 35×10 | 28×12 | 23×14 | 20×16 |
| 20.0 | | | | | | | | 51×8 | 43×9 | 37×10 | 29×12 | 25×14 | 21×16 |

## TABLE A-10

### Various Ductwork Fittings with Equivalent Lengths

**GROUP 1. SUPPLY AND RETURN AIR TAKE-OFF PLENUM FITTINGS**

(These fittings may also be installed on plenums for counter flow units.)

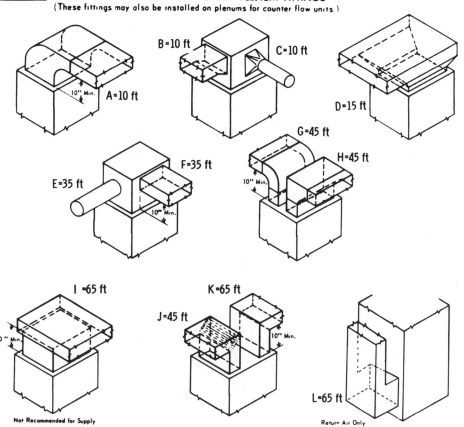

**GROUP 2. REDUCING TRUNK DUCT FITTINGS**

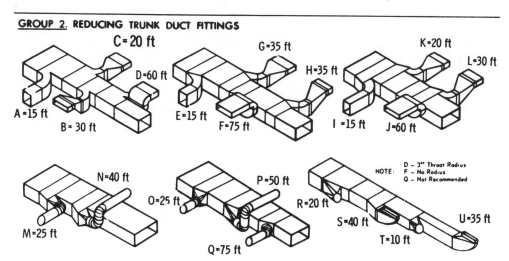

NOTE:  D – 3" Throat Radius
F – No Radius
Q – Not Recommended

## TABLE A–10 *continued*

### GROUP 3. EXTENDED PLENUM FITTINGS

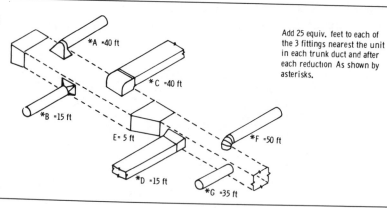

Add 25 equiv. feet to each of the 3 fittings nearest the unit in each trunk duct and after each reduction As shown by asterisks.

*A =40 ft

*C =40 ft

*B =15 ft

E= 5 ft

*F =50 ft

*D =15 ft

*G =35 ft

### GROUP 4    ROUND TRUNK DUCT FITTINGS
(Add 25 equivalent feet to each of the 3 fittings nearest the unit in each Trunk Duct)

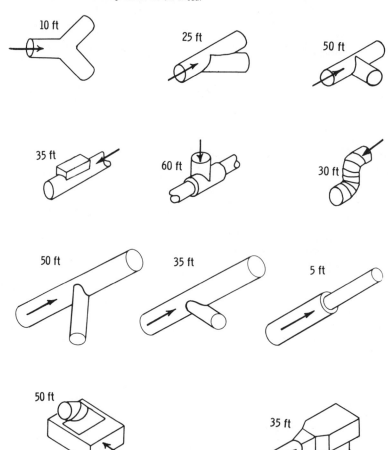

10 ft

25 ft

50 ft

35 ft

60 ft

30 ft

50 ft

35 ft

5 ft

50 ft

35 ft

TABLE A–10 *continued*

### GROUP 5. ANGLES AND ELBOWS FOR TRUNK DUCTS
(Inside Radius = ½ Width of Duct)

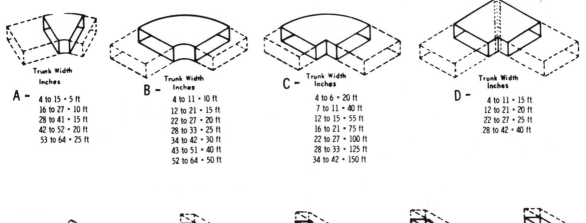

**A –**

| Trunk Width Inches | |
|---|---|
| 4 to 15 | 5 ft |
| 16 to 27 | 10 ft |
| 28 to 41 | 15 ft |
| 42 to 52 | 20 ft |
| 53 to 64 | 25 ft |

**B –**

| Trunk Width Inches | |
|---|---|
| 4 to 11 | 10 ft |
| 12 to 21 | 15 ft |
| 22 to 27 | 20 ft |
| 28 to 33 | 25 ft |
| 34 to 42 | 30 ft |
| 43 to 51 | 40 ft |
| 52 to 64 | 50 ft |

**C –**

| Trunk Width Inches | |
|---|---|
| 4 to 6 | 20 ft |
| 7 to 11 | 40 ft |
| 12 to 15 | 55 ft |
| 16 to 21 | 75 ft |
| 22 to 27 | 100 ft |
| 28 to 33 | 125 ft |
| 34 to 42 | 150 ft |

**D –**

| Trunk Width Inches | |
|---|---|
| 4 to 11 | 15 ft |
| 12 to 21 | 20 ft |
| 22 to 27 | 25 ft |
| 28 to 42 | 40 ft |

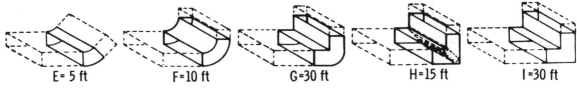

E = 5 ft    F = 10 ft    G = 30 ft    H = 15 ft    I = 30 ft

### GROUP 6. ANGLES AND ELBOWS FOR INDIVIDUAL AND BRANCH DUCTS
(Inside Radius for "A" and "B" = 3 in. and for "F" and "G" = 5 in.)

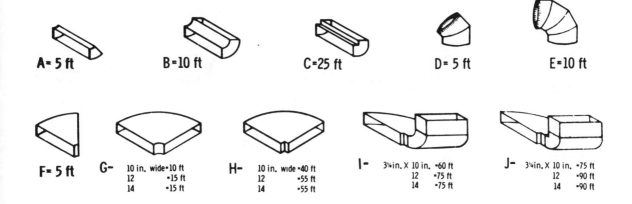

A = 5 ft    B = 10 ft    C = 25 ft    D = 5 ft    E = 10 ft

F = 5 ft

**G –**

| | |
|---|---|
| 10 in. wide | 10 ft |
| 12 | 15 ft |
| 14 | 15 ft |

**H –**

| | |
|---|---|
| 10 in. wide | 40 ft |
| 12 | 55 ft |
| 14 | 55 ft |

**I –**

| | |
|---|---|
| 3¼ in. X 10 in. | 60 ft |
| 12 | 75 ft |
| 14 | 75 ft |

**J –**

| | |
|---|---|
| 3¼ in. X 10 in. | 75 ft |
| 12 | 90 ft |
| 14 | 90 ft |

**TABLE A–10** *continued* _____

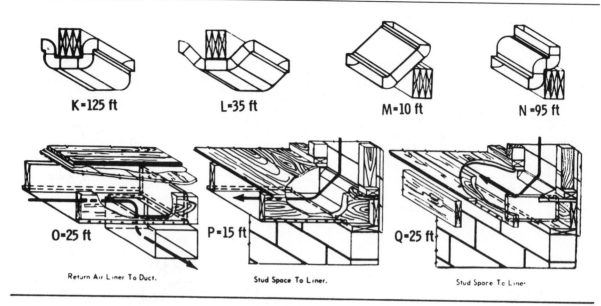

K=125 ft    L=35 ft    M=10 ft    N=95 ft

O=25 ft    P=15 ft    Q=25 ft

Return Air Liner To Duct.    Stud Space To Liner.    Stud Space To Liner

## GROUP 7. BOOT FITTINGS
**(These values may also be used for floor Diffuser Boxes)**

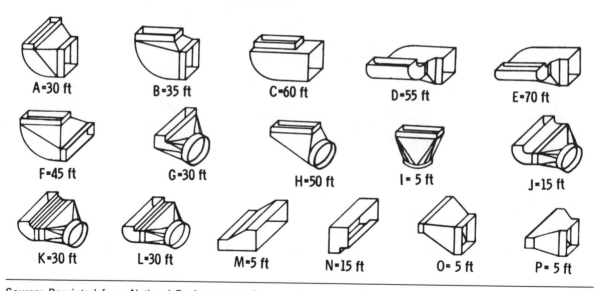

A=30 ft    B=35 ft    C=60 ft    D=55 ft    E=70 ft

F=45 ft    G=30 ft    H=50 ft    I= 5 ft    J=15 ft

K=30 ft    L=30 ft    M=5 ft    N=15 ft    O= 5 ft    P= 5 ft

Source: Reprinted from National Environmental Systems Contractors Association, *Equipment Selection and System Design Procedures* (Arlington, VA, 1973) with permission from ACCA, Air Conditioning Contractors of America

# Index

Note: Boldface numbers in this index refer to the page where the text definition may be found.